ISBN 978-0-266-61204-9
PIBN 10875719

THE

EDINBURGH NEW

PHILOSOPHICAL JOURNAL,

EXHIBITING A VIEW OF THE

PROGRESSIVE DISCOVERIES AND IMPROVEMENTS

IN THE

SCIENCES AND THE ARTS.

EDITORS.

THOMAS ANDERSON, M.D., F.R.S.E.,
REGIUS PROFESSOR OF CHEMISTRY, UNIVERSITY OF GLASGOW;

Sir WILLIAM JARDINE, Bart., F.R.SS.L. and E. ;

JOHN HUTTON BALFOUR, A.M., M.D.,
F.R.S., Sec. R.S. Edin., F.L.S.,
REGIUS KEEPER OF THE ROYAL BOTANIC GARDEN, AND PROFESSOR OF MEDICINE AND BOTANY,
UNIVERSITY OF EDINBURGH.

FOR AMERICA,

HENRY D. ROGERS, LL.D., Hon. F.R.S.E., F.G.S.
STATE GEOLOGIST, PENNSYLVANIA ; PROFESSOR OF NATURAL HISTORY IN THE
UNIVERSITY OF GLASGOW.

JANUARY ...... APRIL 1864.

VOL. XIX. NEW SERIES.

EDINBURGH:
ADAM AND CHARLES BLACK.
LONGMAN, BROWN, GREEN, & LONGMANS, LONDON.
MDCCCLXIV.

PRINTED BY NEILL AND COMPANY, EDINBURGH.

CONTENTS.

REVIEWS:—

CONTENTS.

REVIEWS:—

PROCEEDINGS OF SOCIETIES:—

THE

EDINBURGH NEW

PHILOSOPHICAL JOURNAL.

On some Anomalies in Zoological and Botanical Geography.
By Alfred R. Wallace.

The subject of Geographical Distribution is now generally allowed to be one of the most interesting branches of
Natural History, and owing to the accumulation of much
trustworthy material within the last few years, we are at
length enabled to generalise many of the most important
facts, and to form a tolerably accurate idea of the import
and bearing of the whole inquiry.

In the admirable chapters on this topic in the " Origin
of Species," Mr Darwin has given us a theory as simple as
it is comprehensive, and has besides gone into many of the
details so fully as to render it needless to say another word
here on those parts of the question which he has treated. As
an explanation of the main facts, and of many of the special
difficulties, of geographical distribution, those chapters are
in every respect satisfactory; and I therefore propose now
to consider only the anomalies and discrepancies which so
frequently occur between the distribution of one class or
order and another, and to discuss the possibility of arriving
at a division of the earth into Regions, which shall represent accurately the main facts of distribution in every department of nature.

In doing this I shall consider in detail a few cases of
special difficulty only, and endeavour to establish certain

principles, which, if accepted, will enable us to deal with such
cases for the future, and avoid the confusion into which the
whole question must necessarily fall, if (as has hitherto
been the case) every naturalist proposes a distinct set of
geographical regions for the group to which he pays most
attention.

The entire subject naturally comes under the two heads
of *terrestrial* and *marine* distribution, which may be treated
of independently, but upon similar principles.   I now confine myself entirely to the terrestrial division.   The chief
fault of the Zoological and Botanical regions that have been
hitherto proposed is, that they have generally been too
numerous, and have been more or less artificially bounded
by lines of latitude and longitude.   Those established by
Meyen for plants, and by Woodward for shells, have this
fault, and were, besides, never intended to apply to the whole
organic world.   Swainson's division (Geog. and Class. of
Animals in *Lardner's Cab. Cyc.*), was much more natural,
and was, I believe, the first that took into consideration all
classes of animals, and can lay any claim to rank as a general
system.   But by carrying out even here his favourite quinary
theory, and by following too closely the supposed typical
races of man, he was led into many important errors,—such
as including the northern and southern continents of America
in one region, and placing Northern Asia with India rather
than with Europe.

In June 1857, a paper was read before the Linnean Society by Dr Sclater, entitled, " On the general Geographical
Distribution of the Members of the Class Aves," which marks
an era in this branch of natural history.   The subject was
now for the first time treated in a philosophical manner by
a naturalist well acquainted with the whole class with which
he proposed to deal, and who, by looking chiefly to groups,
—to genera and families rather than to species—and by
taking account of broad contrasts rather than local peculiarities, has succeeded in marking out upon the globe those
divisions, which not only represent accurately the great
facts presented by the distribution of birds, but seem also
well adapted to become the foundation for a general system
of Ontological regions.

The following six Regions are those established by Dr Sclater:—

1st, The Neotropical, comprising South America, Mexico, and the West Indies; 2d, The Nearctic, including the rest of America; 3d, The Palæarctic, composed of Europe, Northern Asia to Japan, and Africa north of the Desert; 4th, The Ethiopian, which contains the rest of Africa and Madagascar; 5th, The Indian, containing Southern Asia and the western half of the Malay Archipelago; and 6th, The Australian, which comprises the eastern half of the Malay Islands, Australia, and most of the Pacific Islands. Each of these regions is characterised by a number of peculiar genera, and even families of birds, which, while found everywhere within the region, do not pass over its boundaries; and by other genera which, though found sparingly in several regions, have their metropolis in one. This scheme of Ornithological distribution has been founded on such an extensive basis of facts, and after having been five years before the world has met with such general acceptance, that it may fairly be taken as established, subject only to modifications of the dividing lines between those regions which gradually merge into each other.

It remains to be shown whether this is not only a true Ornithological, but also a true Zoological and Botanical division of the earth; and if not, to show how it is that what is true for one part of nature should not be equally true for all.

In a paper on the " Geographical Distribution of Reptiles" (*Proc. Zool. Soc.* 1858, p. 373), Dr Gunther has shown that for snakes and batrachians the same divisions will almost exactly apply; the only important discrepancy being that Japan, judging from its snakes, would belong to the Indian region, while its batrachians are decidedly related to those of the Palæarctic region.

In Mammalia the same geographical divisions are very strongly marked, but here again one important discrepancy has been pointed out, namely, that the quadrupeds of North Africa are of Ethiopian, while the birds and reptiles are of European forms. (*Ibis*, vol. i. pp. 93, 157.

In the immense class of Insects, very little has been done

to work out the details of Geographical distribution.   There
is no doubt but that the six regions marked out by Dr
Sclater are generally characterised by distinct forms of insects.   There is one case, however, which has come under
my own observation, in which the entomological would not
correspond with the ornithological regions.   The Moluccas
and New Guinea in their birds and mammals are most decidedly Australian, while the insects show a general correspondence with the Indian type.   It has also been pointed
out that the insects of Chili and of south temperate South
America have little affinity with Neotropical forms.

Land shells, I am informed by the Rev. H. B. Tristram,
generally agree very well with the ornithological regions.
The subdivisions or provinces are, however, often very
strongly marked.

In Plants, I am informed by Dr Hooker, the regions will
in many cases not at all correspond.

In order to arrive at the cause and meaning of these
singular differences in the Geographical distribution of the
various classes, we must inquire how Zoological and Botanical regions are formed, or why organic existences come
to be grouped geographically at all.

It appears to me that this can be explained by a few
simple principles—

1st, The tendency of all species to diffuse themselves over
a wide area, some one or more in each group being actually
found to have so spread, and to have become, as Mr Darwin
terms them, dominant species.

2d, The existence of barriers checking, or absolutely
forbidding that diffusion.

3d, The progressive change or replacement of species,
by allied forms, which has been continually going on in the
organic world.

4th, A corresponding change in the surface, which has
led to the destruction of old and the formation of new
barriers.

5th, Changes of climate and physical conditions, which
will often favour the diffusion and increase of one group,
and lead to the extinction or decrease of another.

By means of these principles we will endeavour to explain the discrepancies already mentioned. And, first, how is it that the snakes in Japan are Indian and the batrachians Palæarctic? Dr Gunther informs us, in the paper already alluded to, that snakes are a pre-eminently tropical group, decreasing rapidly in the temperate regions, and absolutely ceasing at 62° N. Lat. Bactrachians, on the other hand, are almost as fully developed in northern as in tropical regions. They can support intense cold, and are, moreover, more diffusible, geographically, than snakes. These facts furnish a clue to the peculiarities of the Japanese reptile fauna. For let us suppose that Japan once formed a part of northern Asia (with which it is even now almost connected by two chains of islands), it would then have received its birds, mammals, and batrachians from the Palæarctic region; but there could have been few or no snakes, owing to the much lower curve of the isothermal lines in Eastern Asia than Western Europe, giving to Mandtchouria a climate as rigorous as that of Sweden. Now, at a subsequent period, Japan must have been connected with Southern Asia through the line of the Loo-choo and Madjicosima islands, and would then acquire its population of Indian forms of snakes, which would easily establish themselves in an unoccupied region,—whereas the batrachians, as well as the birds and mammals of Southern Asia, would find a firmly established Palæarctic population ready to resist the invasion of intruders, and it is therefore not to be wondered at that but few, if any, Indian forms of these groups should have been able to maintain themselves. Again, the insects of Japan are decidedly Palæarctic in character, except in the case of a few tropical forms of diurnal lepidoptera, which would have been able to establish themselves, like the snakes, on account of the extreme poverty of that group in high latitudes. It would thus appear that the tropical character of the snakes is quite exceptional, depending upon the fact of the whole group being pre-eminently tropical, and can therefore not be held to throw any doubt on the position of Japan in the Palæarctic zoological region.

We have next to consider the supposed discrepancy in the mammals of Algeria compared with the birds, reptiles, insects, and plants—all of which are decidedly of Palæarctic

forms. It will, I think, be found that the facts have here
been somewhat hastily assumed, and that the mammalia
do not differ very much in this respect from other classes,
and in no degree invalidate the position that North Africa
belongs to the Palæarctic region. Leaving out domesticated
animals, I have drawn up a list of the genera of Algerian
mammals (from Captain Loche's Catalogue), and have
divided the species, so as to show how far they correspond
with those of the Palæarctic or other regions.

From an examination of this table, it will be seen that
thirty-three of the Algerian mammals are absolutely identical with European or West Asian species, fourteen more
are representative species of European genera, and ten belong to West Asian and Siberian (and therefore Palæarctic)
genera, giving a total of fifty-seven species and about twenty-
eight genera, as the measure of Palæarctic affinity. Now, to
balance this, what have we to indicate an Ethiopian fauna?
The most important, and what have probably been most
relied on as giving an extra-European character to the country, are the four large felines,—the lion, the leopard, the
serval, and the hunting-leopard,—but as these all range the
whole of Africa, from the Cape to the Mediterranean, and
may very probably have crossed the desert in the tracts of
caravans, they cannot be held to have much weight on the
present question. Then there is the solitary monkey; but
as that actually inhabits Europe, we need hardly have
included it among the representatives of Ethiopian groups,
except to give all the facts that can be fairly claimed on
that side. The antelope is a desert-haunting species, and
therefore may be looked upon as a straggler on the northern
side of the Sahara; and, besides these, we have representatives of two really African genera (Macroscelides and
Zorilla), giving a total of only eight species as the measure of Ethiopian affinity. The remaining species, seven in
number, are true desert-haunters, roaming over North
Africa, Egypt, and Arabia, into the Indian deserts, and have
scarcely any more right to be considered as belonging to
one region than another, since they inhabit the district
which forms the boundary and debateable land of the
Ethiopian, Indian, and Palæarctic regions.

Genera of Algerian Mammals.	Palæarctic Species.	Species of Palæarctic Genera.	Ethiopian Species and Genera.	Asiatic or Desert Species and Genera.
Macacus innuus	1	..	1	...
Ursus	1	...	...	...
Canis	1	3	...	...
Fenecus	...	...	...	1
Hyena	...	1	...	...
Meles	1	...	...	...
Mangusta	...	...	...	1
Genetta	1	1	...	...
Felis	1	...	4	1
Putorius	3	...	...	...
Zorilla	...	...	1	...
Lutra	1	...	...	...
Sus	1	...	...	...
Cervus	1	...	...	...
Dama	1	...	...	...
Antelope	...	...	1	...
Gazella	...	...	...	2
Musimon	...	1	...	...
Vespertilionidæ	8	...	...	...
Sorex	2	...	...	...
Pachyura	...	1	...	...
Crocidura	1	...	...	...
Crossopus	1	...	..	...
Macroscelides	...	...	1	...
Erinaceus	...	1	...	...
Hystrix	1	...	...	...
Myoxus	...	1	...	...
Dipus	1	2	...	...
Alactaga (Siberian Genus)	...	1	...	...
Lepus	1	1	...	...
Gerbillus (N. and W. Asian Genus)	...	7	...	...
Ctenodactylus (Asiatic Group)	...	...	...	1
Sciurus	...	1	...	...
Mus	5	3	...	...
	33	24	8	6

The left margin shows bracketed group labels: **Carnivora.**, **Ruminantia.**, **Insectivora.**, **Rodentia.**

It would seem, therefore, that the supposed discrepancy of the Mammalia, in determining the southern limit of the Palæarctic province, is altogether imaginary. The number of species absolutely identical is not so great as in the birds; but Europe is not the whole Palæarctic province, and if we take *genera* instead of *species*, we shall find the correspondence as complete as possible,—twenty-eight genera being truly Palæarctic, only three Ethiopian, while five are Asiatic, or desert-dwellers. In this case, therefore, the whole of the vertebrata combine with the insects, the land shells, and the plants, to place North Africa in the Palæarctic region.

The case of the insects in the Australian portion of the Malay Archipelago is one of much greater difficulty. Australia itself contains a remarkable assemblage of insects, among which its Lamellicornes, Buprestidæ, and Curculionidæ offer a number of striking forms and genera quite peculiar to it. In New Guinea and the Moluccas, on the other hand, Lamellicornes are comparatively scarce, and with the Buprestidæ and Curculionidæ are of Indian rather than Australian genera; while the great family of the Anthribidæ, which is almost entirely absent in Australia, is here everywhere abundant in genera, species and individuals, though less so than in the Western or Indian region.

To account for this remarkable discrepancy, we must consider,—1*st*, That insects are much more immediately dependent on the character of the vegetation, and therefore on climate, than are vertebrated animals; and, 2*dly*, That water-barriers are much less effective in preventing their dispersion. A narrow strait is an effectual bar to the migration of mammals and of many reptiles and birds, while insects may be transported in the egg and larva state by floating timber, and from their small size and great powers of flight, may be easily carried by the winds from one island to another. Now, the characteristic insects of Australia seem specially adapted to a dry climate and a shrubby flower-bearing vegetation, and could hardly exist in the excessively moist atmosphere and amid the dense flowerless forests of the equatorial islands. If, therefore, we suppose Australia itself to be the most ancient portion of this region (which its great richness in peculiar generic forms seems to indicate), we can easily understand how, when the islands of the Moluccas and New Guinea first rose above the waters and became clothed with dense forests nurtured by tropical heat and perpetual moisture, though the birds and mammals readily adapted themselves to the new conditions, the insects could not do so, but gave way before the immigrants from the islands to the west of them, which having been developed under similar climatal conditions, and thus become specially adapted to them, were enabled, by the enormous powers of multiplication and dispersion possessed by insects, at once to establish themselves in the newly-formed lands, and develop an insect population in many respects at variance with other classes of animals.

There are, however, several instances of groups of insects almost as strictly confined to one-half of the Archipelago as is so remarkably the case with the vertebrata; and when the extensive collections made by myself in most of the islands come to be accurately worked out, no doubt more such instances will be found. Among Coleoptera I may mention the *Tmesisterninæ*, a remarkable sub-family of Longicornes, as being strictly confined to the Australian region, over the whole of which it extends, and has its western limit in Celebes along with the Marsupials and the Trichoglossi. Again, Mr Baly, so well known for his acquaintance with the Phytophagous Coleoptera, finds that one of the principal sub-families of that tribe (Adoxinæ), which he has recently classified, though spread over Europe and the whole of Asia, is only found in the Archipelago in those islands which belong to the Indian region of zoology. This proves that there is an ancient insect-population in the Austro-Malayan Islands, which accords in its distribution with the other classes of animals, but which has been overwhelmed, and in some cases perhaps exterminated, by immigrants from the adjacent countries. The result is a mixture of races, in which the foreign element is in excess; but naturalists need not be bound by the same rule as politicians, and may be permitted to recognise the just claims of the more ancient inhabitants, and to raise up fallen nationalities. The aborigines, and not the invaders, must be looked upon as the rightful owners of the soil, and should determine the position of their country in our system of Zoological geography.

My friend, Mr Bates, has kindly furnished me with some facts as to the entomology of Chili and south temperate America, which would show that the insects of this region have very little connection with those of tropical America.

Out of ten genera of butterflies found in Chili, not one is characteristic of tropical America. Four (Colias, Argynnis, Erebia, and Satyrus) are northern forms, only one of which occurs at all in tropical America, and that high up in the Andes ; three others are peculiar to Chili, but have decided north temperate or Arctic affinities ; and three more (Anthocharis, Lycæna, and Polyommatus) are cosmopolitan, but far more abundant in temperate than tropical regions. Judging, therefore, from butterflies only, we should decidedly have to place south temperate America in the Nearctic region, or form it into a region by itself.

Two important families of Coleoptera, the Geodephaga and the Lamellicornes, furnish different but equally remarkable results. There are 77 genera of these families found in Chili, of which 46 are peculiar to south temperate America, being ⅗ths of the whole ; 17 are cosmopolitan, 2 are north temperate, 10 tropical American, and 1 is African.

But of the 46 peculiar genera, no less than 10 are closely allied to Australian forms, and 3 to South African,—so that the affinities of these groups of coleoptera are almost as strong to Australia as to tropical America ; next comes South Africa, and, lastly, the north temperate zone ; though as the two genera *Carabus* and *Geotrupes* are very extensive and important, and are totally absent from the tropics, but appear again in Chili, the real amount of affinity to northern regions may be taken as somewhat larger.

Here, then, as only 10 genera out of 77 are common to south temperate and tropical America, and as the remainder have wide-spread affinities—to the northern hemisphere, to Australia, and to South Africa—it would seem impossible, from a consideration of these families of Coleoptera alone, not to separate the south temperate zone of South America as a distinct primary region.

Other orders of insects and other families of Coleoptera may very probably give somewhat different results. From Boheman's work on the Cassididæ, I find that the genera of tropical America send representatives into Chili, and even into Patagonia, and that none of the south temperate forms have a direct affinity with those of Australia. But this family is almost exclusively tropical, very few and obscure species inhabiting the colder regions of the earth, while there are no generic forms peculiar to the Australian region.

In many of the preceding facts we have a most interesting correspondence with those furnished by the distribution of plants. Dr Hooker has shown the large amount of resemblance between the flora of southern South America and Australia, especially Tasmania and New Zealand,—one-eighth of the whole New Zealand flora being identical with South American species. Again, the occurrence of northern genera of coleoptera in Chili, and the whole of the butterflies having northern affinities, agrees with the number of northern genera and species of plants in Patagonia and Fuegia, and is an additional proof of the intensity and long continuance of the glacial epoch which sufficed to allow so many generic forms to pass the equator from north to south.

We have here another illustration how much easier of diffusion, and how much more dependent on local conditions are insects than the higher animals. A great part of the southern portion of America is of more recent date than the central tropical mass, and must have had at one time a closer communication than at present with the antarctic lands and Australia, the insects and plants of which, finding a congenial climate, established themselves in the new country, being only feebly opposed by the few northern forms which had already, or soon after, migrated there. And the fact that Tasmania and New Zealand are the poorest countries in the world in butterflies, will enable us to understand how it is that all those found in Chili are northern forms, while the coleoptera of the same countries (Tasmania and New Zealand) being tolerably abundant and varied, and having a shorter journey to perform than the north temperate immigrants, were enabled to get the upper hand in colonizing the new country.

The marsupial Opossums are the most remarkable case of vertebrata in America having Australian affinities. It is very doubtful whether these could have been introduced in the same manner as the plants and insects already alluded to, because the latter have to a considerable extent an ant-arctic character, and do not appear in such numbers as to indicate an actual continuity of land, which would have been almost indispensable for the passage of mammalia, and would at the same time have undoubtedly admitted Australian forms of land birds, which do not exist in South America. It seems more reasonable, therefore, to suppose that these marsupials have inhabited America since the Eocene period, when the same genus existed in Europe, and the marsupial order had probably a universal distribution.

With this one exception, the birds, the mammalia, and the reptiles* of south temperate America have little or no affinity either with north temperate or Australian forms, but are modifications of the true denizens of the Neotropical regions. They appear to have been enabled rapidly to seize hold of the country, and to adapt themselves to its modified climate and physical features—a remarkable instance of which is mentioned by Mr Darwin in the woodpecker of the Pampas, which never climbs a tree. The tropical insects, on the other hand, having become gradually specialized during long periods for a life amid continual verdure and unvarying summer, were totally unfitted for the new conditions presented to them, and only in a very few cases were able to struggle against forms already adapted to a more barren country and a more rigorous climate.

This difference in the adaptive capacity of groups, combined with an unequal power of diffusion, will cause the various kinds of barriers to be sometimes more and sometimes less effective. For example, when a mountain range has attained only a moderate elevation, it will already completely bar the passage of many insects, while mammalia, birds, and reptiles, more capable of sustaining different conditions, will readily pass over it. On the otherhand, an arm of the sea, or even a wide river, will completely

* Except the batrachians, which show some affinities between Australia and South America, a case analogous to that of Japan.

isolate most mammals and many reptiles, while insects have still various means of passing it.

Another consideration which must help to determine the amount of specific peculiarity in a given region, is the average rate at which specific forms have changed. Palæontologists have determined that mammalia have changed much more rapidly than mollusca, from the phenomena of the comparatively recent extinction of so many species of mammals, whose remains are found along with existing species of shells. From the evidence of the distribution of existing species, birds would appear to have changed at least as quickly as mammals, and insects, in some cases, perhaps more so ; owing, no doubt, to their very small diffusibility, and the readiness with which they are affected by local conditions.

Taking the various facts and arguments now brought forward into consideration, it appears evident that no regions (be they few or many in number) can be marked out, which will accurately represent the phenomena of the geographical distribution of all animals and plants. The distribution of the several Classes, Orders, and even Families, will differ, because they differ in their diffusibility, their variability, and their mode of acting and reacting on each other, and on the external world. At the same time, though the details of the distribution of the different groups may differ, there will always be more or less general agreement in this respect, because the great physical features of the earth— those which have longest maintained themselves unchanged —wide oceans, lofty mountains, extensive deserts—will have forbidden the intermingling or migration of all groups alike, during long periods of time. The great primary divisions of the earth for purposes of natural history should, therefore, correspond with the great permanent features of the earth's surface—those that have undergone least change in recent geological periods. Later and less important changes will have led to discrepancies in the actual distribution of the different groups ; but these very discrepancies will enable us to interpret those changes, of which they are the direct effects, and very often the only evidence.

From this examination of the anomalies that occur in

the distribution of different groups, and of the probable causes of such anomalies, it appears that the six regions of Dr Sclater do approximately represent the best primary divisions of the earth for natural history purposes. They agree well with the present distribution of mammalia, birds, reptiles, land shells, and very generally of insects also. The cases in which they do not seem correct are those of isolated groups in restricted localities. The greatest discrepancies occur in groups which have at once great capacities for diffusion, and little adaptability to change of conditions; and, in the case of plants, have probably been much increased by what may be called the adventitious aid of the glacial period and of floating ice.

Of botanical distribution I have said little, from want of knowledge of that branch of the subject, and I can find no detailed information bearing directly upon the questions here discussed, but what I have already mentioned. It is much to be desired that some competent botanist would point out how far these regions agree with, and how far they contradict, the main facts of the distribution of plants. It seems evident that the various modes of glacial action have produced much more effect on the migrations of plants than on those of animals, and also that plants have, on the whole, more varied and more effectual means of dispersal. Still, if the views here advocated are true, the flora of each region should exhibit a characteristic substratum of indigenous forms, though often much modified, and sometimes nearly overwhelmed by successive streams of foreign invasion.

My object in calling attention to the subject by this very partial review of it, is to induce those naturalists, who are working at particular groups, to give more special attention to geographical distribution than has hitherto been done. By carefully working out the distribution of allied genera and closely connected groups of species, they could give the amount of agreement or discrepancy with other groups whose geography is best known, and furnish us with such information on the habits of the species, as might help to explain the anomalies which were found to occur. We should thus soon accumulate a sufficiency of detailed facts to enable us to determine whether these are the best primary divisions of the earth into terrestrial Zoological and Botanical regions, or whether such general divisions are altogether impracticable. Some such simple classification of regions is wanted to enable us readily to exhibit broad results, and to show at a glance the external relations of local faunas and floras. And if we go more into detail, and adopt a larger number of primary divisions, we shall not only lose many of these advantages, but shall probably find insuperable difficulties in harmonising the conflicting distribution of the different groups of organised beings.

On the Slaking of Quicklime. By JOHN DAVY, M.D., F.R.S., Lond. and Edin.*

IN some experiments which I have made on the slaking of quicklime, as its conversion into a hydrate is commonly called, I have noticed certain results new to me, and as I cannot find them noticed in any chemical work I have referred to, I propose to give a brief account of them on the possibility that they may be new to others.

It is well known that as soon as water is added to and absorbed by well-burnt lime fresh from the kiln, an immediate union takes place, the mass becoming broken up and falling into powder, with the production of much heat and steam.† But if the lime has been kept exposed to the air for two or three days, during which time it absorbs a small quantity of water,‡ without at all disintegrating, the same rapid union is not witnessed on the addition of water sufficient to form a hydrate ; on the contrary, some minutes will elapse before the combination takes place, and I find there is a similar retardation of action from other causes as shown by the results of the following experiments :—

1. To a piece of lime taken from a mass, such as that

* Read at the Meeting of the British Association for the Advancement of Science held at Newcastle. (1863.)

† Gunpowder and sulphur have been ignited by it. See Annales de Chem. et de Phys. xxiii. p. 217. About six pounds were slaked.

‡ A little carbonic acid is absorbed at the same time, but this latter is not essential, inasmuch as the lime exhibits the same peculiarity if kept in damp air, excluding carbonic acid.

adverted to, ten grains of water were added. The whole of the water was absorbed. As I held it, a very slight sensation of warmth was perceived by the fingers in contact with it—a sensation that occurred immediately, and for about eight minutes it did not distinctly increase. About that time small cracks began to appear at the surface of the piece; the temperature instantly sensibly rose, and in another minute the heat became too great to be bearable. Now, put down, in a few seconds it became rent, and the hydrate was formed.

2. In a piece of lime of two or three pounds, a hole was bored an inch and half in depth, sufficiently large to admit the bulb of a thermometer. Water, no more than the lime could absorb, was next poured on the mass. The thermometer, from $55°$, immediately rose to $70°$. During about eight minutes little change of temperature was observed; then, in less than a minute, the thermometer rose to $280°$, accompanied with the production of steam and the falling to pieces of the mass.

3. Into a small receiver, $2\frac{1}{2}$ inches high, $1\frac{1}{4}$ inch in diameter, quicklime in fine powder (just pounded) was put in sufficient quantity to fill it to about two-thirds. A thermometer then introduced stood about $60°$. Next added water, more and more, till the whole of the lime appeared to be moistened. This done, in about half a minute, the thermometer stood at $80°$; in about five minutes it had fallen to $78°$; then it began to rise; in one minute it had risen to $100°$, in less than half a minute more to $120°$; then, after a few seconds, an explosion took place, the thermometer was thrown out and the lime was scattered, some even beyond twelve feet.

4. Into the same receiver about the same quantity of pounded lime was introduced with excess of water, and the mixture was immediately stirred. The lime subsided on rest; there was about one-tenth of an inch of superincumbent water. After about twenty minutes the temperature had become a little higher; in about five minutes more steam was produced, but there was no explosion.

5. To a piece of quicklime, weighing 88·3 grs., 9·2 grs. of water were added. A slight increase of temperature was

produced. After about a quarter of an hour it became warm, but quite bearable; portions fell off one after another, but no high temperature was produced. When cool, it was again weighed; its weight was now 93·8 grs.; so it had gained only 5·5 grs. of water. The portions which had fallen off, put into water, were some minutes before they became hot; then they fell to powder, and became a thick paste.

6. A piece of quicklime, weighing 310 grains, was kept an hour and a half in a close vessel with some damp paper. During this time it had increased in weight only half a grain. Now added gradually, in about half a minute, 19·5 grains of water. The water, in minute quantity, was applied successively to different parts of the mass. There was no sensible increase of temperature; 13·5 grains more of water were added; still no increase of temperature. After twenty-four minutes 6 grains more were added, without perceptible effect; and an hour later 6 more, making a total of 45 grains. Shortly after some action had taken place, and a portion had fallen to powder with evolution of heat; next morning the whole mass was found broken up and reduced to the state of powder.

7. To 76 grains of quicklime reduced to fine powder, 11 grains of water were added in two portions, triturating the powder on each addition. I am not aware of any heat having been produced, nor did I see indications of any action. The dry powder was put into a tube and tightly corked. On the day following there was no perceptible change. Now added water in great excess, so as to form a thick paste. After a few minutes there was a slight increase of temperature; after twelve, it had become moderately warm, and it continued so some time, showing the slow formation of the hydrate.

Do not these results warrant the conclusion, that lime is capable of uniting feebly with less water than is required to form the hydrate, that consisting of one proportion of each, the weaker compound containing probably two proportions of lime. In the last-mentioned experiments the quantity of water was nearly in accordance with this composition.

I shall now relate one or two other results which are rather favourable to this conclusion, and at the same time show how unstable is the union (if it be admitted) with the smaller proportion of water.

8. To 39·5 grs. of quicklime added gradually 4 grs. of water. A little heat was produced, but a bearable one. Now plunged the mass into cold water, moving it rapidly for about a minute; it remained cool. Transferred it now to the balance; it had gained 8 grs. In a few seconds action took place, and the hydrate was formed. This experiment has been more than once repeated with the same result.

9. Instead of mere water a mixture of about equal parts of water and sulphuric ether was poured on a piece of quicklime. The immediate effect was the cooling of the little mass by the evaporation of the ether. For many minutes there was a retardation of action, and when it began it went on slowly, the evaporation probably interfering, and preventing rapidity of combination.

10. If, instead of cooling the quicklime, its temperature be at all raised, so much the more rapidly is the hydrate formed on the addition of water. Thus, on pouring a few drops of water on a small piece of quicklime fresh from the fire, allowed to cool, so that its warmth was hardly to be felt, no sooner did the water touch it than the union took place with explosive violence, driving the little fragments to the distance of several feet.

There are other results which I have obtained of like significance.

11. If aqua ammoniæ, or a strong solution of common salt, or of chloride of calcium—compounds having an affinity for water, and not readily parting with it—be added to quicklime, the formation of the hydrate is more or less retarded, but is nowise prevented, and when it takes place it is sudden, with the usual phenomena as to evolution of heat, &c.

12. A similar retardation is witnessed when quicklime in mass is put into a solution of the carbonate or sesquicarbonate of ammonia, or of the carbonate or bicarbonate of potash; but when action commences it is rapid, the hydrate of lime being formed, with the usual production of heat.

But if the lime be reduced to the state of a very fine powder, and mixed (using trituration) with a strong solution of either, a carbonate of lime is formed with a very slight elevation of temperature—that small degree referrible, I believe, to the portion of hydrate at the same time produced.

On the supposition that two of lime can unite with one proportion of water, the instability of the compound is no more than might be expected. There are many analogous examples, such as the sesquicarbonate of ammonia, the neutral carbonate of this alkali, and the bicarbonate of potash and soda. *A priori*, we cannot predicate the chemical relation of one body to another; it may be conjectured, but it can only be determined by experiment. It might be supposed, that because carbonic acid is expelled from lime by a bright red heat, that it would not combine with this acid at a dull red heat. Yet this I find is the case.

Considering the high temperature produced in the act of union of water and lime, and the quantity of steam that may be generated, the idea could hardly fail to occur, that the formation of the hydrate may be applied to some useful purpose, such as the blasting of rocks; and if successful, might be especially useful in collieries as a substitute for gunpowder, which has so often occasioned, by the igniting of gas, terrible accidents with loss of life.

The few trials I have instituted, with a view to this application, have not answered my expectations. I shall mention one or two of the latest I have made. Recently I had a boring made in a block of sandstone, about 15 inches deep and 2 inches in diameter. It was filled with small pieces of quicklime; water was poured in, which, it was inferred, found its way to the bottom in sufficient quantity, and the hole was then firmly closed by a plug of wood. No rending of the rock was produced; yet the hydrate was formed. It must be concluded that the elastic expansive force exerted was not superior to the resistance, and that all the steam was condensed. A second experiment was made, substituting for the boring in rock a strong earthenware jar, capable of holding about a quart. It was similarly charged and tightly corked; the cork bound down

firmly by a cord.   After about fifteen minutes an explosion
took place.   The report was like that of a pistol.   The jar
was broken into several pieces, and some of them were pro-
jected many yards from the spot.

Now, as coal is not nearly so resisting as sandstone,
and as its boring is easily effected, I venture to express the
hope that the experiment may be repeated in a colliery.
It is easily made, at a cost not worth mentioning, is attended with no serious danger ; and should it be successful,
it may conduce to the saving of many valuable lives.

---

*Some Observations on the Blood, chiefly in relation to the
question,* Is Ammonia in its volatile state one of its Normal
Constituents?   By JOHN DAVY, M.D., F.R.S. Lond. and
Edin., &c.

Of the many questions which have been propounded
respecting the blood, there are two in particular which of
late years have excited some interest and have given rise
to much discussion ; one, whether it contains any ammonia
in a volatile state ;—the other, whether the escape of volatile
ammonia is the cause of that characteristic quality of healthy
blood, its coagulation ?

The latter question has been answered, as is well known,
in the affirmative by Dr Benjamin Richardson in a work of
much ability, a successful prize essay, which was published
in 1858.   Shortly after, viz., in the following year, I
endeavoured to show that this is not the case.   The experiments I made were mostly on the blood of the common fowl,
which I selected chiefly on account of the rapid manner in
which the blood of birds coagulates, and its high temperature during the time the phenomenon is in progress.   The
results were all negative.   The coagulation took place
without obvious difference of time whether the blood was
allowed to coagulate in a closed vessel, or exposed to the air,
as, for instance, when received into a vial, the blood completely filling it, and instantly closed by a glass stopper,—
or into a vial of the same kind and the stopper left out.
Further, I found that when ammonia in a notable quantity

was added to the blood, its coagulation was not prevented. These results, all of them well marked, seemed to warrant the conclusion that the volatile alkali is no wise concerned in the coagulation of the blood. I have made many experiments since, and they have all been of a confirmatory kind.[*]

The other question, whether the volatile alkali is a normal constituent of the blood, is not so easily answered, and there is a difference of opinion on the subject among physiologists : thus, Frederichs, who has the reputation of being an accurate observer, thinks that it forms no part of the vital fluid ; whilst Dr Hammond takes the opposite view, and believes with Dr Richardson that it is an integrant part of that fluid, and that he has detected it in no less than fourteen experiments, even in the blood of the common fowl, employing Dr Richardson's test—that is, a slip of glass moistened with hydrochloric acid, and exposed to the vapour rising from the blood.[†]

He does not state the particulars of the trials he made. This, I cannot but hold to be an omission, considering the nature of the fluid, how readily it changes, how apt it is to undergo decomposition,—that of the putrid kind,—and in the act to give rise to the production of ammonia in the form of the volatile carbonate. I believe there is no exaggeration in stating that the instant the blood is taken from the living body a change of this kind commences ; hardly perceptible indeed at first, but with advance of time, especially at a temperature above 60 Fahr., rapidly increasing. In illustration, I shall give the details of an experiment which I have made after the reading of Dr Hammond's statement.

When the thermometer in the open air was 62°, a pullet was killed by dividing the great vessels in the neck. The blood, which was very florid, was collected in three small cups, and each was covered with a plate of glass moistened with hydrochloric acid. In each the blood coagulated in less than two minutes. After an exposure of five minutes the glass from one of them was removed, and the acid

* See Trans. Roy. Soc. of Edin. for 1859 ; and my Physiological Researches, London, 1863.

† Physiological Memoirs, Philadelphia, 1863. I quote this author, being one of the latest and ablest inquirers I can refer to.

evaporated at a low temperature ; now on examination, whilst still warm, using a ⅛-inch power, not a trace of muriate of ammonia was visible. After ten minutes' exposure, another glass was removed ; the result was similar. After fifteen minutes, the third was examined ; now, there were traces of the salt in unmistakable crystals. A fresh plate with acid was now put on the first cup, and left on five minutes ; on evaporation, a distinct formation of the salt was found on the glass.

In further illustration of the rapid manner in which ammonia is formed and evolved in connection with the change which takes place in animal matter after deprivation of life, I shall mention another experiment, made when the temperature of the air was 64°. A portion of muscle, with a little fat (together equal in weight to 90·7 grs.), was taken from the leg of a lamb in less then ten minutes after it had been killed by the butcher ; the flesh was still warm. After fifteen minutes (the time taken in bringing it from the slaughter-house) it was put into a small low cup and covered, without being in contact, with a plate of glass moistened with dilute acid. After an hour, the glass was taken off and the acid evaporated ; a distinct trace of muriate of ammonia was left. The experiment was repeated ; the glass was left on for two hours and twenty minutes : now, the result, as shown by the crystals formed on evaporation, was still more strongly marked. It was again repeated without delay, and the glass was left on from 6.4 P.M., to 12 P.M. The formation of crystals now obtained on evaporation was copious ; they were, as seen with the ⅛-inch power, large and characteristic, and yet the meat was not apparently the least tainted ; it had undergone during the time—altogether about nine hours—no change of colour, and had not acquired the slightest unpleasant smell.

Need I point out the bearing of the results of these experiments on those of Dr Hammond ? If he delayed examining the blood for two or three hours, and more especially if his experiments were made during the summer, his finding ammonia in the vapour of fowl's blood is no more than might be expected.

Still, it may be said, it is open to question whether

ammonia, and in its volatile state, does not exist in the blood of other animals, and even in excessively minute quantity, in the blood of birds.

Reasoning from certain facts this seems probable. It is known in the instance of man that the air expired commonly contains a minute quantity of this alkali, leading to the inference that it is exhaled from the lungs, and previously existed in the blood. Dr Richardson, in the work already quoted, gives an account of many experiments, seemingly conclusive on this point. I may relate a few which I have made, employing the same method which he used. The subjects of the trials were persons of different ages, including infants, all in health. The expiration directed on the moistened glass was made through one nostril, the other and the mouth closed. Forty deep inspirations in my own case were commonly sufficient to afford distinct, though slight, traces of muriate of ammonia, as seen with the high power after the evaporation of the acid, and whilst the glass was still warm. Rarely, in any instance, have no traces of the salt been obtained. The indications, however, have been variable, sometimes stronger, sometimes feebler, as if depending on states of the system at different times of the day and under different conditions,—a variability which Dr Richardson also observed.*

---

* In the experiments described above, the precaution was taken of breathing through the nostril to avoid the risk of error which might occur were the respiration through the mouth : in the latter case, the result might be vitiated, and an erroneous inference made, were there a tooth undergoing decay, or a bit of meat adhering, from which, if in a state of incipient putrefaction, ammonia could not fail to be exhaled. When I have compared the breath expired from the lungs through the nostrils and through the mouth, I have found stronger indications in most instances of ammonia from the latter than from the former. I need hardly remark, it is so obvious to reason, that this test of ammonia may have a useful application in medical practice—*e g.*, in diagnosing the earlier and later stages of pulmonary consumption.

It may be worthy of mention, that the transparent limpid fluid which so often drops from the nose in cold weather (as it were by distillation) has had, as often as I have examined it, an alkaline reaction, and has afforded, after the addition of hydrochloric acid, on evaporation, distinct crystals of muriate of ammonia, mixed with which have been a few of common salt. The healthy saliva, as is well known, has the same reaction, and, similarly treated, affords crystals of the same kind as the preceding—tending to prove that ammonia in each instance is eliminated, and is derived from the blood.

I have obtained also traces of the volatile alkali from
the breath of other animals.   In the instances of the horse
and duck they were unmistakeable.   The trials were made
in the same manner as that described—the mouth and one
nostril closed.   The time occupied was about five minutes.
In the instance of the common fowl, no trace was detected ;
but as its nostril is very much smaller than that of the
duck, it is not so easy to direct the current of expired air
on the spot moistened with the acid.   The crystals obtained
from the breath of the horse were so large as to be distinctly seen with a glass of a quarter-inch power, yet were
hardly appreciable by weight.   These trials were made in
the open air.

Besides an exhalation of ammonia from the lungs, it is
also well ascertained that it is excreted by the skin.   It
has been found in sweat by Berzelius in the form of the
muriate, and has been detected by other inquirers.   Like
Dr Richardson, I have found it evolved even in insensible
perspiration.   Here is an instance :—When the thermometer
in my room was 70°, a slip of glass moistened with the
dilute acid was kept under the palm of the warm hand for
ten minutes, carefully avoiding contact; now, on examination, after evaporation of the acid, a trace of muriate of
ammonia was obtained in minute crystals, sufficiently distinct, and more than I could have expected.   This experiment I have repeated on myself and others with like result.
The warmer the weather, and the higher the temperature
of the surface, the larger commonly has been the proportion
of the salt formed.   It is noteworthy, that when the glass,
without the addition of the acid, had been exposed to
the insensible perspiration during the same length of time,
no trace of salt was detected on it; leading to the inference
that the ammonia exhaled is in the volatile form, and probably in union with carbonic acid.

Reflecting on these facts, it seemed probable that if
ammonia in the volatile state exists in the blood, it is likely
to be in a larger proportion in venous than in arterial
blood, on the supposition that it is exhaled from the lungs
with carbonic acid in the act of expiration.   To test this I
made trial of the two kinds of blood, one from the jugular

vein, the other from the carotid artery of a lamb about to be slaughtered. The weather at the time was warm, the thermometer in the shade 75°. The quantity of blood collected of each kind was about the same, the recipient vessels of like size, capable of holding about three ounces. The instant they were filled they were covered with a plate of glass partially moistened with the acid. After fifteen minutes an examination was made in the manner before described. On the glass exposed to the arterial blood no muriate of ammonia could be detected with the high power of the microscope; but on that exposed to the venous a trace of the salt was observed.

I shall now mention some other instances in which, using the same method, I have examined the blood in special quest of ammonia. The animals from which the blood was obtained were the common fowl, the duck, horse, sheep, heifer, calf, sea-trout (*Salmo trutta*), and toad. In all the experiments, in those already described and those which follow, due precautions were taken to avoid as much as possible error, and this both in relation to the acid used and the place where the trials were made; in many instances, for greater security, a comparative experiment was made under the same circumstances as to time and locality, with the acid alone and the blood alone. The results of these were negative.[*]

1. *Of the Common Fowl.*—When the temperature of the open air was 47°, a hen of about three years old was killed by the division of the great cervical vessels. The blood that first flowed was received into a wine-glass. It coagulated almost instantly, certainly in less than half a minute. A plate of glass moistened with acid was immediately placee over it. After about five minutes it was taken off, and the acid evaporated; no trace of ammonia could be detected on it, on inspecting it with the high power. The blood which flowed last, which also rapidly coagulated, and was nearly

---

[*] In my earlier experiments, those made in 1859, the test employed for the detection of ammonia was a glass rod, moistened with hydrochloric acid of reduced strength, which, as has been well pointed out by Dr Richardson, is less delicate, less to be relied on than that of the crystalline formation of muriate of ammonia, as seen with the microscope.

equal in quantity, was subjected to the same trial, and for about the same time ; it afforded a tolerably distinct trace of muriate of ammonia. Fresh glasses were applied, and it may be generally remarked, that the indications of the evolution of the volatile alkali increased pretty regularly from hour to hour.

2. *Of the Duck.*—When the open air was 40°, a duck of about three months old was killed in the same manner as the fowl. The blood, as it flowed, was received into two wine-glasses, and into a smaller glass, in each of which it was subjected to the same treatment as the preceding. In the first glass, it coagulated in about two minutes ; in the second, there was a slight retardation ; in the third, the retardation reached about ten minutes. The blood which flowed first was brightest. The plates of glass, moistened with acid, were examined after about ten minutes' exposure. On that from the first, not a trace of the salt could be detected under the high power ; on that from the second and third, a slight trace was visible ; on repetition, and four hours' exposure (this in a room the temperature of which was 50°, the first was in the open air), a trace of salt was obtained from each portion, strong from the second, and in proportion to quantity of blood ; also from the third, but very slight from the first. In this, as in the preceding example, the saline formation obtained by repetition of the trials increased, and the more rapidly as putridity advanced.

3. *Of the Horse.*—When the open air was 65°, three ounces of blood were taken from the jugular vein of a carriage-horse. A prepared plate of glass was instantly placed over it. After half an hour, when the blood had coagulated, and when the buffy coat that had formed was equal nearly in thickness to the crassamentum which had subsided, an examination for ammonia was made ; barely a trace of it was discernible, no crystals could be detected, merely numerous minute granules. On a fresh glass, after thirty minutes, a distinct trace was obtained, and in a crystalline form. The trial was repeated as many as four different times in the twenty-four hours. The saline formation was found to increase in quantity with the lapse of time. In the last, the crystals of muriate of ammonia, as

seen with the high power, were numerous and large, and yet the glass had increased in weight only ·01 grain.

4. *Of the Sheep.*—When the temperature of the air was about 70°, two portions of blood were obtained from a sheep of about three years old, one portion just after the great cervical vessels had been divided—this chiefly arterial; another when the flow of blood had become languid—this of a darker hue, and, it may be inferred, chiefly venous. The quantity of each was about three ounces; they were tested in the same manner as the preceding. The first examination was made after an hour. From the arterial blood a very few and minute crystals of muriate of ammonia were obtained; from the venous, more. The difference was well marked. After two hours, and again after sixteen hours, the trial was repeated. Each time there was an increase of the salt, and in a somewhat larger proportion from the venous than from the arterial.

5. *Of the Calf.*—The temperature of the air was about the same as the last mentioned, as was also the quantity of blood, which was the first that flowed. Examined after an hour, a distinct but very minute formation of muriate of ammonia was detected. Examined a second and a third time, after the same intervals as the sheep's blood, the results were very similar. The increase of weight after the last and longest interval did not exceed 0·01 grain.

6. *Of the Bullock.*—The air was about 65°. The quantity of blood collected was five ounces; it had the character of venous blood, having been obtained when the flow had nearly ceased. Examined after eight minutes, distinct crystals of muriate of ammonia were seen on the glass more than in any of the preceding trials. Again, examined at intervals during the twenty-four hours, the results were much the same as the preceding, the quantity of saline matter increasing with the length of time. On the last glass its proportion was greater than in any of the foregoing; the crystals were distinguishable by the naked eye.*

* Of these specimens of blood, the sheep's bore distinct marks of putridity, as indicated by smell and discoloration (reddening of the serum) somewhat earlier than the bullock's; the bullock's than the calf's and horse's. On the

7. *Of the Sea-Trout.*—When in Lewes of the Orkney islands, in the beginning of August, I availed myself of the capture of some fish of this kind in the sea with the net to examine the blood.   About half an ounce was collected from three fish by cutting the gills the instant they were taken out of the water.   After fifteen minutes' exposure, on evaporating the acid, a distinct trace of muriate of ammonia was observed, and rather more on a second trial after four hours' exposure.   The coagulum was soft and dark.

8. *Of the Toad.*—This trial was made in July.   The toad was of ordinary size, and vigorous, as it commonly is in this month.   Though the quantity of blood was small, it afforded a distinct trace of ammonia after an hour's exposure.   As the blood was obtained by decapitation, it was a mixture of venous and arterial.*

9. *Of the Fluid of the Allantoid of the Egg of the Common Fowl.*—In this fluid, when the fœtal chick had nearly reached its full time, I have detected ammonia.   The fluid was of sp. grav. 1016, of alkaline reaction ; on evaporation after the addition of hydrochloric acid, it yielded some minute crystals of muriate of ammonia.   In this stage of existence the chick may be considered as differing but little from the batrachian, the respiratory function being of no greater activity than suffices apparently for the organic changes essential to development, and nowise sufficient to preserve the temperature essential to the life of the fœtus.†

What are the conclusions to be drawn from these results ?   1. Do they not all tend to confirm the inference that the congulation of the blood is not owing to the escape

---

third day, the serum of the calf's blood was still colourless ; that of the horse's only very slightly coloured.

 * When stooping over the toad, a nauseous smell was perceived, and an acrid taste in the pharynx, followed by slight headache and malaise, which at the time I fancied might be owing to vapour from the body of the reptile.   The cutaneous glandular structure was in an active state, and distended with its peculiar acrid fluid.

 † It is very remarkable how rapidly the temperature of the young bird rises when its active respiration is established.   The following is an example :—The temperature of a gosling, in process of hatching, just after the end of the egg had been partially broken, was 94° ; two hours later it had risen to 104° ; after other two hours to 109° ; then its head was out of the shell, and the young bird was making muscular efforts to extricate the body.

of the volatile alkali ? 2. Are they not favourable to the idea that the blood generally contains a very small proportion of ammonia? 3. Do they not also favour the idea that the proportion of ammonia is larger in venous than in arterial blood ? 4. And do they not render it probable that in those animals in which the blood is least thoroughly aërated, such as the Batrachians and other allied genera, the proportion of this alkali is greater than in those animals, such as birds and mammalia, of higher temperature and more complete pulmonary respiration ?

Should these conclusions be admitted, it is not difficult to imagine the source of the volatile alkali, inasmuch as, irrespectively of the metamorphic changes which are in constant progress in the body at large, were we to confine the attention to the stomach and intestines, we might in them, in the ingesta, in the one during the formation of chyme, and in the other in the further changes going on in them, find the production of ammonia. Whenever I have examined either the contents of the stomach of any of the mammalia killed in health,* or the contents of the intestines, I have always detected it ; and that it should pass into the blood, may be easily credited, as the urine is seldom free from it, and often abounds in it.

On the Relative Effects of Acid and Alkaline Solutions on Muscular Action through the Nerve. By H. F. BAXTER, Esq., Cambridge.

The present inquiry originated during an investigation on the subject of Muscular Contraction ; and as some experimental conclusions have been based upon the so-called anelectro-tonic state of the nerve, it became of importance to ascertain whether this state—electro-tonic— depended upon any peculiar electrical condition, or whether the effects which indicate its existence might not be referable to changes, electrolytical, which take place in the nerve ; and the facts which I propose to consider are the

* See Physiological Researches, by the author, p. 339.

following. If a current of electricity be made to traverse a
portion of a nerve connected with the muscles, a difference
will be observed in the effects, as manifested by the muscular contractions, when the current is suspended. The
muscles of the limb whose nerve has been traversed by an
inverse current will contract *tetanically;* whilst the muscles,
if the current be *direct*, will contract but once—the tetanic
contractions being very rarely produced. These effects vary
according to the strength of the current and the time of its
passage. Matteucci,[*] in his papers on the " Physiological
Action of the Electric Current," came to the conclusion that
the passage of the electric current through a mixed nerve
produces a variation in the excitability of the nerve, differing essentially in degree, according to the direction of
the current through the nerve. This excitability being
weakened and destroyed, and that more or less rapidly, according to the intensity of the current, when it circulates
through the nerve from the centre to the periphery (direct
current). The excitability, on the contrary, being preserved and increased by the passage of the same current in
a contrary direction, that is to say, from the periphery
towards the centre (inverse current). But in a subsequent
paper [†] on the " Secondary Electro-Motor Power of Nerves,"
he refers the effects to a secondary electro-motor power, and
says, " the secondary current, the existence of which is
demonstrated by the galvanometer, and which is *direct* for
the nerve that has been traversed by the *inverse* current,
and which is also demonstrated by the contractions of the
galvanoscopic frog [placed upon the nerve], explains, according to the known laws of electro-physiology, the effects
produced by it on the opening of the circuit."

" If any part of a nerve," says Du Bois Reymond,[‡] " be
submitted to the action of a permanent current, the nerve,
in its whole extent, suddenly undergoes a material change in
its internal constitution, which disappears on breaking the
circuit as suddenly as it came on. This change, which is
called the Electro-tonic state, is evidenced by a new electro-
motive power, which every point of the whole length of the

* Phil. Trans., 1846, 1847.   † Ibid., 1861.
‡ On Animal Electricity.   Edited by H. Bence Jones, M.D., p 213.

nerve acquires during the passage of the current, so as to produce, in addition to the nerve-current, a current in the direction of the extrinsic current. As regards this new mode of action, the nerve may be compared to a voltaic pile, and the transverse section loses its essential import. Hence the electric effects of the nerve, when in the electro-tonic state, may also be observed in nerves without previously dividing them." '

On a former occasion* I endeavoured to ascertain whether the electric condition of the nerve might not, under these circumstances, be increased ; the results of my experiments, however, failed to give any evidence in support of that conclusion ; but, on the contrary, the nerve-current was destroyed, and I was led to suppose that the electric current occasioned a disorganisation of the nerve. The later researches of Matteucci clearly point out the cause of my failure, and confirm in a great measure the results I then attained.

In an elaborate article, published in the "British and Foreign Medico-Chirurgical Review" for July 1862, on general Nerve Physiology (German), it would seem that amongst the German physiologists this state of the nerve is still considered as being in a peculiar state (electro-tonic—electrotonus). Pflüger, who has worked at the subject to some extent, speaks of the portion of nerve connected with the cathode as being brought into the state of catelectrotone, and that the irritability of this portion is increased and rendered favourable for conducting ; whilst that portion of the nerve connected with the anode is brought into a state possessing the opposite properties, and is spoken of as being brought into a state of anelectrotone. Reasoning from the facts established by Matteucci, it appears to me that the two portions of the nerve called anelectrotone and catelectrotone, by Pflüger, correspond to the portions where the acid and alkaline compounds are developed by the electric current ; if so, the effects are evidently referable to the secondary electro-motor of the nerve as considered by Matteucci. To ascertain how far this supposition

---

* Edinburgh New Philosophical Journal, New Series, April 1858. Essay on Organic Polarity, chap. ix.

may be correct, it becomes necessary to obtain further evidence than that afforded by the use of the electric current, and for this purpose I employed chemical reagents in the following manner:—

It is reasonable to suppose, that by placing an acid and an alkaline solution on different portions of a nerve connected with the muscles, that some difference might be obtained in the contractions, depending upon the relative position of these solutions in regard to the muscles. In my first experiments the two limbs remained connected with the spinal cord, but the contractions excited in one limb, in consequence of reflex actions, rendered the results doubtful. The following plan was consequently adopted :— The two lower limbs being separated, a long portion of the sciatic nerve was dissected out and placed on two portions of bibulous paper, resting on distinct pieces of glass, and the leg rested on another distinct piece of glass. The bibulous paper, when well soaked, prevented the solutions from running over the surface of the glass: the distances between the place where the solutions were applied and the muscles could be varied at pleasure ; but the piece of glass upon which the limb rested was higher than the others, so as to keep the solutions from running down on to the muscles. To prevent the nerve from being moved during the contractions of the muscles, it was necessary to fasten the limb down, either by a ligature or by placing a weight on the thigh bone.

The limb in which the acid solution was nearest the muscle will be designated by the letter *a* ; that in which the alkaline solution was nearest to the muscle by the letter *b*. I need scarcely add that the nerves should be placed on distinct papers and glasses, otherwise a circuit would be formed between the two nerves, if resting on the same papers.

The acids comprised the sulphuric, nitric, muriatic, and the acetic acid. Three solutions of each were prepared, varying in strength, and consisting—No 1, of 1 part of acid to 10 of water ; No. 2, of equal parts of acid and water ; No. 3, of concentrated acid. The alkaline solutions, comprising those of potash, ammonia, and soda, were formed as follows:—The potash solution, No. 1, consisted of 1 part liq. potassæ to 10 of water; No. 2, of equal parts of liq. potassæ and water; No. 3, of pure liq. potassæ. The ammonia solution—No. 1, of 1 part liq. ammoniæ to 10 of water; No. 2, of equal parts of liq. ammoniæ and water; No. 3, of concentrated liq. ammoniæ. The soda solution— No. 1, of a scruple of bicarbonate of soda to one ounce of water; No. 2, of two drachms of the bicarbonate to one ounce of water. The acids and alkalies were those prepared according to the London Pharmacopœia.

Some care is requisite in applying the solutions. The best plan was to use a glass rod instead of a piece of glass tubing, or a glass pen, which could be easily cleaned; and as it was necessary to apply the solutions simultaneously, two rods were required. Or, instead of dropping the solutions *upon* the nerve, the solutions were placed on the papers and the nerve then dropped upon the solutions.

Having arranged the nerve of *a* and of *b* upon the papers, which were one-eighth of an inch in width, and the distance between the papers a quarter of an inch, whilst that of the muscle from the paper was half an inch, the limbs being firmly secured, the first experiments were for the purpose of ascertaining the effect of placing the solutions, not simultaneously upon the nerve, but one after the other. The nitric acid and potash solutions were employed. Upon placing the alkaline solution, No. 1, upon the nerves, a slight contraction in *a*, but none in *b*; No. 1 acid solution was now placed upon the nerves—no effect. The acid solution, No. 1, was placed upon the nerves—no effect. On placing the alkaline solution on the nerves, a slight contraction in *a*, but none in *b*. The No. 2 solutions were now employed in the same manner: with the alkaline solutions, contractions in both limbs, consisting of two strong contractions in *a* and four in *b*; upon placing the acid solutions on the nerves, three contractions were excited in *a*, but none in *b*. When the acid solutions were first placed upon the nerves, two contractions in *a*, none in *b*; but upon the addition of the alkaline solutions, three strong contractions in *b*, slight fibrillar contractions in *a*. With the solutions No. 3, with the alkaline, three strong contractions in *a*, four in *b*; on the addition of the acid, none in *b*, three strong ones in *a*. With the acid solutions, two in *a*, one in *b*; upon the addition of the alkaline, none in *a*, four strong contractions in *b*.

From the foregoing experiments we see—*First*, that the alkaline solutions excite stronger muscular contractions than the acid solutions; and *Secondly*, that the strong solutions, especially the acids, prevent the transmission of nervous impressions excited at the distal extremities of the nerve, which is no doubt due to the disorganisation of the nervous structure.

In the following experiments, the solutions were placed simultaneously upon the nerves, beginning with the weaker solutions.

EXPERIMENT I.—*With Nitric Acid and Potash.*

With No. 1 *solutions.*—A slight contraction in *a*, none in *b*.

With No. 2 *solutions.*—Two powerful contractions in *b*; slight contractions and more continued in *a*.

With No. 3 *solutions.*—Four powerful contractions in *b*, but only one in *a*.

EXPERIMENT II.—*With Nitric Acid and Ammonia.*

With No. 1 *solutions.*—A slight contraction in *b*, none in *a*.

With No. 2 *solutions.*—Slight fibrillar contractions in *b*, two powerful ones in *a*, and after a short time, slight fibrillar contractions.

With No. 3 *solutions.*—Three powerful contractions in *b*, and two in *a*.

I was much surprised at these results with the ammonia solutions, as Kühne* has stated that ammonia is a strong muscular excitant, and that the vapour of ammonia is sufficient to cause muscular contractions. On applying the strong solution to the surface of the muscle, slight contractions occurred, but with the weak solutions, I could not obtain any effect. I believe that a great many of the results which appear as contradictory arise from neglecting to state the season of the year in which the experiments are performed.

* Report on Muscular Contraction. By A. B. Duffin, M.D., in Beale's Archives of Medicine. Nos. 10 and 11. Lond., April 1862.

My present investigation was undertaken during the months of September and October; the weather at the time being cold, and the frogs were getting apparently into a torpid state, and certainly not so excitable as during the summer months. It is well known that during the spring months, at the time of spawning, the irritability of the frogs is so great, that it is difficult to perform any experiments upon them. The mere division of the spinal cord will throw the animal into a state of tetanus; and I have even found that removing them from the water, and handling them, will be quite sufficient to produce the same state.

Experiment III.—*With Nitric Acid and Soda.*

With No. 1 solutions.—No contractions either in *a* or *b*.
With No. 2 solutions.—Two contractions in *a*, one in *b*.
With No. 3 acid and No. 2 soda.—Two powerful contractions in *a*, and then slight fibrillar contractions; three powerful contractions in *b*.

Experiment IV.—*With Sulphuric Acid and Potash.*

With No. 1 solutions.—Slight contraction in *a*, none in *b*.
With No. 2 solutions.—One strong contraction in *a*, and then slight fibrillar contractions; two strong contractions in *b*, and slight fibrillar contractions, but they did not last so long as those in *a*.
With No. 3 solutions.—Two strong contractions in *b*, three strong contractions in *a*.

Experiment V.—*With Sulphuric Acid and Ammonia.*

With No. 1 solutions.—Slight fibrillar contractions in *b*, none in *a*.
With No. 2 solutions.—Two strong contractions in *b*, three strong contractions in *a*, and slight fibrillar contractions.
With No. 3 solutions.—Two strong contractions in *a*, three strong contractions in *b*, and then fibrillar contractions.

Experiment VI.—*With Sulphuric Acid and Soda.*

With No. 1 solutions.—A slight contraction in *a*, none in *b*.
With No. 2 solutions.—Two strong contractions in *a*, and then slight fibrillar contractions; one strong contraction in *b*.
With No. 3 acid and No. 2 soda.—Two strong contractions in *a*, three strong contractions in *b*.

EXPERIMENT VII.—*With Muriatic Acid and Potash.*

With No. 1 *solutions.*—A slight contraction in *a* and in *b*.

With No. 2 *solutions.*—One strong contraction in *a*, two contractions in *b*, and then fibrillar contractions.

With No. 3 *solutions.*—Two strong contractions in *a*, and then fibrillar contractions; three strong contractions in *b*.

EXPERIMENT VIII.—*With Muriatic Acid and Ammonia.*

With No. 1 *solutions.*—Slight fibrillar contractions both in *a* and *b*.

With No. 2 *solutions.*—One strong contraction in *a*, and then fibrillar; three strong contractions in *b*.

With No. 3 *solutions.*—One strong contraction in *a*, four in *b*.

EXPERIMENT IX.—*With Muriatic Acid and Soda.*

With No. 1 *solutions.*—No contraction either in *a* or *b*.

With No. 2 *solutions.*—Slight contraction in *a*, none in *b*.

With No. 3 *acid, No.* 2 *soda.*—One strong contraction in *a*, and then fibrillar; two in *b*.

EXPERIMENT X.—*With Acetic Acid and Potash.*

With No. 1 *solutions.*—No effect.

With No. 2 *solutions.*—Two strong contractions in *a*, one in *b*.

With No. 3 *solutions.*—One strong contraction in *a*, and then fibrillar; three strong contractions in *b*.

EXPERIMENT XI.—*With Acetic Acid and Ammonia.*

With No. 1 *solutions.*—A slight contraction in *a*, none in *b*.

With No. 2 *solutions.*—Two slight contractions in *b*, three strong contractions in *a*.

With No. 3 *solutions.*—Two strong contractions in *b*, one in *a*.

EXPERIMENT XII.—*With Acetic Acid and Soda.*

With No. 1 *solutions.*—No effect.

With No. 2 *solutions*—Slight contraction in *a*, none in *b*.

With No. 3 *acid, No.* 2 *soda.*—Two slight contractions in *a*, one powerful contraction in *b*.

The first conclusions to be deduced from the results of these experiments, are,—1st, That a difference in the position of the solutions on the nerve produces a difference of effect; 2d, That the contractions which result do not bear any relation, beyond a certain point, to the strength of the solutions, they do not increase with their increased strength ; and, 3d, That a difference is obtained with different solutions, the alkaline solutions producing a greater or more constant effect than those of an acid nature.

The next question which naturally arises for consideration is that respecting the mode of action of chemical reagents upon the nervous tissue, in causing muscular contractions. Do they act by reacting upon the electricity of the nervous tissue, or by reacting upon the compound forming the tissue, and thus indirectly, upon the molecular forces associated with it ; or in other words, upon nerve-force?

We have no reason for supposing that chemical reagents would act upon the nervous tissue in a different manner to that observed during ordinary chemical reactions, much less reason have we for believing that, under these circumstances, the vital property of the tissue, nerve-force, would be increased ; we must therefore consider what are the effects that take place during ordinary chemical actions, and I shall limit myself now to the electrical effects. In ordinary chemical actions, it is well known that during chemical combination the compound which performs the part of an acid takes positive electricity, and that of an alkali negative electricity, and the current of electricity which results therefrom goes from the alkali to the acid. When we apply our acid and alkaline solutions upon a nerve, the same effects are produced, which can be readily proved by the galvanometer. Now, the *direction* of the current, as is well-known, has a most important influence in causing contractions, the *direct* being far more influential than the *inverse*. I will first consider the action of a single solution upon a nerve, and believe it to be as follows:—We have first the chemical changes taking place between the chemical reagent and the nerve, during which changes contractions are excited in the muscle, and an electric current developed at the seat of chemical action; if an acid be employed, the current goes from the nervous tissue to the acid, if an alkali, from the alkali to the nervous tissue. The strength of the current would depend upon that of the solutions. Now, I do not think that when one solution alone is used, that the muscular contractions excited depend upon the electric current that is developed, but rather upon the changes which take place during the chemical reactions ; when the two solutions are placed upon the nerve, then the actions become more complex ; in addition to those acting upon the nerve, we have those arising from the combination between the solutions. If the solutions are weak, there may not be sufficient disorganisation of the nerve produced to prevent it from being able to conduct either nervous impressions, or much less a current of electricity, through the part where the solution has been applied ; but if concentrated, then disorganisation takes place, and the nerve becomes incapable of conveying its own impressions, and even perhaps the electric current, so far as to excite nervous action ; so that the conducting power of a nerve under these circumstances becomes a point of some consideration.

Let us consider for a moment the effects produced by using one solution only, as in my first experiments, and then by adding another. If the weak acid solution be placed near the muscle—no contraction ; let the alkaline solution be now placed on the distal side of the nerve—if of sufficient strength, it will cause contractions, in consequence of the acid not destroying the conducting power of the nerve to its own impressions ; but when the combination of the acid with the alkali takes place, we may then have contractions produced, in consequence of the current thus developed, which being *direct*, going from the alkali to the acid, is favourable for causing contractions. If the acid be too strong, the transmission of the nervous impression is prevented, and very likely the influence of the current also. Let the alkali be placed near the muscle, contractions occur ; now, place the acid upon the distal extremity,—no effect ; the current is now in the reverse direction—unfavourable for producing contractions. Let the two solutions be placed on the nerve, simultaneously, as in Experiment 1, the same reasoning will apply, and it is interesting to observe, that when the No. 2 solutions were applied, a difference in the nature of the contractions were observed, being slightly tetanic in the limb *a*. In other experiments also, when the acid was near the muscle, the fibrillar or slight tetanic contractions, were more frequently produced than in the other limb. It may be said that the acid being a stronger muscular excitant, the effects were due to the acid getting upon the muscle ; to avoid this a small quantity of oil was smeared over the surface of the muscle previous to applying the acid, so as to prevent its action, but the same effects were obtained.

The results of this inquiry go far to confirm the conclusion of Matteucci, that when an electric current has traversed a nerve and is then suspended, the tetanic contractions produced are due to a secondary electro-motor power established in the nerves. If the current be sufficiently strong, there is an electrolyzation of the nervous structure, and consequently, an acid developed at the anode and an alkali at the cathode ; as these tetanic contractions are principally observed in the limb in which the current has been *inverse*, it corresponds to the acid being developed near the muscle, and the alkali at the distal extremity ; and upon the suspension of the current, the current arising from the combination of the compound formed during the electrolyzation of the nervous tissue is then directed towards the muscle, as shown by Matteucci. That the tetanic contractions should be more prolonged after the passage of the electric current, than when the solutions are applied, as in these experiments, is what one would naturally expect, considering that the changes were, in the one case in the substance of the nervous tissue itself, and in the other, merely between the solutions. We have therefore no evidence for believing that the so-called electro-tonic* state of the nerve (electrotonus) is anything more than the secondary electro-motor power of the nerves induced by the electric current. The contractions of the muscles becoming, in fact, a galvanoscopic test of the chemical and electrical changes which take place in its own nerve.

* It is interesting to observe whence the origin of the term *electro-tonic* arose. Faraday employed it in his first series of papers (*Experimental Researches*, vol. i. p. 16) to indicate the peculiar state in which a wire was supposed to be brought, when subject to volta-electric or magneto-electric induction ; but he subsequently found that the supposed effects could be fully explained without admitting the electro-tonic state.

On the Antiquity of Man; a Review of "Lyell" and "Wilson." * By J. W. DAWSON, LL.D., F.R.S., F.G.S., Principal of M'Gill College and University, Montreal. Communicated by the Author.†

Questions of human origins have always been popular, and have been agitated in all sorts of forms. Next to the dread question of the unknown future, the long-buried past is one of the most attractive subjects of inquiry; and while the faith of the Christian rests for both on the statements of Holy Scripture, the imagination of the poetical or the superstitious, and the reason of the philosopher or the sceptic, have found ample scope for exercise. In our day, geological investigation on the one hand, and antiquarian and philological research on the other, have given an exact and scientific character to such researches, which, without detracting from their interest, has fitted them to attract a more sustained and systematic attention; hence the appearance of such works as those above named. One of these works is the summing up of the geological evidence in relation to the origin of man, by one of our greatest masters of inductive reasoning. The other is the effort of a skilful antiquarian and ethnologist to apply to the explanation of the primitive conditions of the old world the facts derived from the study of the more recent primitive state of the western hemisphere. Both books are very valuable. Their methods are quite different, and their results as well; and it may be truly said that the geologist might have profited by the labours of the western antiquarian, had he known of them in time; and that the antiquarian might have found some new problems to solve, and difficulties to remove, had he read the work of the geologist. For this, among other

* The Geological Evidences of the Antiquity of Man; with Remarks on Theories of the Origin of Species by Variation. By Sir Charles Lyell, F.R.S. 8vo, pp. 520, illustrated. London, John Murray; Montreal, Dawson Bros.

Pre-historic Man—Researches into the Origin of Civilisation in the Old and New World. By Daniel Wilson, LL.D., Professor of History in University College, Toronto. 2 vols. 8vo, pp. 488–499, illustrated. London, MacMillan & Co.; Montreal, Dawson Bros.

† From the "Canadian Naturalist," 1863.

reasons, it may be well to consider them together. It will be necessary for us in doing this to summarise the numerous and varied facts adduced, and the reasonings therefrom, and we shall follow the order employed by Sir C. Lyell, bringing in Dr Wilson's antiquarian lore to our aid as we proceed.

The great question to be noticed in this review is that of the connection of human with geological history. How far back in that almost boundless antiquity disclosed by the geologist has man extended? At what precise point of the geological scale was he introduced on the mundane stage; and what his surroundings and condition in his earlier stages? In answer to these questions, negative geological evidence, and some positive considerations, testify, without a dissenting voice, that man is very modern. All the evidences of his existence have, until the last few years, belonged exclusively to the recent or latest period of the geological chronology. Certain late observations would, however, indicate that man may have existed in the latter part of the Post-pliocene period, and may have been contemporary with some animals now extinct. Still the evidences of this, as well as its true significance, are involved in much doubt; partly because many of the facts relied on are open to objection, partly because of the constant accession of new items of information, and partly because the age of the animals whose remains are found with those of man, and the time required by the physical changes involved, are not certain.

To these questions Sir Charles addresses himself, with all his vast knowledge of facts relating to tertiary geology, and his great power of generalisation; and he has, for the first time, enabled those not in the centre of the discussions which have for a few years been carried on upon this subject, to form a definite judgment on the geological evidence of the antiquity of our species.

As a necessary preliminary, Sir Charles inquires as to the *recent* remains of man, including those which are prehistoric in the sense of antedating secular history, but which do not go back to the period of the extinct mammalia. He refers, in the first place, to the detailed researches of the

Danish antiquaries, respecting certain remains in heaps of oyster-shells found on the Danish coast (which appear to be precisely similar to those heaps accumulated by the American Indians on our coasts from Prince Edward Island to Georgia); and respecting similar remains found in peat bogs in that country. These remains show three distinct stages of unrecorded human history in Denmark :—1st, A *stone period*, when the inhabitants were small-sized men, brachykephalous or short-headed, like the modern Lapps, using stone implements, and subsisting by hunting. Then the country, or a considerable part of it, was covered by forests of Scotch fir (*Pinus sylvestris*). 2d, A *bronze period*, in which implements of bronze as well as of stone were used, and the skulls of the people were larger and longer than in the previous period; while the country seems to have been covered with forests of oak (*Quercus Robur*). 3d, An *iron period*, which lasted to the historic times, and in which beech forests replaced those of oak. All of these remains are geologically recent ; and except the changes in the forests, and of some indigenous animals in consequence, and probably a slight elevation of some parts of Denmark, no material changes in organic or inorganic nature have occurred.

The Danish antiquaries have attempted to calculate the age of the oldest of these deposits, by considerations based on the growth of peat and the succession of trees, but these calculations are obviously unreliable. The first forest of pines would, when it attained maturity, naturally be destroyed, as usually happens in America, by forest conflagrations. It might perish in this way in a single summer. The second growth which succeeded would in America be birch, poplar, and similar trees, which would form a new and tall forest in half a century; and in two or three centuries would probably be succeeded by a second permanent forest, which, in the present case, seems to have been of oak.* This would be of longer continuance, and would, independently of human agency, only be replaced by beech,

---

* The details of this process, as it occurs in America, will be found noticed in a paper by the writer in the " Edinburgh Philosophical Journal" for 1847. Such changes are constantly in progress in the American forests.

if, in the course of ages, the latter tree proved itself more suitable to the soil, climate, and other conditions. Both oak and beech are of slow extension, their seeds not being carried by the winds, and only to a limited degree by birds. On the other hand, the changes of forests cannot have been absolute or universal. There must have been oak and beech groves even.in the pine woods ; and the growing and increasing beech woods would be contemporary with the older and decaying oak forest, as this last would probably perish not by fire, but by decay, and by the competition of the beeches. In like manner, the growth of peat is very variable even in the same locality. It goes on very rapidly when moisture and other conditions are favourable, and especially when it is aided by wind-falls, drift-wood, or beaver-dams, impeding drainage and contributing to the accumulation of vegetable matter. It is retarded and finally terminated by the rise of the surface above the drainage level, by the clearing of the country, or by the establishment of natural or artificial drainage. On the one hand, all the changes observed in Denmark may have taken place within a minimum time of two thousand years. On the other hand, no one can affirm that either of the three successive forests may not have flourished for that length of time. A chronology measured by years, and based on such data, is evidently worthless.

Possibly a more accurate measurement of time might be deduced from the introduction of bronze and iron. If the former was, as many antiquarians suppose, a local discovery, and not introduced from abroad, it can give no measurement of time whatever ; since, as the facts so clearly detailed by Dr Wilson show, while a bronze age existed in Peru, it was the copper age in the Mississippi valley, and the stone age elsewhere ; these conditions might have co-existed for any length of time, and could give no indication of relative dates. On the other hand, the iron introduced by European commerce spread at once over the continent, and came into use in the most remote tribes, and its introduction into America clearly marks an historical epoch. With regard to bronze in Europe, we must bear in mind that tin was to be procured only in England and Spain, and in

the latter in very small quantity: the mines of Saxony do
not seem to have been known till the middle ages. We
must further consider that tin ore is a substance not me-
tallic in appearance, and little likely to attract the attention of savages; and that, as we gather from a hint of Pliny,
it was probably first observed, in the west at least, as stream
tin, in the Spanish gold washings. Lastly, when we place
in connection with these considerations, the fact that in
the earliest times of which we have certain knowledge, the
tin trade of Spain and England was monopolised by the
Phœnicians, there seems to be a strong probability that the
extension of the trade of this nation to the western Mediterranean, really inaugurated the bronze period. The only
valid argument against this, is the fact that moulds and
other indications of native bronze casting have been found
in Switzerland, Denmark, and elsewhere; but these show
nothing more than that the natives could recast bronze
articles, just as the American Indians can forge fish-hooks
and knives out of nails and iron hoops. Other considerations might be adduced in proof of this view, but the limits
of our article will not permit us to refer to them. The
important questions still remain: when was this trade commenced, and how rapidly did it extend itself from the sea-
coast across Europe? The British tin trade must have been
in existence in the time of Herodotus, though his notion of
the locality was not more definite than that it was in the
extremity of the earth. The Phœnician settlements in the
western Mediterranean must have existed as early as the
time of Solomon, when "ships of Tarshish" was the general
designation of sea-going ships for long voyages. How long
previously these colonies existed we do not know; but considering the great scarcity and value of tin in those very
ancient times, we may infer that perhaps only the Spanish,
and not the British, deposits were known thus early; or that
the Phœnicians had only indirect access to the latter.
Perhaps we may fix the time when these traders were able
to supply the nations of Europe with abundance of bronze
in exchange for their products, at, say 1000 to 1200 B.C., as
the earliest probable period; and probably from one to two
centuries would be a sufficient allowance for the complete

penetration of the trade throughout Europe; but of course wars or migrations might retard or accelerate the process; and there may have been isolated spots in which a partial stone period extended up to those comparatively modern times, when first the Greek trade, and afterwards the entire overthrow of the Carthaginian power by the Romans, terminated for ever the age of bronze, and substituted the age of iron. This would leave, according to our ordinary chronologies, at least ten or fifteen centuries for the post-diluvian stone period; a time quite sufficient, in our view, for all that part of it represented by such remains as those of the Danish coast, and the still more remarkable platform habitations, whose remains have been found in the Swiss lakes, and which belong properly to the *recent* period of geology. In connection with this, we would advise the reader to study the many converging lines of evidence derived from history, from monuments, and from language, which Dr Wilson shows, in his concluding chapter, to point to the comparatively recent origin of at least post-diluvian man. Let it be observed, also, that the attempts of Bunsen and others to deduce an extraordinarily long chronology from Egyptian monuments, and from the diversity of languages, have signally failed; and that the observations made by Mr Horner in the Nile alluvium are admitted to be open to too many doubts to be relied on.*

Before leaving 'the recent period, it is deserving of note that Sir C. Lyell shows on the best evidence, that in Scotland, since the building of the wall of Antoninus, an elevation of from twenty-five to twenty-seven feet has occurred both on the eastern and western coast, and consequently that the raised sea-bottoms containing canoes, &c., in the valley of the Clyde, supposed by some to be of extremely ancient date, were actually under water in the time of the Romans; a fact of which, but for their occupation of the country, we should have been ignorant.

From the recent period we pass, under the guidance of

* The chronology deduced from the Delta of the Tinière, which would give to the stone period an antiquity of 5000 to 7000 years, appears to us to be similarly defective; and the data assigned to human remains in the valleys of the Mississippi and Ohio, and the old reefs of Florida, still more so.

Sir Charles, to the Post-pliocene, geologically distinguished from the Recent by the fact that its deposits contain the bones of many great extinct quadrupeds ; as for instance the mammoth, *Elephas primigenius*, the woolly rhinoceros, *R. tichorhinus*, and others, heretofore (but it would seem on insufficient evidence) supposed to have disappeared before the advent of man. The evidence now adduced that prim-eval man was really contemporary with these creatures is manifold and apparently conclusive, and in the work before us is carefully sifted and weighed in all its bearings, much being rejected as inapplicable or uncertain. The evidences relied on are chiefly the following :—

1. Human remains found with those of extinct animals in caves in Belgium, in England, and elsewhere, in circum-stances which preclude the probability of their mixture by interments or other modern causes.

2. The finding of flint implements associated with bones of extinct animals in the valley of the Somme, and elsewhere.

3. A supposed sepulchral cave of this period discovered in the south of France. In addition to these there are many minor facts tending to the same conclusion, but with less distinctness.

It is impossible to give extracts which will convey any adequate idea of the facts adduced from the above sources, but the following paragraphs may serve as examples of some of them. They relate to evidence that man was con-temporary with extinct animals, afforded by caverns near Liége, explored by Dr Schmerling, and to the similar evi-dence obtained in the cave of Brixham in England.

" The rock in which the Liége caverns occur belongs generally to the Carboniferous or Mountain Limestone, in some few cases only to the older Devonian formation. Whenever the work of destruction has not gone too far, magnificent sections, sometimes 200 and 300 feet in height, are exposed to view. They confirm Schmerling's doctrine, that most of the materials, organic and inorganic, now filling the caverns, have been washed into them through narrow vertical or oblique fissures, the upper extremities of which are choked up with soil and gravel, and would scarcely ᴇver be discoverable at the surface, especially in so wooded a country.   Among the sections obtained by quarrying, one of the finest which I saw was in the beautiful valley of Fond du Forêt, above Chaudefontaine, not far from the village of Magnée ; where one of the rents communicating with the surface has been filled up to the brim with rounded and half-rounded stones, angular pieces of limestone and shale, besides sand and mud, together with bones, chiefly of the cave bear.   Connected with this main duct, which is from one to two feet in width, are several minor ones, each from one to three inches wide, also extending to the upper country or table-land, and choked up with similar materials. They are inclined at angles of 30° and 40°, their walls being generally coated with stalactite, pieces of which have here and there been broken off and mingled with the contents of the rents, thus helping to explain why we so often meet with detached pieces of that substance in the mud and breccia of the Belgian caves.   It is not easy to conceive that a solid horizontal floor of hard stalagmite should, after its formation, be broken up by running water ; but when the walls of steep and tortuous rents, serving as feeders to the principal fissures, and to inferior vaults and galleries, are en- crusted with stalagmite, some of the incrustation may readily be torn up when heavy fragments of rock are hurried by a flood through passages inclined at angles of 30° or 40°.

" The decay and decomposition of the fossil bones seem to have been arrested in most of the caves by a constant supply of water charged with carbonate of lime, which dripped from the roofs while the caves were becoming gradu- ally filled up.   By similar agency the mud, sand, and pebbles were usually consolidated.

" The following explanation of this phenomenon has been suggested by the eminent chemist Liebig.   On the surface of Franconia, where the limestone abounds in caverns, is a fertile soil in which vegetable matter is continually decay- ing.   This mould or humus, being acted on by moisture and air, evolves carbonic acid, which is dissolved by rain.   The rain-water, thus impregnated, permeates the porous lime- stone, dissolves a portion of it ; and afterwards, when the excess of carbonic acid evaporates in the caverns, parts with the calcareous matter and forms stalactite.   So long as water flows, even occasionally, through a suite of caverns, no layer of pure stalagmite can be produced ; hence the formation of such a layer is generally an event posterior in date to the cessation of the old system of drainage—an event which might be brought about by an earthquake causing new fissures, or by the river wearing its way down to a lower level, and thenceforth running in a new channel.

" In all the subterranean cavities, more than forty in number, explored by Schmerling, he only observed one cave, namely, that of Chokier, where there were two regular layers of stalagmite, divided by fossiliferous cave-mud. In this instance, we may suppose that the stream, after flowing for a long period at one level, cut its way down to an inferior suite of caverns, and, flowing through them for centuries, choked them up with debris ; after which it rose once more to its original higher level : just as in the Mountain Limestone district of Yorkshire some rivers, habitually absorbed by a ' swallow hole,' are occasionally unable to discharge all their water through it ; in which case they rise and rush through a higher subterranean passage, which was at some former period in the regular line of drainage, as is often attested by the fluviatile gravel still contained in it.

" There are now in the basin of the Meuse, not far from Liége, several examples of engulphed brooks and rivers : some of them like that of St. Hadelin, east of Chaudefontaine, which reappears after an underground course of a mile or two ; others, like the Vesdre, which is lost near Goffontaine, and after a time re-emerges ; some, again, like the torrent near Magnée, which, after entering a cave, never again comes to the day. In the season of floods such streams are turbid at their entrance, but clear as a mountain-spring where they issue again ; so that they must be slowly filling up cavities in the interior with mud, sand, pebbles, snail-shells, and the bones of animals which may be carried away during floods.

" The manner in which some of the large thigh and shank bones of the rhinoceros and other pachyderms are rounded, while some of the smaller bones of the same creatures, and of the hyæna, bear, and horse, are reduced to pebbles, shows that they were often transported for some distance in the channels of torrents, before they found a resting-place.

" When we desire to reason or speculate on the probable antiquity of human bones found fossil in such situations as the caverns near Liége, there are two classes of evidence to which we may appeal for our guidance. First, considerations of the time required to allow of many species of carnivorous and herbivorous animals, which flourished in the cave period, becoming first scarce, and then so entirely extinct as we have seen that they had become before the era of the Danish peat and Swiss lake dwellings : secondly, the great number of centuries necessary for the conversion of the physical geography of the Liége district from its ancient to its present configuration ; so many old underground channels, through which brooks and rivers flowed in the cave period, being now laid dry and choked up.

" The great alterations which have taken place in the shape of the valley of the Meuse and some of its tributaries, are often demonstrated by the abrupt manner in which the mouths of fossiliferous caverns open in the face of perpendicular precipices, 200 feet or more in height above the present streams. There appears also, in many cases, to be such a correspondence in the openings of caverns on opposite sides of some of the valleys, both large and small, as to incline one to suspect that they originally belonged to a series of tunnels and galleries, which were continuous before the present system of drainage came into play, or before the existing valleys were scooped out. Other signs of subsequent fluctuations are afforded by gravel containing elephants' bones at slight elevations above the Meuse and several of its tributaries. The loess also, in the suburbs and neighbourhood of Liége, occurring at various heights in patches lying at between 20 and 200 feet above the river, cannot be explained without supposing the filling up and re-excavation of the valleys at a period posterior to the washing in of the animal remains into most of the old caverns. It may be objected that, according to the present rate of change, no lapse of ages would suffice to bring about such revolutions in physical geography as we are here contemplating. This may be true. It is more than probable that the rate of change was once far more active than it is now. Some of the nearest volcanoes, namely, those of the Lower Eifel about sixty miles

to the eastward, seem to have been in eruption in Post-
pliocene times, and may perhaps have been connected and
coeval with repeated risings or sinkings of the land in the
basin of the Meuse.   It might be said, with equal truth,
that according to the present course of events, no series of
ages would suffice to reproduce such an assemblage of cones
and craters as those of the Eifel (near Andernach for example) ; and yet some of them may be of sufficiently modern
date to belong to. the era when man was contemporary with
the mammoth and rhinoceros in the basin of the Meuse.

" But although we may be unable to estimate the minimum of time required for the changes in physical geography
above alluded to, we cannot fail to perceive that the duration of the period must have been very protracted, and that
other ages of comparative inaction may have followed, separating the Post-pliocene from the historical periods, and
constituting an interval no less indefinite in its duration."

*   *   *   *   *   *   *

" As the osseous and other contents of Kent's Hole had,
by repeated diggings, been thrown into much confusion, it
was thought desirable, in 1858, when the entrance of a new
and intact bone-cave was discovered at Brixham, three or
four miles west of Torquay, to have a thorough and systematic examination made of it.   The Royal Society made
two grants towards defraying the expenses,* and a committee
of geologists was charged with the investigations, among
whom Mr Prestwich and Dr Falconer took an active part,
visiting Torquay while the excavations were in progress
under the superintendence of Mr Pengelly.   The last-mentioned geologist had the kindness to conduct me through
the subterranean galleries after they had been cleared out
in 1859; and I saw, in company with Dr Falconer, the
numerous fossils which had been taken from the subterranean fissures and tunnels, all labelled and numbered, with
references to a journal kept during the progress of the work,
and in which the geological position of every specimen was
recorded with scrupulous care.

" The discovery of the existence of this suite of caverns

---

* When these grants failed, Miss Burdett Coutts, then residing at Torquay,
liberally supplied the funds for completing the work.

near the sea at Brixham was made accidentally, by the roof of one of them falling in. None of the five external openings now exposed to view in steep cliffs or the sloping side of a valley, were visible before the breccia and earthy matter which blocked them up were removed during the late exploration. According to a ground-plan drawn up by Professor Ramsay, it appears that some of the passages which run nearly north and south are fissures connected with the vertical dislocation of the rocks, while another set, running nearly east and west, are tunnels, which have the appearance of having been to a great extent hollowed out by the action of running water. The central or main entrance, leading to what is called the Reindeer gallery, because a perfect antler of that animal was found sticking in the stalagmitic floor, is 95 feet above the level of the sea, being also about 60 above the bottom of the adjoining valley. The united length of the five galleries which were cleared out amounted to several hundred feet. Their width never exceeded 8 feet. They were sometimes filled up to the roof with gravel, bones, and mud; but occasionally there was a considerable space between the roof and floor. The latter, in the case of the fissure-caves, was covered with stalagmite, but in the tunnels it was usually free from any such incrustation. The following was the general succession of the deposits forming the contents of the underground passages and channels:—

" 1*st*, At the top, a layer of stalagmite, varying in thickness from 1 to 15 inches, which sometimes contained bones, such as the reindeer's horn, already mentioned, and an entire humerus of the cave-bear.

" 2*dly*, Next below, loam or bone-earth, of an ochreous-red colour, from 1 foot to 15 feet in thickness.

" 3*dly*, At the bottom of all, gravel with many rounded pebbles in it, probed in some places to the depth of 20 feet · without being pierced through, and, as it was barren of fossils, left for the most part unremoved.

" The mammalia obtained from the bone-earth consisted of *Elephas primigenius*, or mammoth ; *Rhinoceros tichorhinus* ; *Ursus spelæus* ; *Hyæna spelæa* ; *Felis spelæa*, or the cave-lion ; *Cervus tarandus*, or the reindeer ; a species

of horse, ox, and several rodents, and others not yet determined.

"No human bones were obtained anywhere during these
excavations, but many flint knives, chiefly from the lowest
part of the bone-earth; and one of the most perfect lay at
the depth of 13 feet from the surface, and was covered with
bone-earth of that thickness.   From a similar position was
taken one of those siliceous nuclei, or cores, from which
flint flakes had been struck off on every side.   Neglecting
the less perfect specimens, some of which were met with
even in the lowest gravel, about fifteen knives, recognised
by the most experienced antiquaries as artificially formed,
were taken from the bone-earth, and usually from near the
bottom.   Such knives, considered apart from the associated
mammalia, afford in themselves no safe criterion of antiquity, as they might belong to any part of the age of stone,
similar tools being sometimes met with in tumuli posterior
in date to the era of the introduction of bronze.   But the
anteriority of those at Brixham to the extinct animals is
demonstrated not only by the occurrence at one point, in
overlying stalagmite, of the bone of a cave-bear, but also by
the discovery at the same level in the bone-earth, and in
close proximity to a very perfect flint tool, of the entire left
hind-leg of a cave-bear.   This specimen, which was shown
me by Dr Falconer and Mr Pengelly, was exhumed from
the earthy deposit in the reindeer gallery, near its junction
with the flint-knife gallery, at the distance of about 65 feet
from the main entrance.   The mass of earth containing it
was removed entire, and the matrix cleared away carefully
by Dr Falconer, in the presence of Mr Pengelly.   Every
bone was in its natural place, the femur, tibia, fibula, ankle-
bone, or astragalus, all in juxtaposition.   Even the patella
or detached bone of the knee-pan was searched for, and not
in vain.   Here, therefore, we have evidence of an entire
limb not having been washed in a fossil state out of an older
alluvium, and then swept afterwards into a cave, so as to be
mingled with flint implements, but having been introduced
when clothed with its flesh, or at least when it had the
separate bones bound together by their natural ligaments,
and in that state buried in mud.

" If they were not all of contemporary date, it is clear from this case, and from the humerus of the *Ursus spelæus*, before cited as found in a floor of stalagmite, that the bear lived after the flint tools were manufactured, or in other words, that man in this district preceded the cave-bear."

Multitudes of questions arise out of these observations, and many of them will probably long remain unanswered; but we may, in the remainder of this article, profitably restrict ourselves to three of them.

1. What style of men were these contemporaries of the mammoth, as compared with those who now walk the earth?

2. How great is their antiquity?

3. What bearing have the conclusions which we must form on these points, on the facts known to us on other evidence than that of geology, as to the origin and early history of man?

The writer of these pages, on a former occasion, ventured to predict that if any osseous remains of antediluvian man should be discovered, they would probably present characters so different from those of modern races that they might be regarded as belonging to a distinct species.* With perhaps one exception, this anticipation has not yet been realized. The skull from the cave of Engis, in Belgium, supposed to be the oldest known, is in the judgment of Professor Huxley, not by any means abnormal, but on the contrary, not unlike some European skulls. Another skull, that of Neanderthal, not found with remains of extinct animals, and therefore of uncertain geological antiquity, has, however, excited more attention than the Engis skull. Its prehistoric antiquity has been assumed by many writers, and its low forehead, prominent superciliary ridges, and general flatness, giving a more ape-like air than that of the heads of any modern tribes, together with the great stoutness and strong muscular impressions of the bones found with it, have been regarded as confirmatory evidence of this supposition. It is quite certain, however, that the characters for which this skeleton is eminent, are found, though perhaps in a less degree, in the rude tribes of

* Archaia, p. 237.

America and Australia. It is also doubtful whether this
skeleton really indicates a race at all. It may have belonged to one of those wild men, half crazed, half idiotic,
cruel and strong, who are always more or less to be found
living on the outskirts of barbarous tribes, and who now
and then appear in civilized communities, to be consigned
perhaps to the penitentiary or to the gallows, when their
murderous propensities manifest themselves. Still, as we
shall show under our third head, this Neanderthal man is
nearer in some respects to our historical idea of antediluvian
man than any other of these very ancient examples ; though,
as Lyell properly suggests, there is no absolutely valid reason
for assuming that he may not even have belonged to the
same nation with the Engis man ; since nearly as great
differences are found in the skulls of individual members of
some unmixed savage races.

One remarkable conclusion, however, deducible from the
answer to this our first question, must not be omitted. Of
all the criteria for the distinction of races of men, the skull
is probably the most certain, and, as any one may perceive
who reads Dr Wilson's book, it affords, in really reliable
hands, the best possible evidence of distinctness or of unity,
except where great mixtures have occurred. Now man is
one of the most variable animals ; and yet it would seem
that, since the Post-pliocene period, he has changed so little
that the skulls of these Post-pliocene men fall within the
limits of modern varieties ; and this, while so great changes
have occurred that multitudes of mammals, once his contemporaries, have utterly perished. Now, if these men are
so ancient as many geologists would assume, nay, if they
are even 6000 years old, surely the human race is very permanent, and Professor Huxley may well say that "the
comparatively large cranial capacity of the Neanderthal
skull, overlaid though it may be with pithecoid bony walls,
and the completely human proportions of the accompanying
limb-bones, together with the very fair development of the
Engis skull, clearly indicate that the first traces of the
primordial stock whence man proceeded need no longer be
sought by those who entertain any form of the doctrine of
in the newest tertiaries, but that they may

be looked for in an epoch more distant from the age of the *Elephas primigenius* than that is from us." They may, in short, spare themselves the trouble of looking for any such transition from apes to men in any period ; for this great lapse of time renders the species practically permanent ; more especially when we hear in mind that of the numerous species whose remains are found with those of these ancient men, some have continued unchanged up to our time, and the rest have become extinct, while not one can be proved to have been transmuted into another species.

Sir Charles devotes no less than five concluding chapters to this doctrine of transmutation, as held by Darwin and others. He does not commit himself to it, but wishes to give it due consideration, as a possible hypothesis, which may at least lead to great truths. We are not disposed to give it quite so high a position. Mr Darwin's book impressed us with the conviction that his hypothesis really explains nothing not otherwise explicable, and requires many assumptions difficult of belief; while the whole argument in its favour is essentially of the nature of reasoning in a circle. The point to be proved is, that variations arising from external influences and "natural selection" may produce specific diversity. Now, in order to begin our proof of this, we require at least one species, with all its powers and properties, to commence with. This being granted, we proceed to show that it may vary into several races, and that these races, if isolated, may be kept distinct and perpetuated. We further proceed to show that these races differ so much, that if wild, and not tampered with, we might suppose them originally distinct. So far all goes well with our demonstration ; but we find that many of the differences of these races are of the nature of mere monstrosities, like the six fingers of some men, which, as far as they go, would exclude the individuals having them, not only from their species, but from their order or class. Further, we find that the differences which do resemble those of species, have not, when tried by the severe test of crossing, that fixity which appertains to true specific differences ; so that with due care all our races can be proved to belong to but one species. Thus our whole argument falls to the ground; unless we are content quietly to assume the thing to be proved, and to say, that after showing that some species are very variable, we have established a certain probability that they may overpass the specific limits; though the fact that with all this variability, no species has been known practically to overpass these limits, should logically bring us to the opposite conclusion, viz., that the laborious investigations of Mr Darwin have more than ever established the fixity of species, though they have shown reason to believe that many so-called species are mere varieties.

Applying this to man, and even admitting, what Sir Charles Lyell very properly declines to admit, that the differences between men and apes are in all respects only differences of degree, and further admitting with Professor Huxley, that the difference between the size of the brain in the highest and lowest races of men is greater than the differences between the latter and the highest apes, nothing would be proved towards the doctrine of transmutation; for all these variations might occur without the ape ever overleaping the dividing line between it and the man; and the one fact to be proved is that this overleap is possible.

Perhaps this question as to man and apes, which some recent transmutationists have started, is one of the most damaging aspects of the doctrine, since it shows better than other cases the essential absurdity of supposing the higher nature to be evolved out of the lower; and thus startles the common sense of ordinary readers, who might detect little that is unreasonable in the transmutation of an oyster into a cockle, or even of a pigeon into a partridge; more especially if the reader or auditor is enabled to perceive the resemblance of type between these creatures, without receiving the further culture necessary to appreciate specific and generic difference, and thus is made ready to believe that similarity of type means something more than similar plan of construction. It is very curious too to observe, that while these theorists seize on occasional instances of degraded individuals in man as evidence of *atavism* reverting to a simian ancestry, they are blind to the similar ex-

planation which those who hold an opposite view may give
to the cases of superior minds appearing in low races, in
which the transmutationists can see nothing but spontaneous elevation.   It is also deserving of a passing remark,
that while, as Dr Gray shows, the doctrine of transmutation
is not subversive of all natural theology—that is, so long as
transmutationists admit the presiding agency of a spiritual
Supreme Being—the application of such views to the human
species attacks leading doctrines of that biblical Christianity which is practically of so much higher importance
to man than mere natural theology.

Still some of our modern naturalists follow with as much
pertinacity these transmutation hypotheses, as did the old
alchemists their attempts to transmute chemical species
into each other.   Perhaps the comparison is hardly fair to
the older school of speculators, for chemical species or
elements tend by their combination to form new substances,
which animal and vegetable species do not ; and by so
much the balance of antecedent probability was on the side
of the alchemists, as compared with the transmutationists,
though their methods and doctrines were very similar.
We may, however, at least hope that, like the researches
of the alchemists, those of their successors may develop
new and important truths. ˙ Leaving then this much vexed
topic, let us proceed to our second inquiry, as to the actual
antiquity of these primitive men.

This antiquity is of course to be measured by the geological scale of time, whose periods are marked not by years or
centuries, but by the extinction of successive faunas and
floras and the progress of physical changes.˙   With respect
to the first of these marks of time, we confess that we have
not regarded the observations of Boucher de Perthes and
others as free from the suspicion that accidental mixtures
of human and fossil bones, or other causes not taken into
the account, may have vitiated their conclusions ; and this
suspicion still applies to some of the cases cited by Sir C.
Lyell, as more or less certain proofs.   After reading the
statements of the present volume, we think the Belgian and
Brixham caves may be taken as good evidence of the probable contemporaneousness of man with the *Elephas primi-*

genius, *Rhinoceros tichorhinus, Ursus spelæus*, and their contemporaries; or rather as evidence that man was beginning to appear in Western Europe before those animals had finally disappeared. In consequence of some flaws in the evidence, as it appears to a reader at a distance, we cannot as yet so implicitly receive the evidence of the Somme flint weapons. The cave of Aurignac described by M. Lartet, and in which seventeen human skeletons were found buried, apparently in a sitting posture, cannot be relied on, owing to the late period at which it was explored. We are sorry to doubt this unique instance of antediluvian sepulchral rites; but all the appearances *actually seen* * by M. Lartet are better explicable on the supposition that a cave, once tenanted by the cave-bear and hyæna, had been partially emptied of its contents by some primitive tribe, who had broken up the bones of the extinct animals, not for their marrow, but to make tools and ornaments of them, and had subsequently used the cave as a place of burial, and the ground in front of it for " feasts for the dead." If the bones are still so perfect as M. Lartet asserts, they must have been quite sound when first disturbed at the early historical time in which the cave may have been ransacked. Further, the skeleton of *Ursus spelæus* found in the interior was below the place of deposit of the human skeletons; and we can suppose it to have been contemporaneous only by the unlikely theory that the earth containing this skeleton was placed in the cave by the aboriginal people.†

We give the above leading cases as examples of the rest which are cited, and all of which may in like manner be divided into those which afford probable, though not absolutely certain evidence of Post-pliocene man, and those which are liable to too grave suspicion to be accepted as evidence. It may be said that we should be more ready to believe, and less critical, but it is not the wont of geologists to be so,

* We refer to M. Lartet's account of his discovery in the " Natural History. Review," as well as the more concise statement given in the book before us.

† It is certainly very curious that the objects and arrangements of these caves, and other ancient European depositories, are so thoroughly American, even to the round stone-hammers, whose use is so oddly misinterpreted by the Danish antiquaries.

when new facts and conclusions are promulgated ; and the present case involves too important consequences, both in relation to history, and to the credibility of geological proof in general, to escape the most searching criticism. Geologists must beware lest their science, at the point where it comes into contact with other lines of investigation, and where its own peculiar methods are most liable to err, should be found wanting, and its reliability fall into discredit.

But when did the fossil mammals named above really become extinct? As a preliminary to our answer to this question, we may state that in Western Europe, in the Post-pliocene period of geologists, these animals were contemporary with many still extant, some of them in Europe, others elsewhere. Pictet even maintains that all, or nearly all of our modern European mammals co-existed with these animals in the Post-pliocene period, and that consequently there has since that time been a progressive diminution of species down to the present day. The mammoth, *Elephas primigenius*, existed, or perhaps began to exist, at a still more ancient period,— the newer Pliocene ; when it was contemporary with *Elephas meridionalis*, and other animals of an older fauna. It continued to survive until the introduction of the modern mammals, and then became extinct along with *Rhinoceros tichorhinus* and several other species, which, however, may have been of younger date than itself. With these species lived the *Megaceros Hibernicus*, or great Irish stag, which lasted longer, but perished before the dawn of history. With them also lived the *Bos primigenius*, or gigantic wild-ox, the aurochs, the musk-ox, and the rein-deer. The first of these existed wild until the time of Cæsar ; the second is still preserved in a forest in Lithuania ; the third exists now in Arctic America ; and the fourth still remains in Lapland. With them also co-existed the wolf, the fox, the hare, the stag, and other creatures still living in Western Europe.

That these creatures have been disappearing at different times seems certain ; some may have been exterminated by man, but the greater part must have perished from other causes. They may have gone *seriatim*, or in considerable numbers at or near the same time ; and there seems some reason to believe in a considerable and rapid decadence at the end of the Post-pliocene and beginning of the recent period. One cause which may be assigned is change of climate. The climate of Europe in the time of the mammoth was very cold, as indicated by the evidence of glaciers, and other forms of ice action, and by the presence of the musk ox. No doubt the extinction of this creature, and of the mammoth and tichorhine rhinoceros as well, would follow from the amelioration in this respect as the recent period approached. This change of climate depended on geographical changes, modifying the distribution of land and water, and the direction of ocean currents. A subsidence in Central America or in Florida might restore the climate of the mammoth by altering the course of the Gulf-stream ; and an elevation of land in these regions may have introduced the climate of the recent period. There is abundant evidence that much subsidence and elevation did occur while these changes in organic life were in progress ; and these may, more directly, by the submergence or elevation of large areas in Europe itself, have tended to extinguish species, or introduce them from other regions. All these points being granted, and abundant evidence of them will be found given by Sir C. Lyell, it remains to ask, can we convert the period required for these changes into solar years ? There is but one way of doing this in consistency with the principles of modern geology, and this is to ascertain how long a time would be occupied by agencies now in operation in effecting the changes of elevation, subsidence, erosion, and deposit, observed. Reasoning on this principle, it is plain that a vast lapse of time will be required, and that we may place the earliest men and the latest mammoths at an almost incredible distance before the oldest historical monuments of the human race.

But can we assume any given rate for such changes ? Not certainly till all the causes which may have influenced them can be ascertained and weighed. We have only recently learned that Scotland has risen 25 feet in 1700 years ; but we do not know that this elevation has been uniform and continuous. There is another older sea-level at 44 feet above the present coast ; and there is a still higher sea-level 524 feet above the sea, which certainly goes back to the time of the mammoth. Now we may calculate that if an elevation of 25 feet requires 1700 years, an elevation of 500 feet will require twenty times that length of time ; but if we should find, on further investigation, that 10 of the 25 feet were raised in the first century of the seventeen, and that the rate had gradually decreased, our calculation must be quite different, and even then might be altogether incorrect, since there may have been periods of rest or of subsidence ; so that " such estimates must be considered in the present state of science as tentative and conjectural."

Again, at the rate in which the Somme, the St. Lawrence, and the Mississippi now cut their channels and deposit alluvium, we can calculate that several tens of thousands of years must have elapsed since the mammoth roamed on their banks ; and we have been accustomed to rest on these calculations as close approximations to the truth : but Sir Charles Lyell has, in his present work, introduced a new and disturbing element, in the strong probability which he establishes that the cold of the glacial period extended to a later time than we have hitherto supposed. If, when the gravels of the Somme were deposited, the climate was of a sub-arctic character, we have to add to modern eroding causes the influence of frost, greater volume of water, spring freshets, and ice-jams, and the whole calculation of time must be revised. So if it can be proved that when the St. Lawrence began to cut the ravine of Niagara, in the Post-pliocene or New Pliocene period, there were great glaciers in the basin of Lake Superior, all our calculations of time would be completely set at nought.

Such are the difficulties which beset the attempt to turn the monumental chronology of geology into years of solar time. The monuments are of undeniable authenticity, and their teachings are most valuable, but they are inscribed with no record of human years ; and we think geologists may wisely leave this matter where the Duke of Argyll, in his address to the Royal Society, lately placed it, as a doubtful point, in so far as geological evidence is concerned, whether the mammoth lived later than we have hitherto supposed, or man lived earlier. Still, as we have already stated, those geologists who hold that we must reason inflexibly on rates of change indicated by modern causes, will necessarily, on the evidence as it now stands, maintain that the human race, though recent geologically, is of very great antiquity, historically.

We must now shortly consider our third question, as to the bearing of these facts and doctrines on our received views of human chronology, derived from the Holy Scriptures and the concurrent testimony of ancient monumental and traditional history. It is certain that many good and well-meaning people will, in this respect, view these late revelations of geology with alarm ; while those self-complacent neophytes in biblical learning who array themselves in the cast-off garments of defeated sceptics, and, when treated with the contempt which they deserve, bemoan themselves as the persecuted representatives of free thought, will rejoice over the powerful allies they have acquired. Both parties may however find themselves mistaken. The truth will in the end vindicate itself ; and it will be found that the results of such careful scrutiny of nature as that to which naturalists now devote themselves, are not destined to rob our race either of its high and noble descent, or its glorious prospects. In the mean time, those who are the true friends of revealed truth will rejoice to give free scope to legitimate scientific investigation, trusting that every new difficulty will disappear with increasing light.

The biblical chronology, though it allows an unlimited time for the prehuman periods of the earth's history; fixes the human period within narrow limits, though it does this not by absolute statement of figures, but rather by inference from chronological lists, with respect to the computation of which there may be and has been some difference, especially in the antediluvian period. Allowing large latitude for these differences, we have, say 2000 to 3000 years for the human antediluvian period, corresponding, it is to be supposed, to the later Post-pliocene of geologists. In this period men may have extended themselves over most of the old continent ; and it has been calculated that they may have been nearly as numerous as at present, but this is probably an exaggeration. They had, locally at least, domesticated animals ; they had discovered the use of the metals, and invented many useful arts, though there must have been a vast, scattered, barbarous population. They had split into two distinct races, some portions of which at least had sunk to a state presumably lower than that of any modern tribe, since these latter are all amenable to the influences of civilisation and Christianity, while the former seem to have been hopelessly depraved and degenerate. At the same time they had much energy for aggression and violence ; and it would seem that these giants of the olden time were in process of extinguishing all of the civilisation of the period when they were overwhelmed with the deluge. This is described in terms which may indicate a great subsidence, of which the Noachian deluge was the culminating point, in so far as Western Asia was concerned. The subsidence, unless wholly miraculous, may have commenced at least at the beginning of the 120 years of Noah's public life, and possibly much earlier, and the re-elevation may have occupied many centuries, and may not have left the distribution of land and water, and consequently climate, in the same state as before. At a very early period of this subsidence, if there were men in Europe, they would be perfectly isolated from the original seats of population in Asia, and so would the land animals, their contemporaries. There is, further, nothing in the Mosaic account to prevent us from supposing that the existence of many species was terminated by this great catastrophe.* These are some of the conditions of the biblical deluge, which we might much further illustrate, were this a proper place for doing so, but those stated will suffice to show precisely in what points the new doctrines of geologists in regard to the antiquity of man appear to conflict with this old narrative.

When we carefully consider the geological facts, in so far as they have been ascertained, it seems to us that the discrepancy may be stated thus. Reasoning on the geological doctrine that all things are to be explained by modern causes, and insisting on a rigid application of that doctrine, we must infer that the date of the introduction of

---

* See "Archaia," pp. 216 *et seq.*, and pp. 238 *et seq.* ; King's " Geology and Religion, " " Deluge."

man was "many ten thousands of years" ago. Adopting
the biblical theory, so to speak, that a great subsidence, of
which modern history affords no example, has occurred
within the human period, we might adopt a very much
shorter chronology. It would seem at present that the
facts can be explained on either view, and that the possibility of reconciling these views must depend on the greater
or less evidence which geologists may find of more rapid
changes than they have heretofore supposed within the
human period. Our own impression, derived from a careful
study of all the facts so well stated by Sir C. Lyell, is, that
the tendency will be in this direction, that the apparent
antiquity of the comparatively insignificant deposits containing remains of man and his works will be reduced, and
that a more complete harmony than heretofore between the
earliest literary monuments of the human race and geological chronology will result. At present the whole inquiry is making rapid progress, and the time may perhaps
be not far distant when its difficulties will receive some
such solution. In the meantime, both of the writers
whose works are noticed in this article deserve careful
study, and will be found to contribute much toward the
solution of these great questions.

---

*The Observed Motions of the Companion of Sirius con-
sidered with Reference to the Disturbing Body indicated
by Theory.* By T. H. Safford, Assistant at the Observatory of Harvard College. Communicated by the
Author.*

It is well known to astronomers that the motions of the
bright star Sirius indicated the presence of a disturbing
body, before the discovery of a companion by Mr Alvan
Clark. It was shown by Bessel,[†] that there were irregularities in the motion of this star in right ascension which
were only to be explained by the presence of an unseen

---

* From the Proceedings of the American Academy of Arts and Sciences,
vol. vi.
† Astronomische Nachrichten, Nos. 514–516.

companion, unless, indeed, we might permit ourselves to doubt the universality of the law of gravitation. C. A. F. Peters,[*] some years later, computed such of the elements of the motion of Sirius around the centre of gravity of the system as could be deduced from the motions in right ascension; and Schubert[†] pointed out that there was some reason to believe that the motion in declination also was irregular, though he seems to have fallen into the error of supposing that the motions in right ascension and declination were not completed in the same period.

Afterwards M. Laugier,[‡] of the French Institute, represented the observations of Sirius in declination from 1690 to 1852 by a formula of interpolation which I fear we must consider erroneous. Laugier gives a certain weight to Flamsteed's position from the "Historia Cœlestis Britannica," which is known to have been reduced (and probably from a single observation), without regard to aberration or nutation; so that it cannot be depended upon within 15″, while the real irregularities of Sirius's motion in declination are less than 2″.

Calandrelli,[§] Director of the Pontifical Observatory [‖] at Rome, has in several places insisted that the Greenwich Twelve-Year Catalogue was in error by about 3″ for the date 1845. This, however, was shown by Main [¶] to be contradicted by the several years' work, and I presume most astronomers would agree in considering Calandrelli's argument as irrelevant.

In No. 28 of Professor Brunnow's valuable "Astronomical Notices," I have shown that, in spite of the misapprehensions to which I have just alluded, the observed motion of Sirius in declination is in fact represented by a formula depending on the previous investigation of Peters, but with four new unknown quantities inserted. The ad-

[*] Astronomische Nachrichten, Nos. 745–747.

[†] Astronomical Journal, vol. i. p. 154.

[‡] Astronomische Nachrichten, No. 1142.

[§] Atti dell' Accademia Pontificia de' Nuovi Lincei, 5 Aprile, 1853, p. 816, and elsewhere.

[‖] This is not to be confounded with the observatory of the Collegio Romano.

[¶] Monthly Notices of the Royal Astronomical Society vol. xx. p. 202.

dition of these four quantities, which I have determined by least squares, enables us to state with a certain degree of accuracy the angle of position of the centre of gravity with respect to the visible mass, and thus the angle of position of the supposed invisible companion.

Closely following the actual publication * of this memoir, came the discovery † of the companion by Mr Clark. The question at once arose, whether this were the disturbing body; the evidence bearing upon this appeared very note-worthy. In the first place, the angle of position agreed (within the uncertainty of observation) with that computed for the disturbing body, assuming my investigation ‡ as the basis. The following table shows the relation for 1862 between computation and observation. To my own computation I have added the similar one of Auwers, published afterwards.§

Computed by Auwers,	1862·1	97°·3	
„ Safford,	1862·1	83·8	{ (yearly diminution 1°·4)
Observed by Bond,‖	1862·2	84·6	
„ Chacornac,¶	1862·2	84·6	
Lassell,**	1862·3	83·8	
„ Rutherfurd,††	1862·2	85·0	

The difference between Dr Auwers's theoretical investigation and my own is perhaps not larger than the uncertainty of all the series of observations on Sirius would explain; as I have before stated, the amount of deviation from which the angle of position was computed is very small.

But that the companion of Sirius may produce the disturbances, it—the faint object barely visible in the largest class of telescopes—must have a mass nearly *two-thirds* that

* The number bears date Dec. 20, 1861; my own communication, Sept. 20th.

† Jan. 31, 1862. First announced by Professor Bond, in No. 1358 of the Astronomische Nachrichten.

‡ This fact was stated by Professor Bond (American Journal of Science for March 1862, p. 287).

§ Astronomische Nachrichten, No. 1371. It is proper for me here to express my sense of the courtesy with which Dr Auwers admitted my priority in the matter.

‖ Astr. Nachr., No. 1374.   ¶ Ibid. No. 1355.   ** Ibid. No. 1360.
†† American Journal of Science, May 1863, p. 407.

of Sirius itself. It is difficult to believe this; but, as the evidence of this year (1863) shows, we may be compelled to do so.

There are three hypotheses logically possible with respect to the new star. It may be either unconnected with the system of Sirius; or, secondly, a satellite, but not the disturbing body; or, thirdly, the disturbing body itself. On the first hypothesis, the proper motion of Sirius itself would put it in the following position, assuming the angle of position 84°·5, for 1862·2, and distance 10″·19 for the same date, the latter being the mean of these results (excluding Lassell's 4″·92, which is quite wrong).

$$10\overset{.}{\cdot}09 \quad \text{Rutherfurd.*}$$
$$10\cdot07 \quad \text{Bond.†}$$
$$10\cdot41 \quad \text{Chacornac,‡}$$

Position and Distance by Hypothesis I.; assuming the little star to be fixed.

1863·0	$79\overset{.}{\cdot}1$	$10\overset{.}{\cdot}80$
1864·0	73·3	11·69

The second hypothesis gives no ground for calculation, and it will be considered further on.

The third hypothesis would give (correcting my own investigation, so as to agree in 1862·2 with observation, by $+0°·9$).

1863·0	$83\overset{.}{\cdot}5$
1864·0	82·1

Observation gives, compared with these hypotheses,

1863·3	Bond,§	$82\overset{.}{\cdot}8$	Hyp. I.	$77\overset{.}{\cdot}4$	Hyp. III.	$83\overset{.}{\cdot}1$
1863·2	Rutherfurd,‖	81·2		77·9		83·2

Computed — Observed.

	I.	III.
Bond,	$-\overset{\circ}{5}\cdot4$	$+\overset{\circ}{0}\cdot3$
Rutherfurd,	$-3\cdot3$	$+2\cdot0$

---

* American Journal of Science for May 1863, p. 407.
† Astronomische Nachrichten, No. 1374.
‡ Ibid. No. 1355.
§ MS. furnished by Professor Bond.
‖ As before, American Journal of Science for May 1863, p. 407.

To which must be added, that the first hypothesis requires an increase of distance between 1862·2 and 1863·2 of 0″·8 ; the third, a very slight diminution ; but observation indicates a diminution of about 0″·55, a quantity, to use Mr Rutherfurd's expression,[*] so small that its existence cannot be asserted with confidence." It is hardly conceivable that the long and careful series of observations of Mr Rutherfurd should be in error 3°·3 ; and also inconceivable that Professor Bond's measures, agreeing as they do within 2° 20′ among themselves, should be in the mean 5°·4 erroneous.

We have therefore nothing to oppose to the hypothesis that the new companion is the disturbing body, but the very improbable supposition that the small star partakes very nearly in the great proper motion of Sirius without physical connection ; or the second hypothesis, that the new star is in the system, but with small mass. If this is the case, the disturbing body must, in lieu of the small light of the companion, have still less, or even be absolutely invisible. *It is consequently highly probable that the disturbing body has been actually found ; that what was predicted by theory has been confirmed by sight.* The importance of continued observations on Sirius cannot be too highly felt. The companion must be measured the coming year, and for several years ; while Sirius itself should be re-observed with meridian instruments. So far as the right ascension element is concerned, a series of observations is now in progress at Cambridge ; while Captain Gilliss has most obligingly consented to make a series of declination-observations at Washington ; and the standard observatories at Greenwich and Paris will doubtless continue their series of fundamental star observations, including, of course, Sirius.

I am much obliged to Mr Rutherfurd for the communication of the details of his observations in 1863, and hope he will publish them, together with similar details of those of 1862, and others to be made hereafter. The subject is one where the co-operation of several observers is desirable. Full certainty here can only be obtained after several years' observations.

---

[*] As before, American Journal of Science for May 1863, p. 407.

Notes on the Fertilisation of Orchids. By WILLIAM RUTHER-FORD, M.D., President of the Royal Medical Society, Resident Physician Royal Infirmary. (Being a portion of a thesis, for which a gold medal was awarded by the Medical Faculty of the University of Edinburgh at the Graduation in 1863.)*

Mr Darwin, in the introduction to his admirable work on " The Fertilisation of Orchids," states, that his chief reason for writing the work was, " to show that the contrivances by which orchids are fertilised, have for their main object the fertilisation of each flower by the pollen of another flower ;" and to show that, in his " Origin of Species," he had good grounds for expressing his belief in what he regards as an apparently universal law—viz., " That no hermaphrodite fertilises itself for a perpetuity of generations, an occasional cross with another individual being required." He, moreover, expresses the hope, that his researches may stimulate others to inquire into the habits of our native species.

During the past summer (1862), I spent some time in the examination of a considerable number of orchids, with a view to ascertain whether or not Mr Darwin's observations were accurate, and the conclusions at which he had arrived correct. The points which I especially wished to test, were, 1*st*, Is insect agency essential for their fertilisation ? 2*d*, Is a flower fertilised by its own pollinia, or by those of other flowers ? As regards the first of these, Mr Darwin says, that in every orchis, with the exception of the bee orchis and *Cephalanthera grandiflora*, insects are required to remove the pollinia, and apply them to the stigma ; and with regard to the second point, he says,—that although in some cases the pollinia may be applied to the stigma of the flower from which they are taken, yet in all they may be—and most generally they are—applied to the stigmas of other flowers ; farther, in some flowers—the marsh Epipactis, for

* Read before the Botanical Society November 12, 1863.

example—the pollinia are removed only when the insect retires from the flower.

Sprengel, in 1795, and Robert Brown, in 1833, though the latter was not without his doubts on the subject, both expressed their belief in the necessity for insect agency; and many others have concurred with the opinion; but Darwin was the first to show that the necessity for insects, which was previously considered to be confined to a few, is almost universal. My observations, so far as they have extended, have most thoroughly convinced me of the truth of Mr Darwin's statement. But I must here mention, to prevent any misunderstanding, that I have examined four species only,—for the district in which I resided contained only these four species, although they were severally represented by large numbers of individuals, so that I was able to make a pretty thorough examination of each species. I was staying in a part of Kent where *Orchis maculata* and *Cephalanthera grandiflora* were especially abundant; and *Gymnadenia conopsea*, and *Orchis pyramidalis*, to a lesser degree. I examined 1175 flowers of *Cephalanthera*, 1000 of *Orchis maculata*, 244 of *Gymnadenia conopsea*, and 60 of *Orchis pyramidalis*, in all 2479 flowers. This number may seem very large; but it must be remembered, that the flowers grew abundantly in the locality; and I had but little difficulty in procuring them. All the plants grew near, or in, woods, so that they were most favourably situated for visitation by insects. Mr Darwin says, that on one occasion only has he seen an insect capable of carrying away the pollinia visit an Orchis. I have been more fortunate; for I have repeatedly observed, especially on warm, cloudy days, lepidopterous insects paying their visits; and on one occasion I actually saw an insect remove the pollinia. Although Mr Darwin thinks that an insect does not confine its visitations to one particular species, but embraces several,— an opinion which he has shown to be true in the case of some one or two insects,—I must say that *Orchis maculata* and *Cephalanthera grandiflora*, although growing together, were visited by totally distinct insects, and either species was only visited by one kind of insect.

This fact is certain regarding the fertilisation of three

out of these four species,—*self-fertilisation is impossible,*—
the pollinia must be removed from the flower and applied to
the stigma of either the same or another flower.   In by far
the greater majority of the flowers, the pollinia, where these
were single, were both removed, and in only a few of these
were the ovaries non-fertilised.   Sometimes I found the
heads of pollinia sticking to the stigmas : this was rare, however ; more frequently I found bundles broken off from the
pollinia adhering to the stigma, and in some of these instances the pollinia remained in the same flower untouched,
showing conclusively, that these flowers had been fertilised
by the pollinia of other flowers.   The flowers I examined were
generally *old*, with the viscid discs and stigmas quite dry, so
that no farther change could take place in the fertilisation
of such flowers.   Out of 1304 flowers, 953 had both pollinia
removed, of which 895 were fertile and 58 were non-fertile.
From this it appears, that although the pollinia may have
been removed from the flowers, these were sometimes non-
fertile.   This is, because the insect has carried away the
pollinia without pushing them against the stigma, and because the flowers have never been visited by insects having
pollinia on their probosces.   If such flowers could ever have
become fertilised (most were old), it must have been by the
pollinia of other flowers.

In 212, both pollinia were still remaining, although the
flowers were mostly dry and shrivelled.   Of these 119 were
fertile, and 96 were non-fertile, so that although these flowers
are incapable of self-fertilisation, the flowers are oftener fertilised than not.   Insects with pollinia attached to their probosces visited the flowers and fertilised them, although they
did not remove the pollinia.   Had the flowers grown in a
less wooded district, where insects are more scarce, many
more of them would have had both pollinia remaining, and
fewer of these would have been fertilised.   Observe (*see* the
Table at the end) how different is the case of *Cephalanthera
grandiflora*, which is capable of self-fertilisation, although to
a small degree: only 39 out of 1175 flowers had both pollinia
remaining, and these, nevertheless, were *all* fertile; while of
the 1128 which had both pollinia removed, only 8 were non-
fertile.   In the two other species which had the pollinia

separate, that is, unattached at the base to one another, the right pollinium was removed rather oftener than the left, a fact which would be difficult to explain. Of the 166 flowers which had only one pollinium removed, 142 were fertile and 24 non-fertile, showing that where only one pollinium is removed, the flower is not so certainly fertilised ; in short, the insects have not visited them so frequently.

It is unnecessary for me to comment further upon the following Table, but I may shortly state, that it fully bears out Mr Darwin's conclusions ; it establishes nothing new, but simply places beyond doubt very important opinions advanced by Darwin, among which the following are the most important :—1*st*, Insect agency is necessary for fertilisation ; 2*d*, Crossing of the individuals of a species is not only permitted, but all the arrangements seem especially adapted to bring about such a result.

One would suppose that hybrids ought to be very common if Mr Darwin's opinion were correct,—that one insect visits several species of orchids,—while it is well known that orchidaceous hybrids are extremely rare. From all that I have observed, I believe it to be the rule that each species has its special visitor, and that the same insect visits several species, to be the exception. I dare not, however, speak too positively on this point, for my observations have not been extensive.

Finally, it may seem superfluous for me to draw attention to the beautiful and laborious investigations contained in Mr Darwin's work on orchids ; but only those who have carried on such researches are able to estimate the severe and prolonged labour which they entail.

Name.	Number of Flowers examined.	Both Pollinia remaining.		Both Pollinia removed.		One Pollinium removed.			
						The Right.		The Left.	
		Fertile.	Non-fertile.	Fertile.	Non-fertile.	Fertile.	Non-fertile.	Fertile.	Non-fertile.
Orchis maculata	1000	50 2 of these young.	28 12 of these young.	765	47 15 of these young.	53	3	44	10
Gymnadenia conopsea	244	54	29	96	9	30	7	15	4
Orchis pyramidalis	60	15	39	34	2	Pollinia attached to one another by their bases.			
Totals,	1304	119 } 215	96	895 } 953	58	83 } 93	10	59 } 73	14
Cephalanthera grandiflora	1175	39	None.	1128	8	Pollinia attached to one another by their bases.			

Remains of Birds' Eggs found at Fisherton, near Salisbury.
By H. P. BLACKMORE, M.D., Salisbury. Communicated
by Sir WILLIAM JARDINE, Bart.

In " The Geologist" for October last, Mr Blackmore of
Salisbury, while giving a list of the fossil mammalia and
flint instruments obtained in the Pleistocene districts of
Fisherton, near Salisbury, states—" Although you ask no
information with regard to birds, it may be interesting to
some of your readers to know that fragments of such fragile
things as birds' eggs have been obtained from the same
deposits ; one in point of size and thickness of shell would
correspond, if entire, to that of a goose, the other to that of
a moor hen." This being the first record, we believe, of
the remains of eggs having been found with those of the
lower animals and others of like time, attracted our attention, and in reply to our inquiry Mr Blackmore has kindly
sent the following particulars to Mrs Strickland :—

" Both fragments of egg-shells are in my possession ; the
larger of the two, which I think is probably part of the egg
of *Grey-Lag Wild Goose,* was found in March 1861, by work-
men digging brick earth in Mr Baker's pit. The clay, sand,
and gravel in this pit are slightly stratified, and at one point
are nearly 30 feet in thickness. The shell was found about
14 feet below the surface, the soil above having evidently
never been disturbed. Within a few feet of the spot where
the shell was discovered the men found two small bones,
one the coracoid, and the other about the upper three-
fourths of the femur of a species of *Anser,* corresponding in
size with similar bones of the Grey-lag. The shell itself is
stained of a pale fawn colour, and both upon the in and outside has many small superficially raised incrustations ; hence
I infer that the shell must have been already broken when
embedded in the clay.

" Since writing to ' The Geologist,' I have joined the fragments of the smaller egg more perfectly together, and find,
from the small size of the fragments, I formed an erroneous
opinion of the size of the egg. The restored fragments I
carefully compared with a collection of recent British

eggs, and find them both in texture and size to correspond closely with the eggs of the *Common Wild Duck.* In colour the shell is rather darker than the larger one, but is in parts similarly incrusted both inside and out. This was found in Mr Harding's pit in November 1862, about 20 feet below the surface, in undisturbed clay. No bones accompanied this specimen."

I. *On Parallel Relations of the Classes of Vertebrates, and on some Characteristics of the Reptilian Birds.*

II. *The Classification of Animals based on the Principle of Cephalization.* No. I. By JAMES D. DANA. Communicated by the Author.*

I. *On certain parallel relations between the classes of Vertebrates, and on the bearing of these relations on the question of the distinctive features of the Reptilian Birds.*

At the close of an article by Professor Hitchcock, a portion of a letter of the writer is quoted, in which a parallelism is drawn between the Oötocoid or semi-oviparous Mammals (*Marsupials* and *Monotremes*), the Ichthyoid Reptiles (*Amphibians* of De Blainville, *Batrachians* of many authors), and the Reptilian Birds. The general fact of this parallelism throws light on (1.) the classification of Mammals, (2.) the distinctive features of the Reptilian birds, and (3.) the geological progress of life.

1. *Classification.*—The Amphibians are made by many zoologists an independent class of Vertebrates, on the ground of the fish-like characteristics of their young. The same systematists, however, leave the Marsupials in the class of Mammals, notwithstanding their divergencies from that type. The number of classes of Vertebrates, usually regarded as four, thus becomes *five*, namely, Mammals, Birds, Reptiles, Amphibians and Fishes. There are some indications that this number will soon be further increased by some zoologists, through the making of another class out of the *Reptilian Birds.*†

The discovery of the Reptilian Birds has brought the general law to view, that, among the four classes of Vertebrates, ordinarily

---

* From the American Journal of Science and Arts, vol. xxxvi., Nov. 1863.

† Professor Agassiz, in vol. i. of his "Contributions to the Natural History of the United States," page 187, subdivides Fishes into four *classes*, namely, 1. Myzonts ; 2. Fishes proper, or Teliosts (Ctenoids and Cycloids); 3. Ganoids ; 4. Selachians ; which would make the total number of classes of Vertebrates *nine.*

received, each, excepting the lowest, consists of, *first* a grand *typical* division, embracing the majority of its species, and *secondly*, an inferior or *hemitypic* division, intermediate between the typical and the class or classes below.

Before proceeding with our illustrations of this point, a word may be added in behalf of these four classes. In order to appreciate their true value, it is necessary to have in view the *type-idea* which is the basis of the fundamental characteristics of each, and which is connected with the existence of *three* distinct habitats for life—the water, the air, and the land : that in Fishes, this idea is that of *swimming aquatic* life ; in Reptiles, that of *creeping terrestrial* life ; in Birds, that of *flying aërial* life ; in Mammals, that of *terrestrial* life, again, but in connection with a higher grade of structure, the Mammalian. The type-idea is expressed in the adults both of the typical and hemitypic groups ; and any attempt to elevate the hemitypic into a separate class tends to obscure these ideal relations of the groups in the natural system of Vertebrates.

The following are the illustrations of the law above mentioned.

(1.) In the classification of Vertebrates, Mammals, the first class, are followed by birds, as the second ; and while the former are viviparous, the latter are, without exception, *oviparous.* The species of the inferior or hemitypic group of Mammals, partake, therefore, in some degree, of an *oviparous* nature, as the term *semioviparous* or *Oötocoid* implies.

In fact, all Vertebrates, excepting Mammals, are typically oviparous, although some cases of viviparous birth occur among both Reptiles and Fishes. In the viviparous Mammals, the embryo, during its development, derives nutriment directly from the body of the parent until birth, and also for a time after birth ; while in the viviparous Fish, the Selachians excepted, there is simply a development of the egg internally, in the same manner essentially as when it takes place externally. Applying, then, the term oviparous to all cases in which the embryo is shut off from any kind of placental nutrition, Reptiles and Fishes, with the exception mentioned, are as essentially oviparous as Birds. Hence, the Oötocoids or non-typical Mammals are actually intermediate in this respect, and in others also, between the typical Mammals, on one side, and the inferior oviparous Vertebrates collectively, on the other.

(2.) Again, the class next below Birds is that of Reptiles. And, correspondingly, the inferior or hemitypic group of Birds is *Reptilian* in some points of structure.

(3.) Again, the class next below Reptiles is that of Fishes ; and therefore the inferior or hemitypic group of Reptiles is the ermediate or *Ichthyoid* one of Amphibians—the young of frogs and salamanders and other included species having gills like fishes, besides some additional fish-like peculiarities.

The parallelism between the three classes, Mammals, Birds, and Reptiles, is thus complete.

(4.) Fishes have no class of Vertebrates below them, so that an *inferior* hemitypic division is not to be looked for. It might be suspected that the intermediate group in this case would be one between Fishes and the lower sub-kingdoms either of Mollusks or of Articulates; but none such exists. The lowest fish, an Amphioxus, is as distinctly a Vertebrate as the highest, and no Mollusk or Articulate exhibits any transition towards a vertebrate structure.

There are, however, *hemitypic* Fishes; but their place is towards the *top* of the class instead of at its bottom. Ganoids constitute one group of this kind, between Fishes and Reptiles, as long since pointed out by Agassiz. Again, Selachians (or Sharks and Rays) constitute another, between Fishes and the higher classes of Vertebrates. This last idea also has, we believe, been suggested by Agassiz (although we cannot refer to the place where published), this author regarding the species as intermediate in character between Fishes and the allantoidian Vertebrates. Moreover, Müller long ago observed the relation of the Sharks to the Mammals in having a vitelline placenta, by which the embryo draws nutriment from the parent, as does the mammalian fœtus by means of its allantoidian placenta.

Ganoids and Selachians are thus two *hemitypic* groups in the class of Fishes.

The scheme of grand divisions is then as follows :— *

I.

A. Typical Mammals.
B. Hemitypic Mammals,
   or Oötocoids.

II.

A. Typical Birds.
B. Hemitypic Birds,
   or Erpetoids.

III.

A. Typical or true Reptiles.
B. Hemitypic Reptiles,
   or Amphibians.

IV.

A. Hemitypic Fishes,   B. Hemitypic Fishes.
   or Selachians.   or Ganoids.
C. Typical Fishes,
   or Teliosts.

One of the groups of hemitypic Fishes looks directly towards

---

* It is here seen that the term *Oötocoid*, applied to Marsupials and Monotremes has great significance; and so likewise, *Erpetoids* and *Amphibians*. *Oötocoid* is simply the Greek form of the term *semi-oviparous*.

Reptiles, and the other towards the three higher classes of Vertebrates collectively, but especially Mammals and Birds.

It is plain from the preceding that the sub-kingdom of Vertebrates, instead of tailing off into the Invertebrates, has well pronounced limits below, and is complete within itself.

2. *Distinctive Features of the Reptilian Division of Birds.*—The skeleton of the fossil Bird, discovered at Solenhofen, has some decided Reptilian peculiarities, as pointed out by Wagner, Owen, and others. But even if perfect, it could not indicate all the Reptilian features present in the *living* animal. It is therefore a question of interest, whether the relations of the hemitypic to the typical species in the two classes Mammals and Reptiles— one superior to that of Birds, and the other inferior—afford any basis for conclusions with regard to characteristics of the hemitypic Birds undiscoverable by direct observation. The following considerations, suggested by analogies from the classes just mentioned, may be regarded as leading to unsatisfactory results ; and yet they deserve attention.

A. *Mammals.*—(1.) It is a fact to be observed that the hemitypic Mammals are as truly and thoroughly *Mammalian,* as regards the fundamental characteristic of the type—the suckling of their young—as the typical species.

(2.) The departure from the typical Mammals is small in the *adult* individuals, especially the adult males. But it is profoundly marked in their *young,* they thus approximating in period of birth and some other respects to oviparous Vertebrates.

B. *Reptiles.*—(1.) The *adult* Amphibians, or hemitypic Reptiles, depart but little from the typical Reptiles, either in structure or habits.

But (2.) the *young,* in their successive stages, from the egg upward, partake strikingly of characters of the inferior class of Fishes.

The law seems, then, to be, that the species of the hemitypic group have their principal or most fundamental resemblance to those of the class or classes below in the *young* state. We should hence conclude that the *young* of the Reptilian Birds or Erpetoids possessed more decided Reptilian peculiarities than the adults.— What these unknown peculiarities, if real, were, we can infer only doubtingly from the analogies of the known cases already considered.

The characteristic of the intermediate type, on which the intermediate character depends, is, in the case of both Mammals and Reptiles, that particular one which is the special distinction of the inferior type. The types inferior to Mammals are *oviparous,* and hence the hemitypic Mammals are semi-oviparous. The type inferior to Reptiles, or that of Fishes, is distinctively *aquatic* and breathes consequently by means of *gills* instead of lungs, and hence the hemitypic Reptiles have gills in the young state.

What then are the characteristics of Reptiles that may have been presented by the inferior or hemitypic Birds? The more prominent distinctions of Reptiles are the following :—

(1.) A covering of scales, or else a naked skin, instead of a covering of feathers.

(2.) A terrestrial creeping mode of life instead of an aerial or flying mode.

(3.) Incomplete circulation, and hence, to some degree, cold-blooded, instead of complete and warm-blooded.

Now, as to the young of the Reptilian Birds, it may be inferred that—

(1.) They were unquestionably unfledged. For this is universal among birds, for a while after leaving the egg. It is quite probable, that they were more completely unfledged, or for a longer time, than is common for the young of ordinary birds ; for even the adult bird, judging from the Solenhofen specimen, was less completely feathered than usual.

(2.) They were unquestionably *walking* chicks. For Birds in the lower division of the class (*Præcoces* of Bonaparte) have the use of their legs immediately after leaving the egg, and seek their own food. A brood of Reptilian bird-chicks, with long tails and nearly naked bodies, creeping over the ground, would have looked exceedingly like young Reptiles—very much, indeed, as if the eggs of a Reptile had been hatched by mistake. Moreover, these Reptilian Birds were probably not only walking birds when young, but as much so as hens and turkeys are, if not more exclusively so, even when adults ; for, in the inferior division of ordinary birds, the species are far inferior as flying animals to those of the superior division, and in some, as is well known, the wings only aid in running.

(3.) But the characteristics which have been mentioned under (1) and (2) are not of fundamental value, like that of the existence of gills in the young of hemitypic Reptiles, or that of the semi-oviparous method of reproduction in Oötocoid Mammals ; and it would seem that there must have been some more profound Reptilian characteristic. It is therefore probable that the third distinction of Reptiles stated belonged also to the young Reptilian Bird ; that is, it had incomplete circulation, and hence, an approximation to the cold-blooded condition of Reptiles. The heart may have had its *four* cavities complete, as in Birds, and in Crocodiles among Reptiles ; but, in addition, there may have been a passage permitting a partial admixture of the venous and arterial blood, such as exists not only in Crocodiles but also in the young Bird during an early stage in its development. This peculiarity in the vascular system of the young Bird of the present day ceases with the beginning of respiration. But in the Reptilian birds it may have continued on through the early part, at least, of the life of the chick, or until it was fledged.

This conclusion is made to appear still more reasonable by the following comparison of the three obvious methods of subdividing Vertebrates, and the connection therewith of the characteristics of the hemitypic groups. These three methods are—

1. Into *viviparous* and *oviparous;* which places the dividing line between Mammals, and the inferior Vertebrates.

2. Into *warm-blooded* and *cold-blooded*, or those having perfect, and those having imperfect, circulation; which places the line between Mammals and Birds, on one side, and Reptiles and Fishes, on the other.

3. Into *pulmonate* and *branchial*, or those with lungs, and those with gills; which places the line between Mammals, Birds, and Reptiles, on one side, and Fishes on the other.

Now the characteristic of the *first* of these methods of subdivision is that on which the hemitypic group of the first class, or that of Mammals, is based. The characteristic of the *third* is that on which the hemitypic group of the third class, or the Reptilian, is based. Hence, the characteristic of the *second* should be, if the analogy holds, that on which the hemitypic group of the second class, or that of Birds, rests for its most fundamental distinction.

3. *Geological History.*—It has been observed, on page 78, that the Vertebrate sub-kingdom has well-drawn limits below, instead of tapering downward into Mollusks or Articulates. This feature of the sub-kingdom is further evident from the fact in geological history that the earliest species of Fishes were not of the *lower* group, that of Teliosts, but of the two higher, or those of Ganoids and Selachians. The Vertebrate type did not originate, therefore, in the sub-kingdom of Mollusks, or of Articulates; neither did it start from what might be considered as its base, that is, the lower limit of the class of Fishes; but in intermediate types, occupying a point between typical Fishes and the classes above.

Moreover, the inferior group did not come into existence until the Cretaceous period, in the latter ar of geological history, when the Reptilian age was commencing its decline.

In the Devonian age, or closing Silurian, appeared the first Ganoids and Selachians. In the Carboniferous, Reptiles were introduced,—first, the inferior Amphibians, and then typical species. Afterward, in the early part of the Reptilian age, as Reptilian life was in course of expansion, there were the first of the Reptilian Birds and the first of the Marsupials or hemitypic Mammals (with probably some typical species of each of these classes). Thus the Vertebrate type, commencing at the point of approximation of Reptiles and Fishes, expanded until each of its higher classes had representative species, before the inferior division of true or typical fishes—Teliosts—came into existence. Afterwards, in the Cenozoic, the true or typical Birds and Mammals had their full expansion.

The Vertebrate type, therefore, not only was not evolved along lines leading up from the lower sub-kingdoms, but was not, as regards its own species, brought out in lineal order from the lowest upward. The sub-kingdom has therefore most evidently a separateness and a roundness below, so to speak, or an entireness in its inferior limits, which belongs only to an independent system.

We find in the facts no support for the Darwinian hypothesis with regard to the origin of the system of life.

II. *The Classification of Animals based on the principle of Cephalization.*

No. I.

As the principle of cephalization is involved in the very foundation of the diverse forms that make up the Animal Kingdom, we may look to it for authoritative guidance with reference to the system that prevails among those forms. Some of its bearings on zoological classification have already been pointed out.* I propose to take up the subject more comprehensively; and, in the present article, to bring the light of the principle to bear on the relations of the Sub-kingdoms, Classes, Orders, and some of the tribes of animal life.

It is essential, first, that the methods or laws of cephalization be systematically set forth, that they may be conveniently studied and compared. The following statement of them is an extension of what has already been presented :—

As an animal is a *cephalized* organism (or one terminating anteriorly in a head), the anterior and posterior extremities have opposite relations. The subdivision of the structure into *anterior* and *posterior* portions has therefore a special importance in this connection. As these terms are used beyond, the *anterior* portion properly includes the head, which is the seat of the senses and mouth, with whatever organs are tributary to its purposes,

---

* Expl. Exp. Report on Crustacea, p. 1412, 1855: Amer. Jour. of Science and Arts [2], xxii. 14, 1856; xxxv. 67, xxxvi. 1, 1863.

anterior in position to the normal locomotive organs; the *poste-rior* portion is the rest of the structure. The anterior is emi-nently the cephalic portion. The digestive viscera from the stomach backward, and the reproductive viscera, belong as char-acteristically to the posterior portion.

It follows, further, from the cephalized nature of an animal, that its *primary centre of force*, or the point from which concentra-tion and the reverse are to be measured, anteriorly and poste-riorly, is in the head, near the anterior extremity of the structure. In an Insect or Crustacean, its position is between the mouth and the organs of the senses—over which part the cephalic mass is located. This is sustained by embryogeny; and also by the fact, that, as the two most fundamental characteristics of an animal are its being sense-bearing and mouth-feeding, the mouth, on de-scending to the simplest of animals, is the last part to become obsolescent. Only in the inferior Invertebrates is the position of the mouth approximately *central* in the structure, as explained on page 90.*

1. *Methods of Cephalization.*

The methods, according to which the grades of cephalization are exhibited, may be arranged under the following heads:

A *Size (force-measured) of life-system:* each type, between Man at one extreme and Protozoans at the other, having its spe-cial range of variation in this respect.

B. *Functional:* or variations as to the distribution of the functions *anteriorly* and *posteriorly*, and as to their condition.

C. *Incremental:* or variations as to vegetative increment, that is, as to amplitude, and multiplicative development.

D. *Structural:* or variations in the conditions of the structure,—whether (1) compacted, or, on the other hand, resolved into nor-mal elements; (2) simple, or complex by specialization; (3) de-fective, or perfect; (4) animal-like, or plant-like.

E. *Postural:* or variations as to posture. (Only in Verte-brates.)

F. *Embryological:* or variations connected with the develop-ment of the young.

G. *Geographical distribution.*

For greater convenience and uniformity, the methods under these heads are mentioned beyond as they appear when viewed along the *descending* line of grade, instead of the ascending, This is, in fact, the more natural way, since the typical form in a

---

* There may also be one or more *secondary* centres of force; but they are, as regards the subject before us, of comparatively small importance. The in-dependent development of the abdomen and cephalothorax in Crustaceans is a case of the kind.

group—the fixed point for reference—holds a position towards the top of the group. The methods, as given, are therefore more strictly methods of *decephalization* than of cephalization; but the former are simply the reverse of the latter.

A. SIZE (OR FORCE) OF LIFE-SYSTEM.

1. *Potential.*—Exhibited in less and less force and size of life-system with decline of grade (and the reverse, with rise of grade); as that in passing from the type of Megasthenes (Quadrumanes, Carnivores, Herbivores, and Mutilates) to that of Microsthenes (Chiropters, Insectivores, Rodents, and Edentates) ; or from that of Decapods to that of Tetradecapods among Crustaceans—in which latter case, unlike the former, there is also *retroferent* decephalization; and so, generally, in passing from a higher to a lower type, it being equivalent to passing to a type of smaller and weaker life-system.

B. FUNCTIONAL.

2. *Retroferent.*—A transfer of functions backward that belong anteriorly in the higher cognate type.

Under this method, there are the following cases :—

a. A transfer of members from the cephalic to the locomotive series ; as the transfer of the forelimbs to the locomotive series in passing from Man to Brute Mammals; that of a pair of maxillipeds or posterior mouth-organs to the locomotive series in passing from Insects to Spiders ; that of two pairs of maxillipeds to the locomotive series in passing from Decapod to Tetradecapod Crustaceans.

b. A transfer of locomotive or prehensile power and function, more or less completely, from the anterior locomotive organs to the posterior.

c. A transfer of the locomotive function, more or less completely, from the limbs (these often becoming obsolete) to the body, and mainly to the caudal extremity.

Under *b* and *c*, the condition may be described as—

(*a*) *Prosthenic* (from the Greek προ, *before*, and σθένος, *strong*), if the anterior locomotive organs have their normal superiority.

(*b*) *Metasthenic* (from μετα, *after*, &c.), if a posterior pair is the more important, and the anterior are weak or obsolete.

(*c*) *Urosthenic* (from ουρα, *tail*, &c.), if the posterior part of the body, or the caudal extremity, is the main organ of locomotion.

Ordinary flying Birds are *prosthenic*, while the *Præcoces* (Gallinaceous Birds, Ostriches, &c.), being poor at flying, or incapable of it, are *metasthenic*, and they thus exhibit their inferiority of grade. Hymenopters, Dipters, Lepidopters, &c., among Insects, are *prosthenic*, while Coleopters, Orthopters, Strepsipters, &c., in which the fore-wings (the *elytra*) do not aid in flight, or but little,

are *metasthenic.* Fleas, which are degradational species, related
to Diptera, have the third or *posterior* pair of legs much the
longest and strongest. Among Macrural Crustaceans, the strongest
legs are, in the higher species, the *first* pair; in others inferior,
the *second;* in others still inferior (the Penæids), the *third* pair.

Viewed on the ascending grade, this method is the *preferent.*

3. *Pervertive.*—A subjection of an organ to any abnormal
function inferior to that normal to it; as in the adaptation of
the nose of the Elephant to prehension; of the antennæ of many
inferior Crustaceans to prehension or locomotion; of the maxillipeds of inferior Macrurans to locomotion; of the forehead in
many Herbivores to purposes of defence.

The perverted nose of the Proboscideans is one of the indications of their inferiority to the Carnivores; but it is not necessarily a mark of Inferiority among Herbivores themselves, as the
faculty of prehension is one of those especially characterising Carnivores and other higher Mammals, and nearly all Herbivores fail
of it.

Viewed on the ascending grade, this method and the following
may be included under the term *perfunctionative.*

4. *Defunctionative.*—Exhibited in the defectiveness or absence
of the normal function of an organ; as in the absence of the
function of prehension from the fore-limbs of Herbivores (this
prehension in the fore-limbs belonging to the Mammalian type);
and that of locomotion mostly from all the limbs in the Mutilates; that of locomotion from the female Bopyrus; that of locomotion from Cirripeds and other attached animals; that of the
sense connected with the *second* pair of antennæ (and probably
also the *first,* these organs being obsolete) in the Lernæas and
Cirripeds, these antennæ being simply prehensile organs in a
Lernæa, and constituting the base of the peduncle in an Anatifa.*

This degradation and loss of functions is connected often with
the *elliptic* and *amplificative* methods of decephalization. It is
connected with the latter in the Bopyrus, and also in Cirripeds
and other attached species.

C. INCREMENTAL.

5. *Amplificative.*—Exhibited in an elongation or general enlargement of the segments or members, and an increased laxness of the
parts. Includes the cases—

a. Lengthening. widening, or laxness in the *anterior* portion of
the body; the same in the *posterior* portion.

---

* See " Expl. Exp. Report on Crustacea," p. 1393, and plate 96, where it is
shown that the antennæ of the young Anatifa have a sucker-like organ for
attachment, and become, in the metamorphosis, the bottom of the peduncle by
which the adult Anatifa is attached.

" *b.* An abnormal enlargement of the general structure.

The elongation or enlargement which takes place with decline of grade is mainly *posterior*, it being small anteriorly, and sometimes none at all. In passing from the Brachyural to the Macrural type of Crustaceans, the change anteriorly is principally in an increased laxness and lengthening of the parts, with little increase in the dimensions of the body anterior to the mouth; while the abdomen (or *posterior* extremity) is enlarged 10 to 50 times beyond the bulk it has in the Crab. Descending from a snail to an oyster, there is diminution anteriorly and great enlargement posteriorly, and the animal is little more than a visceral sac.

In less marked cases of the *amplificative* method, there is only an attenuation or lengthening of the body and limbs, as in many Neuropters, Orthopters, Homopters, wading Birds, &c. The Lepidopters, also, in their very great expanse of wing, exemplify this method. In species that are attached, as the Cirripeds, the young are usually free; and it is only when they begin to outgrow, amplificately, the minute life-system (Entomostracan in the Cirripeds) that they become fixed. As attached animals, they often attain great size.

Viewed on the ascending grade, this method is the *concentrative;* and it is exhibited in the increased abbreviation and condensation of the anterior and posterior members and segments, or of the whole structure.

6. *Multiplicative.*—Exhibited in an abnormal multiplication of segments or members; as in Myriapods, Worms, Phyllopods, Trilobites, &c. There may be—

a. Simple Multiplicative; as in the superior Myriapods, the Chilopods, in which the body-segments, thus multiplied, have each its single or normal pair of members.

b. Compound Multiplicative; as in the Myriapods, of the Iulus division, or Diplopods (Chilognaths), in which there is a duplication of the pair of legs of a body segment. The name *Diplopod*, adopted by Gervais and some other authors, has the advantage of having thus a dynamical value.

The multiplicative method is, in general, a degradational one. When it affects only subordinate parts of the structure, as the length of the tail of Mammals, or of Reptiles, &c., the forms are not necessarily degradational. But when it affects the general structure, and the types are indefinite in segments, like the Myriapods, Worms, and Snakes, the forms are degradational. In Mammals, the tail may be said to have indefiniteness of limit; but, since this part is only an appendage to the body and has little functional importance, its elongation cannot properly be regarded as a mark of degradation, although one of inferiority.

When, however, the posterior extremity is, in magnitude and importance, a part of the main body structure itself, as in Snakes and Fishes, the case is properly an example of multiplicative degradation.

The abnormal number of segments under the multiplicative method may arise from a self-subdivision of enlarging normal segments, or from additions beyond the range of the normal number. The many joints of the antennæ in Crustaceans of the Cyclops group, the writer has shown to result through the former method, and the multiple segments of Phyllopods may be of the same origin : but there are no facts yet ascertained that would refer the multiplication of segments in Myriapods and Worms to this method.

Viewed on the ascending grade, this method is the *limitative*.

D. STRUCTURAL.

7. *Analytic.*—Exhibited in a resolving of the body-structure, or of an organ, more or less completely, into its equal normal elements, or in a tendency to such a resolution.

A relaxed state of the cephalic power leads to a relaxed and elementally-constituted structure. When this method characterizes strongly the general structure, the form is usually degradational ; as in Myriapods, Worms, larves of Insects,—these structures consisting of a series of nearly similar rings (the normal elements of an Articulate), without a subdivision into head, thorax, and abdomen. Fishes, of the Vertebrate type, are, as nearly as may be, in this elementalized condition. An approximation towards analysis or resolution of the body appears in the absence of the constriction between the head and thorax in Spiders and Crustaceans ; and still further, in the absence of the constriction between the thorax and abdomen in the lowest of Spiders,—the Acaroids.

Under this method, there is, in no case, among adults, or larves, a complete analysis or resolution of the head into normal segments ; the closest approximation to it, Insecteans and Crustaceans, occurs in the Gastrurans (Squilla group). But here the mandibular, and one, two, or more maxillary segments are still united. In an Insect, the head contains six normal segments, and the thorax three ; and yet the thorax has 3 to 5 times the bulk of the head ;—showing a condensation in the head-part equal to 6 to 10 times that of the thorax. Concentration in an animal structure is therefore eminently cephalic concentration, or, in a word, *cephalization*,—the head being the part most condensed, and least liable to occur resolved into its elements.

The analytic method, viewed on the ascending grade, is the *synthetic*.

8. *Simplificative.*—Exhibited in increased simplicity of structure, and in an equality of parts that are normally identical. The cases are—

a. Simplicity from diminished number of internal or external organs for carrying on the processes of life; as in the absence of distinct respiratory organs, or of different parts in the digestive system, &c.; or the union of the sexes in one individual, &c.; —a simplification which reaches its extreme limit among Radiates in the Hydra, and among animals, in the Protozoans.

b. Simplicity from equality in parts normally alike; as, equality iu the height of the teeth of some of the earliest of Tertiary Mammals; in the annuli of Worms. This case is related to the analytic.

Viewed on the ascending grade, this method is the *differentiative*, the facts exhibiting which are embraced under the well-known law of differentiation or specialization, which is fundamental in all development.

Differentiation internally, as it multiplies and perfects the means of elaborating the structure, is attended with an increasingly higher grade of chemical change, more perfect nutrition, and more complete decarbonization of the blood; and implies, therefore, improvement in all tissues, a more sensitive nervous system, and greater cephalic power and activity. And from the reverse comes the reverse effect.

9. *Elliptic.*—Exhibited in the defectiveness or absence of segments or members normally pertaining to the type of the order or class containing the species, and arising from *abnormal weakness* in the general system, or in an organ. It is exhibited especially in the degradational or inferior types. The cases are—

Incomplete or deficient (1) segments, or (2) members, in either (*a*) the *anterior*, or (*b*) the *posterior* portion of the body; as in the absence of some or all, of the teeth in Edentates; of the posterior limbs in Whales; of the abnormal appendages and posterior thoracic segments in some Schizopods or degradational Macrurans; of the antennæ, either one or both pairs, in many inferior Entomostracans; of wings in the Flea, &c.

This method of decephalization differs from the defunctionative in implying a deficiency not only of function but also of organ or member.

The incompleteness or deficiency of normal parts referred to above will be better appreciated if contrasted with deficiencies from other causes. The principal other causes are the following :—

(1.) A *high degree* of cephalization or cephalic concentration in the system.—Thus in the crab, the highest of Crustaceans, the abdomen is very small, and *elliptic* both in segments and members, because of the *high degree* of cephalic concentration ; while in the Schizopods referred to above, and in the Limulus and many other inferior Crustaceans, the same deficiency comes from *weakness* of life-system or decephalization.

(2.) High development of one part of an organ, at the expense of other adjoining parts.—This principle may be said to include the preceding, since, in that, there is a high development of the anterior or cephalic portion of the structure at the expense of the posterior or circumferential.   But here, there is reference to special organs rather than to the structure as a whole.   Thus, in the foot of a Horse, there is an enlargement of one toe, normally the third, at the expense of the others, and this enlarged toe has the full normal strength that belongs to the foot under the Herbivore-type.

It is apparent from the facts in paragraphs (1) and (2), that there may be an *elliptic* method of *cephalization* as well as of *decephalization*.   The Crab-type is a striking example of the former. The foot of the Horse, considering separately the *Horse-type*, is a case under the former rather than the latter ; for, in any related species, a lessening of the disparity of the toes would be evidence of weakness and inferiority *under that type*.   Yet, as compared with the higher Carnivore-type, in which the life-system has the strength to develop all the toes in their completeness and fulness of vigour, with great strength of foot, the foot of the horse is *elliptic*, and a mark of inferior cephalization.   In the typical Ruminants, the complete series of teeth is indicated in an embryonic state before birth ; but part of them fail of development, while the others—those specially characteristic of the type—go forward to great size and perfection.   As in the foot of the Horse there is here an enlargement of one portion at the expense of the others. And this, under the Ruminant-type, is progress toward the highest condition of the type, or *cephalization* by an elliptic method.   A Ruminant in which the teeth should be all equally developed would be one of too great feebleness of system to carry the structure to its typical perfection; and such is the Eocene Anoplothere.*
If, however, the Ruminants were referred to the Megasthene-type

---

* "Amongst the varied forms of existing Herbivora we find certain teeth disproportionately developed, sometimes to a monstrous size ; whilst other teeth are reduced to rudimental minuteness, or are wanting altogether : but the number of teeth never exceeds, in any hoofed quadruped, that displayed in the dental formula of the Anoplotherium.   It is likewise most interesting to find that those species with a comparatively defective dentition, as the horned Ruminants for example, manifest transitorily, in the embryo-state, the germs of upper incisors and canines, which disappear before birth, but which were retained and functionally developed in the cloven-footed Anoplothere."— *Goodsir, British Assoc. Rep.* 1838. *Owen's Brit. Mamm.*, 1846, p. 483.

as represented in the Carnivores, the *deficiency* of teeth would be an example of *decephalization* by the elliptic method ; for such a deficiency under the higher type of the Carnivores would be evidence of abnormal weakness.

The same principle is exemplified in Carnivores ; for the size and number of the molar teeth are less the larger the canines. The Machærodus, with its huge tusks and but *three* molars to either side of a jaw, is a remarkable example. Again, in the Elephant, two incisors are developed into the great tusks of the upper jaw at the expense of the other incisors and canines ; and jaws that look as if bearing profoundly the mark of degradation or decephalization, are hence compatible with high *cephalization* under the Herbivore-type.

It is not to be inferred that the enlargement of one part of an organ at the expense of the others, is *necessarily* an indication of *general* elevation of grade. Even in the case of the foot of the Horse, the elevation implied is elevation only under the Horse-type or among Solidungulates, and not elevation above all other Herbivores.

These examples are sufficient to illustrate the contrast between the elliptic method of cephalization and of decephalization ; and also the fact, that a case of the former in one relation may be one of the latter in a higher, that is, if referred to a higher group as the standard type. The cases that would come under the elliptic method of *cephalization* (as that of the Crab) have been already referred to by the writer to the *concentrative*, they being a result of concentration in the life-system.

(3.) That simplicity of structure which is opposed to the specialized or differentiated condition of superiority of type.—It is evident that the examples of elliptic decephalization, taking this term in its most comprehensive sense, may include the various simplifications which mark unspecialised structures of inferior types. Yet we propose to restrict the term to those examples of deficiencies which are obviously connected with degradational or hypotypic conditions under any type.

Viewed on the ascending grade, this method is the *completive.*

10. *Phytozoic.*—Exhibited in a departure from the Animal-type through a participation in structural features of the Plant-type, that is, through a plant-like arrangement of the organs. The cases are—

a. A radiate arrangement of external organs ; as in the Bryozoans and inferior Tunicates.

b. A radiate arrangement of internal as well as external organs; as in Radiates.

c. Perfect, or nearly perfect, symmetry in the radiation, instead of eccentric or irregular forms. Perfect symmetry is most general where the number of rays is based on the numbers 4 or 6 (which, it is to be noted, are multiples of 2 and 3), 4 being the number for the class of Medusæ, and both 4 and 6 occurring in that of Polyps. But if the number of rays is 5, as in the highest of Radiates, the Echinoderms, while examples of perfect symmetry occur, there are many cases of unsymmetrical forms (as in the Spatangi) in which the Radiate type seems to tend to emerge from phytoid towards true animal-like forms. In the regularly radiate, the mouth is central or very nearly so, while in the Spatangi, there is something of the fore-and-aft form of the animal.

Among species under the true animal-type there are forms showing an approximation to the central position which the mouth has in Radiates. In a Limulus, for example, the mouth-aperture is only one-half less remote from the anterior margin of the body than from the posterior (base of caudal spine). The Limuli are extreme in *amplificative* docephalization and in lowness of grade. Under the *multiplicative* method also, there is something similar in Worms and Myriapods. The head is here strictly at the anterior extremity; but the cephalic force has so feeble control, that joints multiply behind; and in the lowest of Worms, each separate segment is nearly equal in all functions to the cephalic segment. Moreover, in the embryological development of an Annelid, the first segment (with its pair of appendages) that is formed after the appearance of the head is not the anterior one close to the head, but the *eighth* (or one near this); and from this point the rings form in succession posteriorly, and also towards it from the head; as if, in these *multiplicate* species, there was a *secondary centre of force* distant from the front which preponderates over the *primary* one.

This method viewed on the ascending grade is the *holozoic* (from ὅλος all, and ζωον animal); it is exhibited in a rise from the plant-like type to the true animal-like type.

E. Postural.

11. *Postural.*—Exhibited in an increasing proneness in the position of the nervous system—the extremes being *verticality* in Man, and *horizontality* in the Fish.

F. Embryological.

12. *Prematurative.*—Exhibited in precocity of young or larves. Thus, the chicken, as soon as born, runs about and seeks its own food, while the young of those Birds which belong to the superior group,—the true flying Birds—remain helpless until able to fly; a fact recognised in Bonaparte's classification of Birds. So the young colt or calf (Herbivorous) is on its legs almost as soon as born; but the young kitten (Carnivorous, and higher in type) is for a considerable time helpless.

Prematurity has often been recognised as evidence of low

development and low rank; and the following is the explanation of it:—

When an animal has reached the condition required for locomotion and for the care of itself, it has already the essential faculties of an adult; and although these faculties of locomotion and self-feeding are of comparatively low grade, the animal possessing them is approximately mature in its cephalic forces, and afterwards rises but little with growth. Prematurity hence involves inferiority. The pupa-state of an Insect is a means of higher development the more perfect its inactivity. For this complete rest allows all the forces of the individual to be concentrated on the internal processes, and favours, therefore, that cephalic growth which makes a special demand on these forces; while in an *active* pupa (or rather the larve that passes through no pupa-state), activity, whether that of locomotion, or of digestion, constantly exhausts force; and only the balance, not thus run away with, goes towards the maturing process. With such an open outlet of force, the animal may mature physically, that is, grow and perfect its outer structure; but cephalically, or, in all those points of structure, as well as psychical powers, that are connected with superior cephalic development, it makes little advance.

Hence (*a*), those insects whose larves are essentially like the adults, and undergo no metamorphosis, are inferior in type,—as generally so recognised.

Again (*b*), those Insects (as most Hymenopterous) whose larves are footless grubs are superior in type to those (as the Lepidopterous) whose larves are most highly developed and active.

Viewed on the ascending grade, this method is the *permaturative.*

13. *Gemmative.*—Exhibited in multiplication by buds. Budding may produce—

a. Perfect individuals, capable of egg-production.

b. Individuals capable only of budding, and giving origin to a perfect egg-producing individual, as the last of a series of buddings.

c. Caducous, or persistent buds; the *latter* leading to compound forms, either branching, lamellar, or massive.

This power of reproduction by buds occurs in many Worms, both superior and inferior; in Bryozoan and many Ascidian Mollusks; in Polyps and many other Radiates. The production of persistent buds is the lowest grade, and is common in the budding Mollusks and Radiates, but not the Articulates. Among budding Articulates, case *b* appears to be of lower grade than case *a*.

This method is allied to the *multiplicative*, p. 85. It is also *phytozoic* (p. 89), or a plant-like feature in animal life.

14. *Genetic.—Number of young or eggs.*—As is well known, there is a mark of grade in the number of eggs or young produced at a single period or in a given time—the number, other things equal, being inversely as the rank or grade of the species.

15. *Thermotic.— Temperature required for embryonic development.*—Another mark of grade is afforded by the temperature required for egg-development :—for, in general, the higher the temperature, the higher the grade. Thus, the eggs of Birds require heat above ordinary summer heat, while those of Reptiles do not. The embryos of Mammals require still higher and more uniformly continued heat until their maturity, the Oötocoids alone excepted, in which birth is premature. The eggs of some Hymenopterous Insects mature inside of the larves of other Insects, where they are never exposed to a temperature of 32° Fahr.; while those of ordinary Lepidopters and many other species mature in the summer heat, and may stand a temperature below 0° Fahr.

The necessity of a higher temperature indicates, ordinarily, that the chemical processes in the vital economy are of a higher or more delicate character, or those required for a higher grade of cephalization.

G. Geographical Distribution.

16. *Habitational.*—(1.) *Terrestrial species higher than Aquatic.*— This law, announced by Agassiz, is also directly dependent on the conditions determining the grade of cephalization.

a. In the case of *aquatic* species, the ova, as well as the adult animals, are bathed in a liquid that penetrates to the interior, and dilutes, to some degree, the nutrient or developing fluids ; and, under such circumstances, the grade of chemical or vital evolution cannot be as high as in the atmosphere. The germ must therefore be one of an inferior kind. Aquatic animals are, in an important sense, *diluted* animals.

b. Again, *terrestrial* species whose ova are hatched in water, or whose young are aquatic, are for the same reason inferior, as a general rule, to those whose ova are hatched on the land.

Aquatic development or life is one of the most important marks of low grade. Among embryological characteristics, it has often a profounder value than prematurity. The *inferior division* of a *class, order, tribe,* and even *subordinate group,* is often one consisting either of *aquatic* species, or those that are *semi-aquatic* (aquatic in habit though not strictly so in mode of life, or aquatic in the young state when not in the adult).

(2.) *Living* (a) *in impure waters, or those abnormal in condition; or* (b) *in deficient light, as in shaded places, or the ocean's depths, mark of inferiority.*—Muddy waters, or salt waters excessively saline as in some inland lakes, or waters only brackish, are here included.

But *oceanic waters*, although saline, are not properly impure. Of the sub-kingdoms and the classes containing aquatic animals, the *highest* groups are those of *marine* waters. Thus, the highest of Mollusks, the Cephalopods, are marine; the highest of Radiates, the Echinoderms; the highest of Fishes, the Selachians; of Crustaceans, or the Maioid or Triangular Crabs; of Worms, the Dorsibranchs; of Acalephs, all but the Hydroids are marine; while *all* species of Echinoderms and Polyps are marine. Among the subordinate groups there are some fitted particularly for fresh water. Types that belong to fresh water sometimes have inferior species in brackish or salt water; and those that belong to salt water sometimes have inferior species in brackish or fresh water.

(3.) *Species of cold climates inferior to those of warm.*—According to the 15th canon, the highest oviparous animals should be tropical species; but not necessarily so the viviparous Mammals, since, with them, the requisite temperature for embryonic development is obtained within the parent.

An exception to this, as regards oviparous species, is afforded by Crustaceans; for, as shown by the writer, the highest kinds, the Maioid or Triangular Crabs, have their fullest development in the cooler temperate zone.

(4). *Having a wide range with regard to any of the earth's physical conditions, as (a) climate, (b) height, (c) oceanic temperature, (d) oceanic depth, (e) hygrometric conditions, &c., commonly a mark of inferiority.*—For if the development of a high order of cephalized life requires rest for a while in the young, as, for example, the nursing time in the higher Mammals and Birds, and the Pupa state in Insects, and also an absence from diluting or impure waters and the presence of the full light of the sun, it should also equally demand precise or narrowly restricted limits in all physical conditions, these being essential to the more refined or delicate chemical or vital processes. Man is the chief exception to this law,—and for the reason that he is not simply in and of nature, but also above nature, and has the will and power to bring her forces under subjection, overcoming the rigours of climate, and subjugating other inimical agencies by his art. Protophytes and Man are the only species that have the range of the world—the one because so low, the other so high. The Dog accompanies Man in his wide wanderings: but only through the virtue which is in Man, who provides the artificial heat, protection, and food his brute attendant needs. Even the human race dwindles in extremes of climate, either hot or cold.

Recapitulation.—The following are the names of the several methods of cephalization pointed out, both those based on the descending and ascending lines of grade.

		Descending.	Ascending.
A. Size of Life-system,	1.	Potential.	1. Potential.
B. Functional, . .	2.	Retroferent.	2. Preferent.
" . .	3.	Pervertive.	3. ⎫
" : .	4.	Defunctionative.	4. ⎬ Perfunctionative.
C. Incremental, . .	5.	Amplificative.	5. Concentrative.
" . .	6.	Multiplicative.	6. Limitative.
D. Structural, . .	7.	Analytic.	7. Synthetic.
" . . .	8.	Simplificative.	8. Differentiative.
" . . .	9.	Elliptic.	9. Completive.
" . . .	10.	Phytozoic.	10. Holozoic.
E Postural, . . .	11.	Postural.	11. Postural.
F. Embryological, .	12.	Prematurative.	12. Prematurative.

The remaining terms fall into both columns.

With *ascending* grade, the changes are mostly *concentrative;* with *descending*, they are *diffusive* or *decentrative.*

2. *Additional Observations.*

1. *Typical, Degradational and Hemitypic forms.* — Typical species are those within type-limits, and *degradational* those outside of the same.[*] But, as groups of all grades have each their own type and type-limits, species may be typical in one relation, and degradational in another; as Fishes, for example, while degradational Vertebrates, have still their own type and type-limits, the Teliosts being the typical Fishes, or those within these limits.

The characteristics of a type, in any case, are those fundamentally distinctive of the group. As to that of the Animal Kingdom at large, we observe that an animal is (1) a fore-and-aft, (2) cephalized, (3) forward-moving organism. The type-idea is hence expressed in a structure having (1) fore-and-aft and dorsoventral polarity; (2) a head at the forward extremity containing the seats or organs of the senses, as well as the mouth and mouth organs; and (3) the power of locomotion, if not also limbs for the purpose. Consequently Radiates, as they fail in the first criterion, are not within type-limits; neither are any *attached* species of animal, and only in a partial degree species without limbs for locomotion.

Again, the Vertebrate-type, in addition to having the characteristics of the animal type and the vertebrate structure, is essentially terrestrial, and therefore the requisite limbs and structure

---

[*] The term *degradational* has no reference to any method of origin by degradation: it implies only that the forms so called represent or correspond to a *degraded* condition of the type.

for terrestrial life are in the type-idea. Fishes are therefore outside of type-limits, or are degradational species.

The Mammal-type, the highest under Vertebrates, in addition to the characteristics of the Vertebrate type, has that of being viviparous in its births, embracing under this quality that of sustaining the embryo by placental nutrition until its maturity (as is not true of the oviparous); and with this there is also that of sustaining the young for a while after birth, by suckling. Hence the Oötocoids, in which there is only imperfect placental nutrition, and birth is premature, and there is an approximation thus to oviparous species, constitute a degradational type.

The Megasthene-type, under Mammals, has its degradational group in the Cetaceans or Mutilates, which fail mostly of limbs, and are aquatic species; and the Carnivore, its degradational group in the Seal and related Pinnipeds. The latter have the type-structure of the Carnivores, while the Mutilates have the type-structure of neither Carnivores nor Herbivores, and are therefore an independent type under the division of Megasthenes.

Again, the Bird-type, in addition to the characteristics of the Vertebrate-type, embraces features adapting the animal to flying, as feathers and wings; perfect circulation; and also a vertebral column which is posteriorly limitate, instead of one admitting of a caudal elongation—somewhat as Insects and Spiders are *closed* types behind, in contrast with the *multiplicate* Myriapods. Hence the Reptilian Birds, having *indefinite* posterior elongation, and some other Reptilian characteristics, are outside of type-limits. So, again, under the subdivisions of Birds, species that have the wings unfledged or but half-fledged, and which, therefore, cannot lead an *aerial* life, are degradational; and species that have the feet imperfectly digitate by their being web-footed, and which therefore lead a *semi-aquatic* life, are semi-degradational in the group to which they may belong.

These examples are sufficient to illustrate the uses of the words typical and degradational.

It is of the highest importance, for the correct classification of species, that in all cases it should be rightly determined whether a degradational genus is degradational to the *family* to which it belongs, or to the *tribe* or *order*, or to a still higher division. Although Seals and Whales are similarly adapted to the water, it is plain to one familiar with the species that the former are degradational Carnivores, and the latter degradational Megasthenes, as stated above. But like cases come up in every part of the Animal Kingdom, and close study is necessary for a true decision. The first preliminary towards such a decision is a clear idea of the class-type, order-type, tribe-type, or subordinate type under which the *genus or group* falls.

The term *hemitypic* has been shown in the preceding paper to imply in general a grade of the degradational. But, in some groups, as in the class of Fishes among Vertebrates, it is applicable to cases which are not typical because of their being intermediate between the type of the group and a *superior* type or types (p. 77).

Typical groups, or more properly, the groups above the degradational, may be of several grades. Thus under Vertebrates, the classes of Mammals, Birds, and Reptiles, represent different grades of Vertebrate types, and the grades may be designated in order, *Alphatypic, Betatypic, Gammatypic* (from the first three Greek letters, *a, β, γ*). Under Mammals also, there are three grades, those of Man, Megasthenes, and Microsthenes ; then, below these, the hemitypic or degradational Oötocoids. Under tribes, families, and genera, the number of grades may be large.

Degradational subdivisions are strictly *hypotypic*, or *below* the typical range.

Typical subdivisions, or those above the degradational, are not, in all cases, *true* typical, as well exemplified by the orders of Fishes : the Teliosts alone being true typical, and the Ganoids and Selachians, called *hemitypic* above, being properly *hypertypic*, or *above* the typical range. Another example of this is afforded by the subdivisions of Megasthenes. Carnivores and Herbivores are different grades of the *true typical*, the former the more perfect, or *eutypic ;* while the Quadrumanes or Monkeys are *hypertypic*, being an *intermediate* type between the typical Megasthenes and Man ; and the Mutilates (Cetaceans, &c.) are *hypotypic.* Among the Microsthenes, the Chiropters or Bats are *hypertypic*, the Insectivores and Rodents *true typical* of two grades, and the Edentates *hypotypic.*

Among the subdivisions of Mammals there are *three* grades of true typical ; and of them man is *archetypic*, as he has been styled, being the *one perfect* type.

Degradational forms may be classed under three heads, as follows :—

1. *Degenerative ;* in which the forms are thoroughly animal in type. The methods of decephalization which lead most commonly to degenerative forms are the analytic, multiplicative, elliptic, and defunctionative.

2. *Hemiphytoid ;* when, without an *internal* radiate structure, the species are (*a*) attached to a support, like plants (see *defunctionative* method, p. 84) ; (*b*), budding (*gemmative*, p. 91) ; (*c*), radiate externally (*phytozoic*, case *a*, p. 89).

The externally radiate structure is a lower grade of hemiphytoid degradation than either, being attached, or gemmate.

3. *Phytoid* (from φυτον, *a plant*) ; when the structural arrange-

ments are *internally*, as well as externally, radiate (Phytozoic, case *b*).

As Radiates have no limbs, and but imperfect senses, the higher grades among them are manifested most prominently in the conditions of the nutritive system. Some of them (the Echinoderms) are superior, as animals, to the lower *hemiphytoid* species, such as the Bryozoans.

2. *Further exemplifications of the preceding methods of Cephalization.*—In order to give greater clearness to the explanations which have been made on the preceding pages, the application of the terms expressing the methods of cephalization to grades of species may here be further illustrated.

In the class of Crustaceans, the distinction between the 1st and 2d orders, or Decapods and Tetradecapods, depends on case *a* under the *retroferent* method—a transfer of members from the cephalic to the locomotive series. In connection with it, there is also an exhibition, to some extent, of the *analytic* method, more of the segments of the body in the latter being free, and all, more regular or normal in form.

Under Decapods, the difference between the 1st and 2d tribes, the Brachyural and Macrural, depends mainly on the *amplificative* method—there being in the latter, by an abrupt transition, greater length and laxness before and behind. Under the *analytic*, also, the lengthened abdomen in the Macruran has its normal number of segments and members.

Among the subdivisions of *Macrurans*, the *retroferent* method appears prominently in the transfer of force from the *first* pair of legs to the *second*, and, among the lower genera, to the *third* pair (see p. 83); the *amplificative*, in the length of antennæ in some families, and in the length of abdomen as compared with that of the cephalothorax in others; the *elliptic*, in the absence of posterior cephalothoracic members, and also the obsolescence of the abdominal members in many Schizopods or degradational Macrurans; the *pervertive*, in the outer maxillipeds taking the form and functions of feet, as in many inferior Macrurans.

Under *Tetradecapods*, the difference between the 1st and 2d tribes, or Isopods and Amphipods, depends on the very same methods as that between the 1st and 2d under the Decapods; that is, on the *amplificative*, as shown in the greater length of cephalothorax and the elongated abdomen, and on the *analytic*, the lengthened abdomen having its normal segments and approximately normal members.

Under the Amphipods, the *amplificative* method is variously illustrated; the *elliptic* in the obsolescent abdomen of the Caprellids, as well as in the absence or obsolescence in many species of two pairs of thoracic legs.

Again, in the class of Insecteans, the distinction between the 1st and 2d orders, or Insects and Spiders, depends on *case* (*a*) under the *retroferent* method (p. 83); and, in connection, there is an exhibition of an incipient stage of the *analytic*, the head and thorax in Spiders constituting a single mass (p. 86).

Under Insects, the difference between the two highest divisions, *Prosthenics* and *Metasthenics*, depends on *case* (*b*) under the *retroferent* method, or a transfer of the flying function mainly or wholly to the posterior pair of wings. And the third is a degradational group, in which, by the *amplificative, analytic,* and *elliptic* methods, the species (Lepismæ, &c.) are wingless and larve-like.

Among Herbivores, the Elephant shows superiority (1) in having, as in Carnivores, the teeth (its tusks) for defensive weapons; (2) in having, as in Carnivores, the power of prehension, a quality, however, transferred from the teeth to one of the organs of sense, the nose; this organ of prehension also aids in defence; (3) in having the normal number of toes; (4) in having pectoral mammæ, as in the highest Megasthenes or Quadrumanes, the highest Microsthenes or Bats, and also in Man. The great size is not a mark of overgrowth and inferiority, for the animal is neither stupid nor sluggish. The Ruminants are inferior to the Elephant in having, not an *inferior* organ of sense, but the forehead, or typically the most important part of the head, perverted to use for self-defence; and also in other ways. Among Ruminants, the Stag or Elk-type shows superiority to the Ox-type, in (1) its more compact and smaller head; (2) its less magnitude *posteriorly*; (3) its limbs adapted to fleet motion; (4) its fore-limbs adapted for climbing and clinging, giving them a special *prosthenic* character and great superiority to those of the Ox. The Horse-type shows inferiority to the Elephant-type, in (1) its long head and neck (amplificate); (2) its one-hoofed foot; (3) its being *metasthenic*, the hind legs serving as the principal organs of defence; and also in the characters mentioned above.

The discussion of the subject of classification farther on, will be found to be a continued exemplification of the laws of cephalization, and we refer to it for additional elucidation.

3. *The forms, resulting from the expression of the same law of cephalization in diverse groups, often similar; and hence come some of the analogies between groups, or their osculations.*—It is apparent that the grades of cephalization may have expression in *any* division of the animal kingdom, and that hence may come *parallel* results as to form. For example, there may be cases of *amplificative* decephalization—or of long-bodied or long-legged species— in the different orders or tribes of Insects; and, when so, the *ies, in these* different groups thus characterized, will be, in a , *representatives* of one another, and the groups will " *osculate*"

at such points. One example is that of Orthopters and Neuropters through the Mantids in the former, and the Mantispids in the latter; also, that of Dipters and Neuropters, through the slender Tipulids of the former. The same may be exemplified among the orders of Birds. The degradational feature, for example, of webbed feet, or that of defective wings, may characterise the inferior species of different subdivisions, and so produce osculant groups; so may the *amplificative* feature of great length of limb and neck, the Herons among the Altrices, thus representing the Grallatores among the Præcoces.

The osculations or close approximations of classes, orders, tribes, &c., are thus often connected with like expressions of the methods of cephalization.

4. *Forms resulting from high and low cephalization sometimes similar.*—High and low cephalization often lead to similar forms, the former through cephalic *concentration*, the latter through cephalic and general feebleness; just as a thing may be small, when the material is condensed or concentrated, and equally small when dilute and there is little of it. Thus the Crab has a very small memberless abdomen, from a contracting of the sphere of growth through concentrative cephalization; on the other hand, the Schizopod has a memberless abdomen, through a limitation of the sphere of growth resulting from mere feebleness in the life-system. The abbreviated memberless abdomen of the Caprellid and the obsolescent spine-like abdomen of the Limulus are other examples among Crustaceans of this *elliptic* decephalization. The Butterflies have very large wings through the amplificative method; but some inferior nocturnal species have the wings narrow, through inferiority of grade, on the above principle, and not properly through concentration and elevation.

There is, in general, no danger of confounding the two cases, because the accompaniments in the structure of the superior species, as well as those of the inferior, commonly indicate their true relations, at once, to the mind that is well versed in the department of zoology to which the species belong; but there are many cases in which it is not safe to make a hasty decision.

5. *Uniformity of shape and size in any group greater among the higher typical species than among the lower typical or degradational species.*—On the higher typical level in any class, order, tribe, &c, the type is represented generally in its greatest number of species, and always under the least extravagance of form and size. Thus, Insects, the higher typical division of Insecteans, are vastly more numerous in species, and less diversified in size, form, and structure, than Crustaceans or Worms. And, under Insects,

the Hymenopters have little variety of form of body, and form or size of wings, compared with the Neuropters, Lepidopters, Homopters, and even the Coleopters; and the Coleopters, little compared with the Orthopters. The fantastic shapes, in all cases, occur in the inferior typical or the degradational groups. In these, cephalization is of low grade, and as a consequence of this relaxing of the system, or its inferior concentration, the forms run off into varied extravagances.

6. *Classification hereby placed on a dynamical or sthenic basis.*— The laws of celaphalization, as is apparent from the explanations which have been made, are based upon the idea that an animal is centralized force; and that the degree of concentration of this force may be exhibited in the structure; that, consequently, the various grades of species or groups become apparent, to some extent, through size and form, and their determination is thus, in part, a matter of simple measurement. Dimensions or spatial conditions have a relation to force in the animal kingdom as well as in that of the celestial spheres.

Rank or grade are thus brought to the rule and plummet, and classification, thereby, has a dynamical basis. The distinctions between groups have a dynamical or sthenic character, and all subdivisions in classification, when thoroughly understood, will have recognised sthenic relations.

It must, however, be kept in mind that the element of *size*, when used in the application of the principle, or as a mark of superiority, is not *absolute* size. For it is one of the laws of life that vegetative growth may enlarge a weak life-system to gigantic dimensions. Thus, the life-system of an Entomostracan takes great magnitude in a Limulus; of a Tretradecapod, in a female Bopyrus; of an Edéntate, in a Megathere; of a Mutilate, in a Whale. The body of a Crab has fifty times the dimensions of that of an Insect; and its head probably 100 times that of the head of an Insect, although an Insect is the superior species.

Neither is mere muscular strength an indication of grade; for there is force used in sustaining the structure which is greater the higher the organism; and superior to this, there is sensorial and other cephalic force. Were we to base our comparison between the grade of life-system in a Crab and that of a Bee on the ground of muscular strength, we should go far astray; and still wider from the mark, were we to rely on the relative sizes of the cephalic nervous masses; for this nervous mass in a common Crab (*Maia squinado* of European seas) has twenty-five to thirty times the bulk of that in a Bee. Man yields in size and muscular strength not only to the higher Megasthenes, but to the *Whales, or lowest;* and the brain in the Elephant and the Whale outweighs his. The Megathere, although much more powerful than a Rodent, has not, on this account, as his structure and habits show, any claims to a place above the lowest of Microsthenes.

The terms *Megasthenes* and *Microsthenes* are not to be understood as signifying large Mammals and small Mammals, but Mammals of *strong life-system* and *weak life-system*. Comparing the typical species of Megasthenes * with those of Microsthenes, there is some correspondence between average size of structure and strength of life-system. But a comparison of the typical of the former with the degradational of the latter leads to very false results.

An approximation to the right ratio is obtained from a comparison of the degradational species of each ; but this is of no importance in its bearing on the question, since vegetative growth is apt to give the greatest proportional enlargement to the *lowest* species.

These facts teach that relative size of body, or of brain, is no necessary test of relative rank. The ratio, in *bulk*, of 1 : 3 between the brain of an average Man and that of a gorilla tells nothing of the actual difference of life-system, or of brain-power. The relative *lineal* dimensions of Microsthenes and Megasthenes has been estimated at 1 : 4, which gives, for the relative *bulk*, 1 : 64. If this be the typical ratio between the life systems of the highest Microsthenes and highest Megasthenes, surely that between the highest Megasthenes and normal man—he constituting a *distinct order*—must be at least as great.

The same ratio of 1 : 4, as shown by the writer, is that for the mean size, lineally, of Tetradecapods and Decapods, under Crustaceans. In two cases, then, consecutive orders differ by a like ratio, or approximately so, in dimensions. As has been remarked, deductions from mere size may be very erroneous; yet there is no reason, in either of the above cases, to suppose the ratio of life-systems less than that thus indicated. May not, therefore, some similar ratio exist between other analogous consecutive orders, where size does not manifest it,—as, for example, between Spiders and Insects ? And is not the ratio a much greater one between the highest of Insecteans and highest of Crustaceans, since these subdivisions of Articulates are not orders but *classes ?* Important results may flow from following out the idea here touched upon.

* These orders of Mammals, make parallel series—the Chiropters or Bats of the Microsthenes representing the Quadrumanes of the Megasthenes, the Insectivores representing the Carnivores, the Rodents the Herbivores, and the Edentates the Mutilates.

After the preceding explanations, I proceed to exhibit some of the relations of the higher groups in zoological classification, as they appear in the light of this subject of cephalization.

(*To be continued in next number.*)

Synopsis of Canadian Ferns and Filicoid Plants. By George Lawson, Ph.D., LL.D., Professor of Chemistry and Natural History in Dalhousie College, Halifax, Nova Scotia.

The following Synopsis embraces a concise statement of what is known respecting Canadian ferns and filicoid plants. Imperfect as it is, I trust that it will prove useful to botanists and fern fanciers, and stimulate to renewed diligence in investigation. The whole number of species enumerated is 74. Of these 11 are doubtful. Farther investigation will probably lead to the elimination of several of the doubtful species, which are retained for the present with a view to promote inquiry; but a few additional species, as yet unknown within the boundaries of Canada, may be discovered. The above number (74) may be regarded, then, as a fair estimate—perhaps slightly in excess—of the actual number of ferns and filicoid plants existing in Canada. The number certainly known to exist, after deducting the species of doubtful occurrence, is 63.

The number of species described in Professor Asa Gray's exhaustive "Manual," as actually known to inhabit the northern United States, that is to say, the country lying to the south of the St Lawrence River and great lakes, stretching to and including Virginia and Kentucky in the south, and extending westward to the Mississippi River, is 75. This number does not include any doubtful species.

The number described in Dr Chapman's "Flora," as inhabiting the Southern States, that is, all the states south of Virginia and Kentucky and east of the Mississippi, is 69.*

---

* *Mr D. C. Eaton, M.A., is author of that portion of Dr Chapman's "Flora" ~h relates to the ferns.*

From these statements it will be seen that we have our due share of ferns in Canada.

The whole number of ferns in all the American States, and the British North American Provinces, is estimated, in a recent letter from Mr Eaton, as probably over 100.

In the British Islands there are about 60 ferns and filicoid plants. In islands of warmer regions the number is greatly increased. Thus Mr Eaton's Enumeration of the true ferns collected by Wright, Scott, and Hayes, in Cuba, embraces 357 species. The proportions of ferns to phanerogamous plants in the floras of different countries are thus indicated by Professor Balfour, in the " Class Book of Botany," page 998, § 1604 :—" In the low plains of the great continents within the tropics ferns are to phanerogamous plants as 1 to 20 ; on the mountainous parts of the great continents, in the same latitudes as 1 to 8 or 1 to 6 ; in Congo as 1 to 27 ; in New Holland as 1 to 26. In small islands, dispersed over a wide ocean, the proportion of ferns increases ; thus, while in Jamaica the proportion is 1 to 8, in Otaheite it is 1 to 4, and in St Helena and Ascension nearly 1 to 2. In the temperate Zone, Humboldt gives the proportion of ferns to phanerogamous plants as 1 to 70. In North America the proportion is 1 to 35 ; in France 1 to 58 ; in Germany 1 to 52; in the dry parts of South Italy as 1 to 74; and in Greece 1 to 84. In colder regions the proportion increases ; that is to say, ferns decrease more slowly in number than phanerogamous plants. Thus, in Lapland, the proportion is 1 to 25 ; in Iceland 1 to 18 ; and in Greenland 1 to 12. The proportion is least in the middle temperate zone, and it increases both towards the equator and towards the poles ; at the same time it must be remarked, that ferns reach their absolute maximum in the torrid zone, and their absolute minimum in the arctic zone."

Canada consists of a belt of land, lying to the north of the St Lawrence River and the great lakes. By these it is separated, along nearly the whole extent of its south-eastern and western boundaries, from the northern United States, which thus enclose Canada on two sides. A striking resemblance, amounting almost to identity, is therefore to be looked for in the floras of the two countries. Yet species *appear in* each that are absent in the other.

The species of ferns and filicoid plants which are certainly Canadian, number . . . 63
Of these there inhabit the Northern States, 58
   Do.   do.   Southern States, . 38
   Do.   do.   Europe, ' . . 36

The following table is designed to show some of the geographical relations of our Canadian ferns. The first column (I.) refers exclusively to the occurrence of the species within the Canadian boundary. The plus sign (+) indicates that the species is general, or at least does not show any decided tendency towards the extreme eastern or western, or northern or southern parts of the province. The letters N, S, E, W, &c., variously combined, indicate that the species is so limited to the corresponding northern, southern, eastern, or western parts of the province, or at least has a well-defined tendency to such limitation. The mark of interrogation (?) signifies doubt as to the occurrence of the species. The second column (II.) shows what Canadian species occur also in the Northern States, that is the region embraced by A. Gray's Manual ; and the third column (III.) those that extend down south into Chapman's territory. The fourth column (IV.) shows the occurrence of our species in Europe ; C in this column indicating Continental Europe, and B the British Islands. The fifth or last column (V.) shows the species that extend northwards into the Arctic circle—35 in all, of which, however, only 14, or perhaps 15, are known to be arctic in America. Am, As, Eu, and G indicate respectively Arctic America, Arctic Asia, Arctic Europe, and Arctic Greenland. The information contained in the last column has been chiefly derived from Dr Hooker's able Memoir in the Linnean Transactions (vol. xxiii. p. 251).

Hitherto no attention whatever has been paid, in Canada, to the study of those remarkable variations in form to which the species of ferns are so peculiarly liable. In Britain, the study of varieties has now been pursued by botanists so fully as to show that the phenomena which they present have a most important bearing upon many physiological and taxological questions of the greatest scientific interest. The varieties are studied in a systematic manner, and the laws of variation have been to a certain extent ascertained. And as the astronomer can point out the existence of a planet before it has been seen, and the chemist can construct formulæ for organic compounds—members of homologous series—in anticipation of their actual discovery, so, in like manner the pteridologist now studies the variations of species by a comparative system, which enables him to look for equivalent forms in the corresponding species of different groups. Studies so pursued are calculated to evolve more accurate and definite notions as to the real nature of species, and the laws of divergence in form of which they are ·capable. I would therefore earnestly invite Canadian botanists to a more careful study of the *varieties* of the Canadian ferns, after the manner of Moore and other European leaders in this comparatively new path. The elasticity, or proneness to variation, of the species in certain groups of animals and plants has been somewhat rashly used to account for the origin of species, by what is called the process of variation. It seems to tell all the other way. Innumerable as are the grotesque variations of ferns, in forkings, and frillings, and tassellings, and abnormal veinings, &c. (see the figures in Moore's works), we do not know of a single species in which *such* peculiarities have become permanent or general, that is *specific*, so that the species can be traced back to such an origin ; surely something of the kind would have happened had all species originated by a process of variation.

Tabular View of the Distribution of Canadian Ferns and Allied Plants over certain parts of the Northern Hemisphere.*

NAME.	I. Canada.	II. Northern States.	III. Southern States.	IV. Europe.	V. Arctic Circle.
POLYPODIACEÆ.					
1. Polypodium vulgare, . .	+	+	+	C.B.	Eu.
2. P. hexagonopterum, . . .	+	+	+	...	...
3. P. Phegopteris, ...	+	+	...	C.B.	Eu. G.
4. P. Dryopteris, ...	+	+	...	C.B.	Eu.Am.G.

* In the above Table, the doubtful species are included ; but all reference to varieties is omitted.

NAME.	I. Canada.	II. Northern States.	III. Southern States.	IV. Europe.	V. Arctic Circle.
5. P. Robertianum, ...	+	+	...	C.B.	..←
6. Adiantum pedatum, . . .	+	+	+	...	...
7. Pteris aquilina, ...	+	+	+	C.B.	Eu.
8. Pellæa atropurpurea, . . .	S.	+	+	...	...
9. Allosorus Stelleri, . . .	+	+	...	...	...
10. Cryptogramma acrostichoides,	W.W.	...	...	?	Am.
11. Struthiopteris germanica, .	+	+	...	C	Eu.
12. Onoclea sensibilis, . . .	+	+	+	...	...
13. Asplenium Trichomanes, .	+	+	+	C.B.	...
14. A. viride, ...	N.E.	...	...	C.B.	Eu. G.
15. A. angustifolium, ...	S.W.	+	+	...	...
16. A. ebeneum, ...	+	+	+	...	...
17. A. marinum, ...	E. ?	...	...	C.B.	...
18. A. thelypteroides, . . .	+	+	+	...	...
19. A. montanum, ...	?	+	+	...	...
20. A. Ruta-muraria, ...	?	+	+	C.B.	Eu.
21. Athyrium Filix-fœmina, .	+	+	+	C.B	Eu.
22. Woodwardia virginica, . .	S.W.	+	+	...	...
23. Scolopendrium vulgare, . .	W.W.	+	...	C.B.	...
24. Camptosorus rhizophyllus, .	W.	+	+	...	...
25. Lastrea dilatata, ...	+	+	+	C.B.	Eu. Am.
26. L. marginalis, ...	+	+	+	...	...
27. L. Filix-mas, ...	? ?	...	...	C.B.	Eu. G.
28. L. cristata, ...	+	+	...	C.B.	...
29. L. Goldieana, ...	W.	+	...	...	...
30. L. fragrans, ...	N.W. ?	+	...	...	As. Am. G
31. L. Thelypteris, ...	+	+	+	C.B.	...
32. L. Nov-Eboracensis, . .	+	+	+	...	...
33. Polystichum angulare, . .	+	+	...	C.B.	Eu.
34. P. Lonchitis, ...	N.W.	+	...	C.B.	Eu. Am.G.
35. P. acrostichoides, ...	+	+	+	...	...
36. Cystopteris fragilis, . . .	+	+	+	C.B.	Eu.Am.G.
37. C. bulbifera, ...	+	+	+	...	...
38. Dennstædtia punctilobula, .	+	+	+	...	...
39. Woodsia Ilvensis, . . .	+	+	+	C.B.	{Eu. As. Am. G.
40. W. alpina, ...	+	...	...	C.B.	Eu. G.
41. W. glabella, ...	+	+	...	...	Am.
42. W. obtusa, ...	?	+	+	...	...
43. Osmunda regalis, ...	+	+	+	C.B.	...
44. O. cinnamomea, ...	+	+	+	...	...
45. O. Claytoniana, ...	+	+	+	...	...
46. Schizæa pusilla, ...	?	+	...	...	...

NAME.	I. Canada.	II. Northern States.	III. Southern States.	IV. Europe.	V. Arctic Circle.
OPHIOGLOSSACEÆ.					
47. Botrychium virginicum,	+	+	+	...	Eu. G.
48. B. lunarioides,	+	+	+	?	...
49. B. Lunaria,	N.	...	...	C.B	Eu. G.
50. Ophioglossum vulgatum,	?	+	+	C.B.	Eu.
LYCOPODIACEÆ.					
51. Plananthus Selago,	N.?	+	+	C.B.	Eu. As. Am. G.
52. P. lucidulus,	+	+	+	C.	...
53. P. alopecuroides,	? ?	+	+	...	...
54. P. inundatus,	+	+	+	C.B.	...
55. Lycopodium clavatum,	+	+	+	C.B.	Eu. G.
56. L. annotinum,	+	+	+	C.B.	Eu.Am.G.
57. L. dendroideum,	+	+	+	...	...
58. L. complanatum,	+	+	+	C.	Eu. As.
59. Selaginella spinulosa,	N.E.	+	+	C.B.	Eu. G.
60. Stachygynandrum rupestre,	+	+	+	...	...
61. Diplostachyum apodum,	+	+	+	...	...
MARSILEACEÆ.					
62. Azolla caroliniana,	S.	+	+	...	...
63. Salvinia natans,	? ?	...	+	C.	...
64. Isoetes lacustris,	+	+	+	C.B.	Eu. G.
EQUISETACEÆ.					
65. Equisetum sylvaticum,	+	+	...	C.B.	Eu.Am.G.
66. E. umbrosum,	+	+	...	C.B.	Eu.
67. E. arvense,	+	+	...	C.B.	Eu. As. Am. G.
68. E. Telmateja,	W.	+	...	C.B.	...
69. E. limosum,	+	+	...	C.B.	Eu.
70. E. hyemale,	+	+	...	C.B.	Eu.
71. E. robustum,	+	+	...	...	...
72. E. variegatum,	N.E.	+	...	C.B.	Eu. Am.? G.
73. E. scirpoides,	+	+	...	C.	Eu. As. Am. G.
74. E. palustre,	N.	...	...	C.B.	Eu. Am.

Nat. Ord. POLYPODIACEÆ.

POLYPODIUM.

P. vulgare, Linn.—Frond linear-oblong or somewhat lanceolate, more or less acuminate, deeply pinnatifid, in some forms almost pinnate; lobes (or pinnæ) linear-oblong, obtuse, often acute, rarely acuminate, entire or crenate or serrate; sori large; very variable as regards outline of the frond, form, &c., of the lobes, and serrature. *P. vulgare*, Linn., A. Gray, Moore, &c. *P. virginianum* of English gardens. *P. vulgare*, var. *americanum*, Hook., Torrey Fl. N. Y., ii. 480.—On rocks in the woods, not rare around the city of Kingston; abundant on the rocky banks of the St Lawrence, in Pittsburg; in the woods at Collins's Bay; and on Judge Malloch's farm, a mile west from Brockville; Gananoque lakes and rivers; Farmersville; Newboro-on-the-Rideau; Toronto; on the great boulder of the Trent Valley, near Trenton; on rocks west from Brockville, outcrop of Potsdam Sandstone at Oxford, and Hull mountains near Chelsea, C.E., B. Billings, jr.; near Gatineau Mills, D. M'Gillivray, M.D.; Mount Johnson, C.E., and Niagara River, P. W. Maclagan, M.D.; Brighton, in the crevice of a rock in a field, and abundant on rocky banks, right bank of the Moira, above Belleville, J. Macoun; Ramsay, Rev. J. K. M'Morine, M.A.; north-west from Granite Point, Lake Superior, R. Bell, jr.; mountain top, near Mr Brydge's house, Hamilton, C.W., Judge Logie; River Rouge and lower end of Gut Lake, W. S. M. D'Urban; Cape Haldimand, Gaspé, John Bell, B.A.; Red River Settlement, Governor M'Tavish; Pied du cap Tourmente, M. L'Abbé Provancher; L'Orignal and Grenville, C.E., J. Bell, B.A. The habitats above cited show that although this fern is not so common in Canada as in Britain, it is nevertheless widely distributed. It is common in New York State, according to Professor Torrey; and in the Northern States generally, according to Professor Asa Gray; rarer in the South, according to Dr Chapman.

P. hexagonopterum, Mich.—Frond triangular in outline, acuminate, pinnate, hairy throughout; pinnæ broadly lanceolate, pinnatifid; lowest pair of pinnæ larger than the others, not deflexed; lobes of the pinnæ linear-oblong or lanceolate, strongly toothed, or almost pinnatifid. The decurrent pinnæ have a tendency to form conspicuous irregular angled wings along the rachis. Stipe not scaly except at the base. Rhizome long, slender, ramifying. Whole plant much larger than *P. Phegopteris*, and quite a different species. *P. hexagonopterum*, Michx., A. Gray, &c. The figure in Lowe's Ferns, vol. i. p. 143, tab. 49, is a little too much like Phegopteris. *P. Phegopteris γ. majus*, Hook. Fl. Bor. Amer., ii. p. 258. Hooker's *β. intermedia* of Phegopteris is *connectile*, Willd., which A. Gray refers to *P. Phegopteris*, L. *Phegopteris hexagonoptera*, J. Sm. Cat., p. 17.—Canada, Goldie in Hook. Fl. B. Amer.; Chippawa, C. W., P. W. Maclagan, M.D.; Mirwin's Woods, near Prescott, rare, B. Billings, jr.; near Westminster Pond, London, W. Saunders. Not by any means so general in Canada as in New York State, where, Professor Torrey states, it is common.

P. Phegopteris, Linn.—Frond acutely triangular in outline, acuminate, pinnate ; the pinnæ linear-lanceolate, pinnatifid, lowest pair deflexed ; lobes of the pinnæ oblong, scythe-shaped, obtuse, approximate,
entire ; rachis hairy and minutely scaly to the apex of the frond, as well
as the mid-ribs of the pinnæ. *P. Phegopteris,* Linn., A. Gray, Moore,
&c. *Phegopteris vulgaris,* J. Sm. *P. connectile,* Michx., Pursh Fl.
Am. Sept., ed. 2, vol. ii. p. 659.—Canada, Hooker ; Black Lead Falls
and De Salaberry, west line, W. S. M. D'Urban ; Ramsay, Rev. J. K.
M'Morine, M.A. ; Nicolet, P. W. Maclagan, M.D. ; Prescott, damp
woods, not common, Osgood Station of the Ottawa and Prescott Railway, also Gloucester, near Ottawa, growing on the side of a ravine, and
Chelsea, C.E., B. Billings, Jr ; opposite Grand Island, Lake Superior,
R. Bell, jr. ; L'Orignal and Harrington, J. Bell, B.A.

P. Dryopteris, Linn.—Frond thin, light-green, pentangular in outline,
consisting of three divaricate triangular subdivisions, each of which is
pinnate, with its pinnæ more or less deeply pinnatifid ; pinnules oblong,
obtuse, nearly entire ; stipe slender and weak, not glandulose. *P. Dryopteris,* Linn., A. Gray, Moore, &c. *Phegopteris Dryopteris,* J. Sm.—
Abundant in the woods around Kingston ; Ramsay, Rev. J. K. M'Morine,
M.A. ; very common in woods about Prescott, B. Billings, jr. ; Montreal
and Nicolet Rivers, C.E., P. W. Maclagan, M.D. ; Belleville, common
in the woods, J. Macoun ; opposite Grand Island, Lake Superior, R.
Bell, jr. ; River Rouge, Round Lake, Montreal, De Salaberry, west
line, and Black Lead Falls, W. S. M. D'Urban ; Newfoundland, Labrador ; Somerset and St Joachim, M. L'Abbé Provancher ; L'Orignal,
J. Bell, B.A.

Var. β. erectum.—Frond erect, rigid, with a very stout and very long
glabrous stipe (18 inches long) ; beech woods at Collins's Bay, near Kingston, with the normal form. This variety resembles *P. Robertianum* in
general aspect, but is not at all glandulose.

P. Robertianum, Hoffman.—A stouter plant than *P. Dryopteris ;*
fronds more rigid and erect ; rachis, &c., closely beset with minute-
stalked glands. *P. Robertianum,* Hoffman, Moore, &c. *P. calcareum,*
Sm. *P. Dryopteris,* var. *calcareum,* A. Gray.—Canada, Moore and
other authors ; United States, Gray and others. This species is commonly spoken and written of as a Canadian Fern. Not having had an
opportunity of seeing Canadian specimens, I cannot cite special habitats.
The minutely glandulose rachis serves at once to distinguish it.

ADIANTUM.

A. pedatum, Linn.—Stipe black and shining, erect, forked at top,
the forks secundly branched, the branches bearing oblique triangular-
oblong pinnules. *A. pedatum,* Linn., A. Gray, &c., Lowe's Ferns,
vol. iii. pl. 14. Abundant in vegetable soil in the woods around Kingston ; woods around the iron mines at Newboro-on-the-Rideau ; Farmersville ; Toronto ; Montreal, Chippawa, Wolfe Island, and Malden, P. W.
Maclagan, M.D. ; Belleville, in rich woods, abundant, J. Macoun ;
Ramsay, Rev. J. K. M'Morine, M.A. ; Ke-we-naw Point, R. Bell, jr. ;
at the Sulphur Spring, and common everywhere about Hamilton, Judge

Logie ; Lake Huron, Hook. Fl. B. A. ; De Salaberry, west line, W. S. M. D'Urban ; on the Gatineau, near Gilmour's rafting ground, D. M'Gillivray, M.D. ; London, W. Saunders ; St Joachim and Isle St Paul, Montreal, M. L'Abbé Provancher ; West Hawkesbury and Grenville, C.E., J. Bell, B.A. Apparently common everywhere in Upper Canada. I cannot speak so definitely of the Lower Province. This is one of our finest Canadian ferns ; "the most graceful and delicate of North American ferns," says Torrey. It is easily cultivated. Fine as it is in the Canadian woods, I have specimens even more handsome from Schooley's Mountains (A. O. Brodie, Ceylon Civil Service) ; their fan-like fronds spread out in a semicircle, with a radius of 2½ feet. It is not a variable species in Canada. T. Moore, in " Index Filicum," gives its distribution as N. and N.W. America, California to Sitka, North India, Sikkim, Nepal, Gurwhal, Simla, Kumaon, Japan. There is a var. β. *aleuticum*, Rupr., in the Aleutian Islands.

PTERIS.

Pt. aquilina, Linn.—Stipe stout, 1 to 3 feet high, frond ternate, branches bipinnate, pinnules oblong lanceolate, sori continuous under their recurved margins. *Pt. aquilina*, Linn., A. Gray, Moore, &c.—Abundant on Dr Yates's farm in Pittsburg, and elsewhere about Kingston ; Water-down Road, Hamilton, common, Judge Logie ; Chippawa and Malden, C.W., P. W. Maclagan, M.D. ; Ramsay, Rev. J. K. M'Morine, M.A., Prescott, common, B. Billings. jr. ; Belleville, very common on barren ridges, J. Macoun ; Grand Island, Lake Superior, R. Bell, jr. ; Red Lake River, also between Wild Rice and Red Lake Rivers, and Otter Tail Lake and River, between Snake Hill River and Pembina, &c., J. C. Schultz, M.D. ; Black Lead Falls, and Portage to Bark Lake, W. S. M. D'Urban ; Gatineau Mills, very common, D. M'Gillivray, M.D. ; Lakefield, North Douro, Mrs Traill ; New Brunswick, Hook. Fl. Bor. Amer. ; L'Orignal, J. Bell, B.A. ; London, W. Saunders.

α. *vera*.—Pinnules pinnatifid (the normal or typical form of Moore), Dr Yates's farm, Kingston.

β. *integerrima*.—Pinnules entire (a sub-variety), common in Canada and westward. There are various other sub-varieties, differing in size, pubescence, &c.

γ. *decipiens*.—Frond bipinnate, thin and membranous, lanuginose, pinnules pinnatifidly toothed, or, in small forms, entire, barren ; L'Anse à Cabièlle, Gaspé, John Bell, B.A. This is a very remarkable fern, resembling a Lastrea, and in the absence of fructification, it is doubt-fully referred to *Pteris aquilina*, yet the venation seems to indicate that it belongs to that species, which is remarkable for its puzzling forms. Being at a loss what to make of this fern, I sent it to Mr D. C. Eaton, M.A., who is justly looked up to by American botanists as our best authority on American ferns, and he likewise failed to recognise it. I hope some visitor to Gaspé will endeavour to obtain it in a fertile state, and thus relieve the doubt.*

* Since the above was written, I have had an opportunity of studying the *forms* and development of *Pteris aquilina*, and am quite satisfied that the *doubtful plant is a state* of that species, not old enough to be fertile.

[Var. *δ. caudata* appears occasionally in lists. I have as yet no satisfactory evidence of its occurrence in Canada proper. The nearest approach to it is a specimen from the Hudson's Bay territories, probably from the Red River District (Governor M'Tavish). In the South it is a very distinct form, of which there are beautiful specimens in Wright's Cuban Plants (No. 872), and is very close to the *Pteris esculenta* of Australia.]

PELLÆA.

P. atropurpurea, Link.—Stipe and rachis almost black, shining, 6 to 12 inches high, frond coriaceous, pinnate, divisions opposite, linear-oblong or somewhat oval. *Pteris atropurpurea,* Linn. *Platyloma atrop.,* J. Sm., Torr Fl. N. Y., ii. p. 488. *Allosorus atropurpureus,* A. Gray. *Pellæa atropurpurea,* Link., Fée, J. Sm. in Cat., Eaton.— Niagara River, at the Whirlpool, three miles below the Falls. This fern seems to retain its fronds all winter, for I have fertile specimens, in a fine state, collected at the Whirlpool at the end of February 1859 by A. O. Brodie. Dr P. W. Maclagan has also collected it there. It is not common anywhere on the American Continent so far as I can learn. Mr Lowe speaks of it as in cultivation in Britain, " an evergreen frame or greenhouse species, not sufficiently hardy to stand over winter's cold." There must be some other reason for want of success in its cultivation in Britain.

ALLOSORUS.

A. Stelleri, Ruprecht.—Fronds pale-green, thin and papery, 3 to 9 inches long, bipinnate or tripinnate, some of the smaller barren fronds scarcely more than pinnate; pinnæ five or six pairs; lobes of the barren frond, rounded, oval, veiny; of the fertile frond, much narrower, linear-lanceolate, firmer; sori at the tips of the forked veins along the margins, stipe red, whole plant glabrous. A beautiful and delicate fern, growing in the crevices of rocks, rare. *Allosorus Stelleri,* Ledeb. Fl. Rossica. *Allosorus gracilis,* Presl., A. Gray, Torrey Fl. N. Y. ii. p. 487. In a letter from Mr T. Moore (1857), he mentioned to me that he had learned from specimens from Dr Regel, St Petersburg, that " the North American *Allosorus gracilis* was the old *Pteris Stelleri* of Amman, so that it spreads from North America through Siberia to India, whence Dr Hooker has it." *Allosorus minutus,* Turcz. Pl. Exs. *Cheilanthes gracilis,* Klf. *Cryptogramma gracilis,* Torrey. *Pteris Stelleri,* Gmelin. *Pteris minuta,* Turcz. Cat. Pl. Baik. Dah. *Pt. gracilis,* Michaux.— Near Lakefield, North Douro, C.W., on rocks, Mrs Traill; abundant in crevices of limestone rocks, on the rocky banks of the Moira, Belleville, Co. Hastings, J. Macoun; Lake of Three Mountains, W. S. M. D'Urban; Canada to the Saskatchewan, Hook. Fl. Bor. Am.; Dartmouth, Gaspé, John Bell, B.A. This is a northern species, and rare in the United States.

CRYPTOGRAMMA.

C. acrostichoides, R. Br.—" Remarkable for its sporangia extending far down on the oblique veins, so as to form linear lines of fruit." λ have not seen the plant. It is referred by Sir William Hooker to *Allosorus crispus* (A. Gr. in Enum. of Dr Parry's Rky. Mtn. Plants). *Cryptogramma acrostichoides*, R. Br., Moore. *Allosorus acrostichoides*, A. Gr.—Isle Royale, Lake Superior. Placed in Dr Hooker's Table as a Canadian species that does not extend into the United States. It has recently been found on the Rocky Mountains. *Allosorus crispus* is general throughout Europe, and occurs at Sitka, in North-West America. Mr Moore observes that the Eastern (Indian) species, *A. Brunoniana*, is very doubtfully distinct from the European plant.

STRUTHIOPTERIS.

S. germanica var. *β pennsylvanica.*—Rhizome stout, erect; fronds tufted; sterile ones large pinnate, erect-spreading, deeply pinnatifid; the fertile ones erect, rigid, with revolute contracted divisions, wholly covered on the back by sporangia. A very graceful fern, well suited for cultivation in gardens. *Struthiopteris pennsylvanica*, Willd., Pursh, J. Sm. Cat. *S. germanica*, Hooker, Torrey Fl. N. Y. ii. p. 486, Gray. *Osmunda Struthiopteris*, Linn.; *Onoclea Struthiopteris*, Schkr.; *Onoclea nodulosa*, Schkr., according to Hooker. Torrey refers *O. nodulosa*, Michx., to *Woodwardia angustifolia.*—Frankville, Kitley; Longpoint; Lansdowne; Hardwood Creek; usually found along the margins of creeks, &c.; common in rich, wet woods near Prescott, and abundant around Ottawa, B. Billings, jr.; low rich grounds, Belleville, abundant along Cold Creek, J. Macoun; Re-we-naw Point, Lake Superior, in low ground, at times under water, R. Bell, jr.; Ramsay, Rev. J. K. M'Morine, M.A.; near Lakefield, North Douro, Mrs Traill; field beyond Waterdown, Hamilton, Judge Logie; Osnabruck and Prescott Junction, Rev. E. M. Epstein; near Montreal, W. S. M. D'Urban; Assiniboine River, John C. Schultz, M.D.; Canada, to the Saskatchewan, Hook. Fl. Bor. A.; Pied du Tourmente, M. L'Abbé Provancher. This is the commonest plant in the Bedford Swamps; Gaspé and L'Orignal, J. Bell, B.A.; London, W. Saunders. Found in the western part of New York State, but rare according to Torrey.

ONOCLEA.

O. sensibilis, Linn.—Rhizome creeping; barren frond broad, leafy, deeply pinnatifid; fertile ones erect, spicate, contracted, doubly pinnate, with small revolute pinnules, enclosing the sporangia, not at all leafy. *Onoclea sensibilis*, Linn., A. Gr., J. Sm., &c. Lowe's Ferns, vol. vi. pl. 1.—In woods along the banks of the Little Cataraqui Creek in great abundance, and in moist swampy places in the woods in various other places about Kingston; west end of Loborough Lake; Becancour, M. L'Abbé Provancher; London, W. Saunders; common in marshy ground at Hamilton, Judge Logie; Lakefield, North Douro, Mrs Traill; St John's, C. E., Niagara and Malden, P. W. Maclagan, M.D.; Belleville, in low marshy places, abundant, J. Macoun; Ramsay, Rev. J. K. M'Morine, M.A.; Amagos Creek, Lake Superior, R. Bell, jr.; Prescott, common, B. Billings, jr.; on the river shore, Gatineau Mills, D. M'Gillivray, M.D.; L'Anse au Cousin, Gaspé and L'Orignal, J. Bell; Nova Scotia. This curious fern has been cultivated in England since

1699; at Kew, since 1793. It is very variable as regards the outline and subdivision of the barren frond.

Var. *β. bipinnata.*—Fronds bipinnate; perhaps not a constant form. Fertile fronds of this variety originated the *O. obtusilobata*, Schkr. Pêche River, and near Cantley, Hull, D. M'Gillivray, M.D.

ASPLENIUM.

A. Trichomanes, Linn.—Frond small, narrow, linear, pinnate; pinnæ roundish-oblong or oval, oblique, almost sessile, crenate: rachis blackish brown, shining, margined; sori distant from the midrib. *Asplenium Trichomanes*, Linn., Moore, Gray, &c., Lowe's Ferns, vol. v. pl. 22. *Asp. melanocaulon*, Willd., Pursh. Fl. Sept. Americ. ii. p. 666. *Asp. anceps*, Lowe.—Inhabits rocky river banks, &c., but is not common in Canada. On rocky banks, at Marble Rock, on the Gananoque River; Namainse, dry ground on the top of a mountain, R. Bell, jr.; rocky woodlands west from Brockville, rare, B. Billings, jr.; Montreal, Jones's Falls and Niagara, P. W. Maclagan, M.D.; Lake Medad, Hamilton, Judge Logie; Pittsburg, near Kingston, John Bell, B.A.; Pied du cap Tourmente, M. L'Abbé Provancher; near Belleville, J. Macoun.

β. delicatulum.—Frond narrower, pinnæ much smaller, thinner, and wider apart than in the normal form. This is a sub-variety, passing by intermediate states into the typical plant, which is the common form of northern Europe. The variety is the prevalent form in Canada, but also occurs farther south in the United States, for I have specimens from Catskill (A. O. Brodie), and is not confined to the American continent, for Professor Caruel, the acute author of " Flora Italiana," sends specimens of a similar form from Florence. There is an *Asp. Trich.* var. *majus* in Cuba (according to Mr Eaton's Enumeration of Wright's Cuban ferns). *A. anceps* is a Madeiran form, not distinguishable, so far as I can see, from common European states of *A. Trichomanes*.

A. viride, Hudson.—Frond small, linear, pinnate; pinnæ roundish-oblong or oval, more or less cuneate at base, slightly stalked, crenate or slightly lobed; rachis bright green; sori approximate to the midrib; in outline of frond and general aspect resembles the preceding species. *A. viride*, Hudson, Flora Anglica, 385 ; Sm., Bab., Moore, &c. *A. Trichomanes, β ramosum*, Linn.—This beautiful alpine fern was found in Canada for the first time last summer, having been collected in considerable quantity at Gaspé, C.E., by John Bell, B.A., who formed one of a party of the Provincial Geological Survey. It was previously known to occur sparingly in N.W. America, at one spot on the Rocky Mountains, and in Greenland. Mr Bell's discovery of its occurrence in Gaspé is therefore extremely interesting in a geographical point of view. The Gaspé specimens although young, agree perfectly with the typical European form of *A. viride*, of which I have a full series of Scotch examples, as well as others collected in Norway by T. Anderson, M.D. In young specimens the pinnæ are usually large, thin, and more cuneate and lobed than in the mature plant, in which they are roundish-ovate.

A. angustifolium, Michx.—Frond large (1 to 3 feet high), annual, lanceolate, pinnate; pinnæ long, linear-lanceolate, acute ; fertile fronds more contracted than the barren ones, "bearing sixty to eighty curved fruit dots on the upper branches of the pinnate forking veins," (Eaton). *A. angustifolium,* Michaux, A. Gray, Eaton, J. Smith, Lowe's Ferns, vol. v. pl. 24.—In Canada this fern appears to be confined to the extreme south-western point of the province;* Malden, P. W. Maclagan, M D.; at the Oil Wells, township of Enniskillen, Lady Alexander Russell. For information of the latter station I am indebted to the kindness of Judge Logie of Hamilton. This fern appears to be still rare in cultivation among the fern fanciers of Europe. It was introduced to Britain in 1812 by Mr John Lyon of Dundee.

A. ebeneum, Aiton.—Frond erect, lance-linear, pinnate; pinnæ numerous, lanceolate (the lower oblong), sessile, slightly auricled at base and finely serrate; rachis blackish-brown, shining. *Asplenium ebeneum,* Aiton, Hortus Kewensis, ed. 2, vol. v. p. 516, Gray, Eaton, J. Smith, Lowe's Ferns, vol. v. pl. 2. *A. polypodioides,* Schkr.— Rocky woods, Brockville, B. Billings, jr.; the only locality in Canada from which I have seen specimens.† Although so rare with us, this species appears to be not uncommon in the United States. Gray speaks of it as "rather common;" I have specimens from Schooley's Mountains, West Point, N. Y., Providence, Philadelphia, &c. Judging from Mr Eaton's indication in Chapman's Flora, it again seems to decrease in the south, so that its present headquarters are in the Northern States.

[*A. marinum,* Linn.—Frond broad and leafy, linear-lanceolate, tapered above, pinnate; pinnæ ovate-oblong or linear, oblique, shortly stalked, rarely pinnatifid, the upper ones confluent, stipe brownish, rachis brown below, green and winged above, sori large, linear, oblique; grows on rocks. *Asplenium marinum,* Linn., Moore, J. Smith, &c. *A. latum,* Hort.—New Brunswick, E. N. Kendal, in Hook. Fl Bor. Am. I cannot learn that this fern has been subsequently found in North America, and hope, therefore, that botanists will look for it on the rocky shores of New Brunswick. It usually grows out of the crevices of shore cliffs, and is very limited in its geographical range, growing, according to Moore, only in the western part of Europe, crossing from Spain to Tangiers on the African coast, and being again met with in Madeira, the Azores, and Canary Isles.]

A. thelypteroides, Michaux.—Fronds large oblong-ovate, pinnate; pinnæ lanceolate, acuminate, from a broad sessile base, and deeply pinnatifid, the lobes oblong, minutely toothed. *Asplenium thelypteroides* Michaux, Pursh, Bigelow, Torrey, Beck, Darlington, Gray, Eaton. *Diplazium thelypteroides,* Presl, J. Sm.—In rich woods, De Salaberry, west line, W. S. M. D'Urban; Mirwin's woods, &c., Prescott, B. Billings, jr.; Beloeil Mountain, P. W. Maclagan, M.D.; moist woods near the Hop Garden, Belleville, rare, J. Macoun (a deeply serrated, leafy form); Ramsay, J. K. M'Morine, M.A.; St Joachim, M. L'Abbé Provancher; London, W. Saunders. Not a common fern in Canada; perhaps more plentiful in the United States. I have a fine series of specimens from Schooley's Mountains (A. O. Brodie), and others from Providence.

* Subsequently found in the Belleville district by Mr Macoun.
† *Subsequently* found near Belleville by Mr Macoun.

β. *serratum.*—Lobes of the pinnæ ovate-oblong, approximate, strongly and incisely serrate. This may be regarded as a sub-variety.—Belleville, J. Macoun.

[*A. montanum*, Willd., which extends along the Alleghanies, has not yet been found in Canada, but may possibly occur. It grows on cliffs.]

[*A. Ruta-muraria*, Linn.—The wall-rue, a small species, which grows in the crevices of limestone cliffs in the Northern States, and is common on stone walls and old buildings in Britain, is to be looked for in Canada.]

ATHYRIUM.

A. Filix-fœmina, R. Br.—Frond ample (1–3 feet long), broadly oblong-lanceolate, bipinnate; pinnæ also lanceolate; pinnules ovate-lanceolate or oblong, incisely toothed. Grows in large tufts, the fronds delicate, of a bright green hue. Lady Fern of the poets. *Athyrium Filix-fœmina*, R. Br., Spreng., Roth., Hook., Moore, &c. *Aspidium Filix-fœmina*, Swartz, Pursh, Beck. *Aspidium asplenioides*, Swartz, Willd., Pursh. *Asplenium Athyrium*, Schkr. *Asplenium Michauxii*, Spreng. *Asplenium Filix-fœmina*, A. Gray Man., p. 595. *Nephrodium asplenioides* and *Filix-fœmina*, Michx. *Asplenium angustum*, Willd., Pursh.—Common in the woods, as near Kingston, Toronto, Trenton, &c.; Pêche River, Ottawa, Dr M'Gillivray; Temiscouata, Chippawa and Malden, P. W. Maclagan, M.D.; Belleville, moist woods, very common, several varieties, J. Macoun; Ramsay, Rev. J. K. M'Morine, M.A.; mouth of the Awaganissis Brook, Gulf of St Lawrence, C.E., and Schibwah River, Lake Superior, R. Bell, jr.; Cemetery grounds, Hamilton, and on Princes Island, Judge Logie; Hamilton's Farm and base of Silver Mt, W. S. M. D'Urban; Mountain Fall, H. B. T., Governor M'Tavish; Snake Hill River, John C. Schültz, M.D.; L'Anse à la Barbe, Gaspé and L'Orignal, John Bell, B.A.; St Tite, M. L'Abbé Provancher; London, W. Saunders.

β. *angustum.* — Frond narrow, linear-lanceolate; pinnæ rather crowded; pinnules not pinnatifid, but incisely toothed, with recurved margins; sori short, curved (*Aspidium angustum*, Willd.?)—Farmersville; Delta; Belleville, J. Macoun.

γ. *rhæticum.*—Frond rather small, firm, narrowly lanceolate in outline; pinnæ more or less distant, and narrowly lanceolate; pinnules incisely toothed or deeply pinnatifid, linear, or more frequently lanceolate-acute, and acquiring a linear aspect from the reflection of the lobes, often crowded with confluent sori.—Dr Yates's farm, on the banks of the St Lawrence, near Kingston; near Montreal, Rev. E. M. Epstein, M.D.; near Lakefield, North Douro, Mrs Traill.

δ. *rigidum.*—Frond small, rigid; pinnules approximate, connected at the base by a broad decurrent membrane, sori confined to the lower part of each pinnule.—Lakefield, North Douro, Mrs Traill.

There are other forms of this species, dependent in many cases, no doubt, upon situation; some with thin veiny fronds of great size, bearing few scattered sori. One form, very like the British var. *molle*, was gathered at Belleville by Mr Macoun. I know no fern more variable than this. Our Canadian forms require careful examination.

WOODWARDIA.

W. virginica, Willd.—Frond pinnate ; pinnæ lanceolate, pinnatifid ; sori arranged in line on either side of the midribs of pinnæ and pinnules. *Woodwardia virginica*, Willd.; A. Gray Man. p. 593. (*Doodia*, R. Br.)—Millgrove Marsh, C.W., Judge Logie ; sphagnous swamp near Heck's Mills, ten miles from Prescott, Augusta, C.W., B Billings, jr.; Pelham, C.W., P. W. Maclagan, M.D.; Belleville, J. Macoun.

(*To be continued in next number.*)

REVIEWS AND NOTICES OF BOOKS.

1. *Victoria Toto Cœlo ; or, Modern Astronomy Recast.* By JAMES REDDIE, F.A.S.L., Hon. Mem. Dial. Soc. Edin. Univ. London, 1863. Hardwicke.
2. *Quadrature du Cercle.* Par un Membre de l'Association Britannique pour l'avancement de la Science (JAMES SMITH). Bordeaux, 1863.

Well said the poet, though he did not anticipate the present application of his lines—

> " The times have been,
> That, when the brains were out, the man would die,
> And there an end : but now they rise again,
> With twenty mortal murders on their crowns,
> And push us from our stools."

Quietly ignored by the British Association, of which they are members, or, worse, informed politely that their lucubrations are of that order which the French Academy (" in consequence of a resolution already ancient") so wisely refuses to entertain, Messrs Reddie and Smith rush into print in Latin and French, the so-called languages, *par excellence*, of true science, with the barely concealed hope that at last the world will do them justice. So far their cases are parallel ; but in respect of patient endurance of well-merited neglect, or still more richly-deserved castigation, at the hands of common-sense, Mr Smith is far in advance of his fellow-martyr. He deserves, therefore, some little consideration (which we can hardly extend to Mr Reddie, for a reason presently to be shown), and we shall endeavour in a word or two to show where and how he errs.

It is certain, even to Mr Smith, that the area of a circle is half the product of its radius and circumference.

It is also certain that the circumference is proportional to the radius. In scientific works this ratio (whatever it may be) is 'ed by $2\pi : 1$ where π is some definite number.

From these two *facts* it follows that the area of a circle is proportional to the square of its radius, and that the proportion is $\pi : 1$. These form the starting-point of Mr Smith's work.

Now, if we merely compare one circle with another, as to area or circumference, we shall have continual verifications of the foregoing facts, whether we assume π to be $3\frac{1}{8}$, as Mr Smith does, or 20,000,000,000,000, which we would advise him to try. All the numerical work, (with a slight exception) in Mr Smith's pamphlet, consists of deductions of one of the above facts from the other two, and would be equally successful whatever number we assume for π.

Mr Smith, of course, breaks down when he comes to apply his supposed value to compare the circumference of a circle with that of an inscribed polygon (which we cannot but suppose he will allow to be shorter). But if π be $3\frac{1}{8}$, the circumference of a circle whose radius is 1 is 6·25, and that of the inscribed polygon of twenty equal sides (which can be constructed by pure geometry, and is therefore *not* liable to suspicion, as results of logarithmic tables are supposed to be by squarers of the circle) is 6·2573... as Mr Smith may easily assure himself by calculation. How does he get over this? But we need not inquire, nothing is impossible to a squarer of the circle.

Mr Reddie's former difficulty lay in his not being able to conceive the idea of mass or inertia as distinct from, and capable of existence without, weight. In this he was, of course, entirely wrong and unutterably silly.

He has now discovered that the only motions we are cognisant of are *relative*, and here he is indisputably right, though a little late in the field. But when he asserts, as a consequence, that Newton's view of the solar system, or rather that of Copernicus and Kepler, falls to the ground, he has completely ignored the fact that Newton deals with *relative* motion only. He will find his main argument (that the moon's path is throughout concave towards the sun) in most good treatises on astronomy; and will find, if he would inquire into what has been done before proceeding to act for himself, that all this is perfectly consistent with Kepler, gravitation, laws of motion, &c., &c. So much for his latest performance; where on earth or in heaven will he *next* break ground?

But a word with him before we close. We have done no more than hold up to deserved, but *kindly*, ridicule his preposterous ideas, which, since he not only published them, but *sent* them to us for review, were public property, so far as our honest but discerning criticism was concerned. He retorts upon us as " a shabby hack" (p. 36). We were " mean enough deliberately to make a mis-statement, finding nothing we dared gainsay" (p. 52). We are " a mendacious and impertinent writer" (p. 52). We have

" put ourselves out of the category of gentlemanly writers" (the same pregnant p. 52), because we animadverted on the conduct of certain journals in praising his curiously absurd essay, instead of pointing out its errors, or passing it over in silence. *Tantæne animis cœlestibus iræ?*

To the last of the above charges we are afraid we must again expose ourselves. Mr Worms, of whom we had hoped we had heard the last, has permitted the publication, by Mr Reddic, of a note in which he states that he considers our estimate of his book (*ante*, vol. xvii. p. 104) "quite beneath his notice." Messrs Worms and Reddic, though *both* in egregious error, do not very cordially agree—except in their opinion of us. Now to every word of our remarks on Mr Worms we most positively adhere, and we are profoundly afflicted (for the sake of British science) that a journal of such deservedly high character as the "Athenæum" has in this instance (though it *did* detect Mr Reddic) unguardedly given its sanction to a most absurd mixture of original nonsense and spoiled extracts. Let us add in conclusion, that our good-humour and impartiality are not to be soured or distorted by abuse, and that, in spite of the terrible epithets Mr Reddic has applied to us, we shall keep on the gloves as usual in our next encounter with him, which, unless appearances are deceitful, is not likely to be long deferred.

Sketch of Elementary Dynamics. By W. Thomson and P. G. Tait. Edinburgh : Maclachlan and Stewart.

We do not remember ever to have seen a work on dynamics in which the fundamental principles of the science were more clearly and beautifully explained than they are in the pamphlet which we have just perused.

Any one who has glanced over the volumes on elementary mechanics (and their number is legion) which form the text-books of our schools, must, we think, have perceived in most of these a want of philosophical method and clearness of conception displayed by the author in his treatment of the subject.

We appeal especially to the student who has read one of these text-books for information, if it did not require a very great mental effort on his part to recognise the different branches as really forming one science. We think that such an one must have often wondered at the unnatural separation maintained between the parallelogram of forces and that of velocities, and sometimes also speculated as to what becomes of the motion when two non-elastic bodies impinge against each other in opposite directions.

Then, again, in works of a somewhat more ambitious character, in which the subject is treated analytically, the mode by which you are led to a result is very much the same as that by which the railway traveller who journeys by night arrives at his destination—the process is not invigorating, and there is no appreciation of the beauties of the way. So when you have once formed your differential equation, and abandoned yourself to the labour of solving it, you have quite lost sight of the physics and the geometry of the problem.

In the present pamphlet, which we are content to view only as an instalment, the authors have, we think very successfully, introduced much of the beauty of the geometrical method with very little of its cumbrousness. They have also availed themselves to the fullest extent of that great discovery of modern times—the conservation of energy; and we shall afterwards adduce an example of their happy employment of this principle. We must add to this a philosophical arrangement and a unity of conception in order to characterise the book. So novel are many of its solutions, that our first emotion is that of surprise, as we recognise some very old friend in his new dress; while our second is altogether one of pleasure from his improved appearance.

We now quote a few lines from the commencement of the volume:—

" Dynamics is the science which investigates the action of force. Force is recognised as acting in two ways,—1*st*, So as to compel rest or to prevent change of motion; and, 2*d*, So as to produce or to change motion. Dynamics, therefore, is divided into two parts —Statics and Kinetics.

In Kinetics it is not mere *motion* which is investigated, but the relation of *force* to motion. The circumstances of mere motion, considered without reference to the bodies moved or to the forces producing the motion, or to the forces called into action by the motion, constitute the subject of a branch of pure mathematics, which is called ' Kinematics.' "

The first part of the pamphlet is therefore devoted to Kinematics. Here, after defining velocity, it is shown, that if two velocities be represented by the sides of a parallelogram, their resultant is represented by its diagonal. Acceleration, in its most general sense, is defined as the " *rate of change of velocity, whether this change take place in the direction of motion or not;*" also, " the *moment* of a velocity or a force about any point is the rectangle under its magnitude, and the perpendicular from the point upon its direction." It is then shown, that the moment of the resultant velocity of a particle about any point in the plane of the components is equal to the algebraic sum of the moments of the components.

We have entered thus into detail in order to introduce to our readers the following very beautiful proof of the " equable description of areas :"—

" When the acceleration is directed to a fixed point, the path is in a plane passing through that point ; and in this plane the areas traced out by the radius-vector are proportional to the time employed. . . . Evidently there is no acceleration perpendicular to the plane containing the fixed and moving points, and the direction of motion of .the second at any instant ; and there being no velocity perpendicular to this plane at starting, there is therefore none throughout the motion ; thus the point moves in the plane. Again, if one of the components (of the resultant velocity of a particle) always passes through the point (about which the moments are reckoned), its moment vanishes. This is the case of a motion in which the acceleration is directed to a fixed point, and we thus prove, that in the case supposed, the areas described by the radius-vector are proportional to the times ; for the moment of velocity, which in this case is constant, is evidently double the rate at which the area is traced out by the radius-vector."

We have only time, before leaving this division of the subject, to request our readers' attention to the novelty and beauty of the authors' proof that the evolute of a cycloid is a similar and equal cycloid ; and with this remark we pass on to the second part of the pamphlet, in which dynamical laws and principles are discussed. And first, we note that the term *centre of inertia* is advantageously substituted for *centre of gravity*.

" The *centre of inertia* of a system of equal material points (whether connected with one another or not), is this point whose distance is equal to their average distance from any plane whatever. The centre of inertia of any system of material points whatever (whether rigidly connected with one another, or connected in any way, or quite detached), is a point whose distance from any plane is equal to the sum of the products of each mass into its distance from the same plane divided by the sum of the masses.

" The sum of the momenta of the parts of the system in any direction is equal to the momentum, in that direction, of the whole mass collected at the centre of inertia."

The following is the authors' statement of the second law of motion, and its corollary the parallelogram of forces :—

" *When any forces whatever act on a body, then, whether the body be originally at rest or moving with any velocity and in any direction, each force produces in the body the exact change of motion which it would have produced if it had acted singly on the ⁓‵‾ originally at rest.*

" A remarkable consequence follows immediately from this view of the second law. Since forces are measured by the changes of motion they produce, and their directions assigned by the directions in which these changes are produced; and since the changes of motion of one and the same body are in the directions of, and proportional to, the changes of velocity, a single force, measured by the resultant change of velocity, and in its direction, will be the equivalent of any number of simultaneously acting forces. Hence,

" *The resultant of any number of forces (applied at one point) is to be found by the same geometrical process as the resultant of any number of simultaneous velocities.*"

After stating the third law of motion, which asserts the equality of action and reaction, the authors then deduce, as immediate consequences of the second and third laws, the following amongst other important propositions :—

" (*a*) The centre of inertia of a rigid body moving in any manner, but free from external forces, moves uniformly in a straight line.

" (*b*) When any forces whatever act on the body, the motion of the centre of inertia is the same as it would have been had these forces been applied, with their proper magnitudes and directions, at that point itself."

We present the following as the authors' view of the *conservation of energy* :—

" A limited system of bodies is said to be *dynamically conservative*, if the mutual forces between its parts always perform or always consume the same amount of work during any motion whatever by which it can pass from one particular configuration to another. The whole theory of energy in physical science is founded on the following proposition :—

" If the mutual forces between the parts of a material system are independent of their velocities, whether relative to one another or relative to any external matter, the system must be dynamically conservative.

" For if more work is done by the mutual forces on the different parts of the system in passing from one particular configuration to another, by one set of paths than by another set of paths, let the system be directed, by frictionless constraint, to pass from the first configuration to the second by one set of paths and return by the other, over and over again for ever. It will be a continual source of energy without any consumption of material, which is impossible."

We have only time to direct the reader's attention to the beautiful proof given in page 40, that " a particle attracted to a fixed point, with a force inversely as the square of the distance, describes a conic of which the point is a focus ;" and

We conclude with the following ingenious application of the law of energy :—

"*When an incompressible liquid escapes from an orifice, the velocity is the same as would be acquired by falling from the free surface to the level of the orifice.*

" For we may neglect (provided the vessel is large compared with the orifice) the kinetic energy (energy of motion) of the bulk of the liquid ; the kinetic energy of the escaping liquid is due to the loss of potential energy of the whole by the depression of the free surface. Thus the proposition at once."

It is hardly necessary, after these quotations, to make any observation, except to recommend this little pamphlet, as one full of beautiful and ingenious solutions, and as well worthy of being read by all who feel an interest in the subject, but especially by those who mean to study the more complete treatise which the authors have promised to the public.

Climate : An Inquiry into the Causes of its Differences, and into its Influence on Vegetable Life. By C. Daubeny, M.D., F.R.S., &c., Professor of Botany and of Rural Economy in the University of Oxford.

This small volume contains four lectures delivered by Professor Daubeny before the Natural History Society of Torquay in February last, on Climate and its influence on vegetation. Whilst the subject is handled by the learned professor in its more popular and practical bearings, there is at the same time contained in these able lectures a large amount of valuable scientific information, embodying the recent results arrived at by meteorologists with respect to the atmosphere and its phenomena. In the discussion of the question, the author defines the climate of a country to be its relations to temperature, light, moisture, winds, atmospheric pressure, and electricity ; but amongst these the first place must be conceded to the intensity of the heat, and to its distribution over different portions of the year, as it is this which in a great degree regulates the other conditions, and is also itself of all others the one most indispensable for the exercise of the functions of vegetable and animal life. This important element of heat is derived from the living bodies scattered over the globe, from the various processes carried on by natural agencies or through the *instrumentality of* man, from the interior of the earth, and from solar radiation. Of these sources whence heat is derived, it is only necessary for the meteorologist to consider the heat received from the sun's rays, because that received from the other sources is too inconsiderable in amount to be appreciable, except in rare and exceptional cases

Professor Dove has calculated the normal temperatures of the different parts of the globe ; that is, what would be the temperature of each place provided the earth presented a solid uniform surface everywhere to the sun's rays. These normal lines are of necessity parallel to the equator, or coincident with the parallels of latitude— the temperature increasing on each side of the equator to 10° of latitude, and thence decreasing proportionally on either side toward the poles. But, on account of the unequal distribution of land and water, the irregularities of the surface, the prevailing atmospheric and oceanic currents, and other concurrent causes modifying the climates of particular localities, few places possess what may be called their normal temperatures, some falling short of it, but a still greater number exceeding the temperature assigned. Thus London, in latitude 51°·30, ought to have a mean temperature of 39°·0, whereas observation assigns to it one of 50°·8, or 11°·8 higher. Everywhere on the Atlantic shore of North America the temperature is lower than in Europe. The respective mean temperatures of Nantes in France and St John's in Newfoundland, both in corresponding latitudes, are 54°·9 and 38°·4,—the difference in favour of the former being 16°·5. If the temperature of different places be compared for the seasons, it will be found that the discrepancy is still more striking. Thus, whilst the winter temperature of Drontheim, lat. 63°, is + 23°·3, that of Yakoutzk, lat. 62°, is − 36°·4, showing an enormous difference of 60°·0 ; whilst the winter temperature of Quebec is 14°·2, that of Penzance, 4° nearer the pole, is 44°·2, or 30°·0 higher than that of Quebec. Again, the summer temperature of Edinburgh is 57°·2, while in Moscow, in the same parallel of latitude, it is 65°·4, or about 8°·0 higher.

" Climates, then, may be divided into equable and excessive, according to the degree in which the mean temperature of the summer and winter differs from that of the entire year ; and with reference to the growth of vegetables, far more importance must be attached to the heat prevailing during summer than to the mean temperature of the climate collectively taken.

" Thus even in Russia and Siberia fine crops of wheat and other kinds of corn are obtained, because the summer temperature rises to the requisite point for ripening the seed, whilst in the north of Scotland, the Orkneys, and the Faroe Islands, although the mean temperature of the year is higher, these crops do not succeed.

" It may be doubted, however, whether even the mean temperature of the summer season affords a sufficient clue to all the variations in the character of the vegetation which are attributable to heat. A plant is not like a spring which is pushed forwards a certain number of degrees by the application of a definite force, and when that pressure is removed, returns again to its original position; for when the stimulus of heat is applied to it, its organs undergo a degree of development, which they retain even although the temperature should afterwards be reduced. Hence it is necessary to note the extremes of temperature to which a country is liable, as well as the mean of its summer and winter climate.

" Should it happen, for instance, that the cold in a low or subtropical latitude ever approaches even for a single night to the zero of Fahrenheit, certain trees, such as the Orange, would infallibly perish; and hence they can never be indigenous in countries subject to such contingencies.

" The general climate of the British Isles is so exceptionally mild, that we have introduced the plants of warmer regions generally into cultivation, and begun to consider them as in a manner naturalized; but that they are not so, and could never have established themselves in the soil without the aid of man, became evident from the effects of the rigorous winter of 1860-61,—one of those seasons of unusual severity which, however, are sure to occur within a certain cycle of years, and to entail the destruction of all these denizens of a more temperate climate, which, rashly presuming upon the mildness of many preceding winters, had begun to regard this as their home."

Among the causes which affect climate are the relative distribution of land and water; the proximity to a large continent or to a wide expanse of sea; the chilling influence of mountains which retain the winter snow during the greater part of the year, or push down glaciers into the valleys; and the shelter afforded by forests or hills of moderate height. Of these, the relative power of land and water, with respect to the reception, radiation, and transmission of heat is the most important influence. The following extracts show the changes produced by different arrangements of sea and land in different countries :—

"Thus it is only on extensive plains, at a distance from snow-capped mountains, that a high range of temperature can maintain itself even during the day in a northern latitude. On the other hand, over such level tracts as those of Russia and Siberia, great heats prevail in summer even in comparatively northern regions.

" Moscow, for instance, in north latitude 55° 48', where the mean temperature of the coldest month is only 13° of Fahr., enjoys in July a heat of 66° 4'; and at Yakoutzk, a Siberian *own, in latitude 62°,* at which the mean temperature of January is 36° 37′ below zero, the thermometer in July rises to nearly 69° of Fahr.

" In islands, on the contrary, situated in northern parallels, the radiation of heat in summer exerts a much feebler influence; so that Stromness, in lat. 58° 57′, has a summer temperature of only 54°; and Unst, the most northern of the Shetland Islands, in lat. 60°, one only of 52°.

"Hence whilst at Christiania, in lat. 59° 55′, fine timber abounds, and crops of wheat and other grain ripen, inasmuch as the mean summer temperature rises nearly to 60°, none but the hardiest kind of barley will grow in the Hebrides, between the parallels of 56° and 58°; and the trees are there reduced to a few of the robuster species, which are both stunted and uncommon.

" In winter, however, the case is reversed. The extreme cold of extensive continents, such as Russia and Siberia, is caused in part by the radiation of heat from their surface during the long nights of these northern latitudes, and still more through their participation in the climate of regions more northerly than themselves, owing to the winds which commonly come from that quarter in the winter season, and which, bringing with them the temperature of the Arctic circle, first condense the moisture into snow, and afterwards impart to the countries they pass over the dry and cutting cold which characterises them.

" But on the sea the circumstances are different, owing to a property peculiar to water, which would seem specially designed as a provision for mitigating the intensity of cold.

" This is its arriving at its greatest density, not at the point at which it freezes, but 8° above, so that whilst it goes on progressively contracting in volume down to 40°, it afterwards again expands, until it falls to the temperature of 32°. . . . .

" Hence owing to this constant circulation of the lighter and heavier portions of the water, the whole must attain the temperature of 40° before any ice is formed upon its surface, and accordingly it is hardly possible that a deep lake should be frozen over even by the longest and most intense frost that can occur.

" The effect of this constant circulation throughout all portions of a body of water in mitigating the severity of insular climates is sufficiently apparent.

" The ocean may, in fact, be regarded as a store-house of heat, which it dispenses to the air passing over its surface, thus rendering it impossible that the latter should ever attain the same extreme degree of cold which it acquires on a continent.

" Hence the equable character of the climate in insular situations, which has been pointed out as prevailing during the summer, holds good also, for the reasons just given, in the winter likewise."

Again, the currents of the atmosphere and of the ocean exercise a powerful influence over the climate of a country; and it is to the influence of two such currents that the British Isles owe their singularly mild, equable, and healthy climate. The Florida Gulf stream impinges on our western shores, bringing with it the warmer temperature of southern latitudes; and since the prevailing direction of the wind is south west, a higher temperature is imparted to the British Isles in common with the whole of Western Europe. The south-western parts of Ireland and Great Britain receive the greatest benefit, and these places accordingly are the favourite residences of invalids during winter, and they may also be noticed as famous for the early growth of vegetables for the London market. The western coasts of South America may be referred to as illustrating the effects of a cold oceanic current flowing from the Polar Seas, and thus reducing the temperature of those parts by which it passes. Thus the mean temperature of Callao in 12° S. Lat. is only 68°·9, while that of Rio Janeiro in 23° S. Lat. is 73°·8.

The following important practical remark is made in discussing the temperature of the soil:—

" It is known that every plant requires a certain amount of heat, varying in the case of each species, for the renewal of its growth at the commencement of the season.

" Now when this degree of heat has spurred into activity those parts that are above ground, and caused them to elaborate the sap, it is necessary that the subterranean portions should at the same time be excited by the heat of the ground to absorb the materials which are to supply the plant with nourishment. Unless the latter function is provided for, the aërial portions of the plant will languish from want of food to assimilate. Indeed, it is even advisable that the roots should take the start of the leaves, in order to have in readiness a store of food for the latter to draw upon."

It is only in well-drained well-pulverised soils that plants enjoy the fullest benefit of this consideration; for, as has been shown in the Scottish Meteorological Society's Report for the quarter ending December 1862, the effect of draining is,—by drawing off the water, and thus filling the interstices of the soil with air which is about the worst conductor of heat,—to prevent the penetration of frosts into such soils to any great depth.

The next most important element of climate is the moisture of the air.

" We must, however, carefully distinguish between the humidity present in the atmosphere in the form of vesicular vapour, and that existing in it in an invisible or aëriform condition.

" All liquids have a tendency to pass into vapour, until checked *by the pressure* of their own atmosphere; and this tendency, which goes by the name of the tension of vapour, increases in each case in proportion to the advance of its temperature.

" Whenever, therefore, the soil and subsoil are not entirely destitute of humidity, the atmosphere above must contain a certain amount of water, not as vesicular vapour, but in a perfectly aëriform condition.

" Now the effects produced upon living beings by the presence of moisture in the atmosphere are entirely different, according as it exists in one or other of these states.

" Vesicular vapour, manifesting itself in the form of fog or mist, causes, as every one knows, a sensation of chill, owing to the abstraction of heat from our persons, caused by the moisture which attaches itself to them, and likewise, for the same reason, interferes with the healthy functions of the skin, and even of the lungs.

" But in an aëriform condition the very opposite effect takes place.

" Professor Tyndall has lately pointed out that humid air, or air containing much moisture in a transparent or an aëriform condition, exerts a remarkable influence both in absorbing and in radiating heat. Owing to the former property, aqueous vapour acts as a kind of blanket upon the ground, and contributes in a very striking manner to the retention of its heat.

" Hence when the air is perfectly divested of moisture, as in the sandy deserts of Africa, in Siberia, and even in Australia, the cold at night is almost insupportable, owing to the absence of that protection which is afforded by aqueous vapour when present in the atmosphere ; whilst during the day the rapid abstraction of moisture from the surface of plants and animals, caused by the dryness, is equally deleterious to both.

" And as the radiation of heat from a body is always equal to its power of absorbing it, it follows that air containing much moisture will, when it rises into the higher regions, sink rapidly in temperature, in consequence of the heat it sends forth into space ; and, indeed, according to Tyndall, the amount radiated from air saturated with moisture is 16,000 times as great as that of air perfectly dry.

" One cause, therefore, of the profuse rains that occur in the tropics may be the cooling of the heated air, which rises from the earth into the higher regions of the atmosphere, and which, when it arrives there, radiates its heat freely into space, and thus has its capacity for moisture reduced.

" Professor Tyndal calculates, that 10 per cent. of the heat radiated from the earth in this country is stopped by 10 feet of the air which lies nearest to the ground.

" *It would* appear from the recent investigations of M. Duchartre in France that the refreshing influence of rain and dew upon plants does not arise from the moisture deposited upon their surfaces being absorbed,—for not a particle of water enters into them through the leaves or stem,—but from its supplying the ground with water for the roots to absorb; and likewise, as I should infer, from the check which its presence affords to a too rapid radiation of heat from the parts above ground, by which the prejudicial effects of a too rapid cooling are provided against."

The last two lectures are devoted to the discussion of the effect of these influences on vegetation, and the way on which climate determines the vegetation of a country. Palms and most other monocotyledonous trees are ill suited to the climates of temperate regions, because, from the liquid nature of the sap with which they are filled, they are liable to freeze,—an act which, by the expansion it occasions, proves of most fatal consequence at all times to the vegetable organisation. On the other hand, dicotyledonous trees, which belong to temperate countries, are protected from frosts either by numerous layers of bark, abounding in air-vessels, or are provided with essential oils and other juices not susceptible of freezing, such as are found in the bark and wood of conifers. The valuable rule on this subject laid down by Decandolle may be briefly stated. The power each plant possesses to resist the extremes of temperature is in direct proportion to the oils contained in the juices, the viscidity of the juices, the smallness of the cells and vessels, and the quantity of air entangled between the parts of the vegetable tissues, or the protection afforded by hairs, down, or air-vessels; whilst the liability to suffer from great heat or from frost is proportional to the quantity of water in the tissues, the mobility of the juices, the size of the cells and vessels, and the absence of air in the external layers.

But the summer and winter climates of a country are what chiefly determine its vegetation.

" There are certain plants, like the Vine, which require an intense heat in summer to bring their fruit to maturity, but yet are capable of resisting a severe cold during winter; they thrive, for instance, on the borders of the Rhine, or in Switzerland, where the winters are very severe; but scarcely even ripen their fruit in England, even in Devonshire or Cornwall, where the thermometer rarely falls to the freezing point of water.

" Other plants, on the contrary, like the Myrtle, cannot resist cold, but do not demand during any part of the year an exalted temperature.

" Thus they luxuriate near the southern coasts of England, but do not show themselves on the Continent, until we reach a much lower latitude than that of this country.

" *The former* class of plants, therefore, may be said to be adapted for an excessive climate, and therefore thrive best on continents ; the latter for an equable one, and consequently succeed most upon islands.

" It may also be remarked, that annuals are adapted for continental climates, as they require heat for the ripening of their seeds, but die away in winter; whilst perennials are better calculated for islands, as it is essential that the winter should not be so rigorous as to destroy them, but not equally necessary that the summers should be always hot enough to ripen their seeds, a failure in this respect for several successive years not entailing the destruction of the species."

Here the extreme temperatures of a country, as distinguished from the mean temperature both of its seasons and its year, must be carefully kept in view. For that which determines the flora of a country, or the plants which will thrive and live in its climate, is not whether the plants will live through such winters as ordinarily occur, but whether they will survive the severest frosts that have happened, and may therefore be expected to recur in a succession of years. But the number of species which admit of cultivation is greater than that of those which grow wild, inasmuch as the former can be maintained by the art of man protecting them during winter, or replacing them with new plants when the old ones happen to be destroyed by frost. This is particularly the case with those whose cultivation is remunerative. As man has little or no power to change the climate, it is only by studying the climate of his neighbourhood in all its details that he can take every advantage of it, and thus rise superior to it, and in some sense overcome it. The subject of local climate is one deserving more attention than it has yet received. The exposure of a piece of ground, its sloping or level character, its relatively low or high position, the nature of its subsoil or its drainage, its sheltered or exposed situation, and its relation to moisture or dryness, may all conspire to give it as mild and genial a climate as is enjoyed some degrees southward.

We conclude this notice with the following extract, showing the effect different seasons have on agricultural produce :—

" It may be useful for the farmer to possess the data for estimating the influence which a summer, warmer or colder, wetter or drier than ordinary, has exerted upon the production of his farm, so as not to be misled in his calculations as to the advantages or disadvantages of any novel plan of cultivation.

" These data have in part been supplied by Mr Lawes in an elaborate paper published in the eighth volume of the Royal Agricultural Society's Journal for 1848, in which he shows, that in 1844, 1845, and 1846, the difference in the amount of produce was in accordance with the general character of their respective seasons.

"Thus it will be seen by turning to the table referred to in the note,* that, in 1844, when there were only 81 rainy days, and when the mean summer temperature was 57° 5', the farm produced 16 bushels of corn to the acre, and the weight of the bushel was 60½ lb., whilst the grain bore to the straw as high a ratio as about 82 to 100.

"In 1846, when there were 93 rainy days, and when the mean temperature was as high as 59°, the yield was about 17 bushels per acre, and the weight of a bushel 68 lb.; but the proportion of grain to straw was lower, namely 76 to 100, showing that the yield of corn had been influenced by temperature, but that the quantity of straw had been increased by the amount of rain.

"Lastly, in 1845, when the number of rainy days was greater, though the temperature had been lower than in either of the two other years,—the former being 110, the latter 55,—the yield was much greater, amounting to 23 bushels per acre, but the weight of the corn per bushel less—namely, 56 lb.; and the increase in the produce of straw such, that the grain only bore to it the proportion of 49 to 100.

"These figures appear to show that, although the greater quantity of rain was favourable to the amount of grain, yet that it tended to increase in a still greater ratio that of straw; and that the higher the temperature of the year was, the heavier the grain would prove, so as to make up in quality for its deficiency in actual quantity.

"Thus it would appear, that, under the same treatment, the produce may vary in the proportion of 7 bushels per acre, according to the difference of season, which is equivalent to one quarter of the normal produce; or, calculating this at 30 million quarters in a good year, may be as much as 7½ million deficient."

* The following table indicates the effect of climate upon the quantity and quality of the produce of the unmanured piece of the experimental wheat-field, (during three seasons); the average results of the variously manured, &c., are also given:—

	1844.	1845.	1846.
Corn per acre in bushels, ...	16	28	17·25
Straw per acre in pounds, ...	1120	2712	1513
Weight of Corn per bushel in pounds, . .	58½	56¼	68½
Percentage of Corn to straw (straw 1000), .	821	534	797
Mean of all the Plots.			
Weight of Corn per bushel in pounds, . .	60½	56¼	68
Percentage of Corn in straw (straw 1000), .	868	490	765
Mean temperature, ...	57° 5'	55° 2'	59° 1'
Rainy days in 30½ weeks, ...	81	110	93

Manual of the Metalloids.   By JAMES APJOHN, M.D., F.R.S.,
M.R.I.A., Professor of Chemistry in the University of
Dublin.   London: Longman & Co.   1864.   12mo.   Pp.
596.

In his preface the author states that the work has been written
as one of Galbraith and Haughton's series of Manuals on different
branches of science.   The object has been to produce a condensed,
but at the same time tolerably comprehensive treatise, in which
no topic of importance should be omitted, while all would be discussed with as much brevity as is consistent with clearness.   It
is intended as a handbook in Chemistry for students in medicine
and engineering.

In the Introduction, the laws of Combination, Chemical Notation
and Nomenclature, the Relations of Atomic Weights, the Law of
Volume, Atomic Volume, the Unitary System of Atomic Weights,
Isomerism, Chemical Formulæ, Isomorphism, Dimorphism, Reaction of Bodies on each other, the causes which determine Decompositions, and the Division of Simple Bodies are considered.   The
following are the fifteen substances which he considers as metalloids :—Oxygen, Hydrogen, Nitrogen, Sulphur, Selenium, Tel-
lurium, Chlorine, Bromine, Iodine, Fluorine, Phosphorus, Arsenic,
Boron, Silicon, and Carbon.   The use of the simpler forms of
algebraic calculation has enabled the author to give precision and
brevity to many of his statements.   At the present day, it is
clear that any one who wishes to study chemistry, must be properly instructed in arithmetic and the elements of algebra.

In speaking of coal gas, Dr Apjohn remarks :—

" *Coal Gas.*—A mixture of inflammable gases, to be burned for
the light, and sometimes for the heat it yields, has been derived
from different sources, viz., from turf, vegetable and animal oils,
and from rosin.   The gas from turf has a low illuminating power,
and is now seldom made; and that yielded by rosin and the oils,
though of good quality, is relatively so costly, that its manufacture has also been abandoned.   Our streets and houses are now
illumined exclusively by gas derived from bituminous coal; and
the subject, therefore, of the gaseous hydrocarbons cannot be considered as sufficiently discussed without some remarks upon the
preparation, composition, properties, and uses of this most important combustible substance.   As an introduction to this subject,
attention is directed to the following table, which gives the ultimate
composition in 100 parts of a Welsh anthracite, of five specimens
of English bituminous coal, and of a variety of brown coal, or

lignite. The last two columns of the table give, the one the Ash, the other the Coke, yielded by each specimen.

	Carbon.	Hydrogen.	Nitrogen.	Sulphur.	Oxygen.	Ash.	Coke.
Anthracite, Wales . . .	91·44	3·36	0·21	0·79	2·58	1·52	92·20
Coking Coal, Newcastle .	81·41	5·83	2·05	0·75	7·90	2·07	66·70
Cannel Coal, Wigan . .	80·07	5·53	2·12	1·50	8·09	2·70	60·36
Coal, Wolverhampton . .	78·57	5·29	1·84	0·39	12·88	10·30	57·21
Wallsend Coal, Elgin . .	76·09	5·22	1·41	1·53	5·05	10·70	58·40
St Helen's Coal, Lancashire	75·80	5·21	1·92	0·90	11·89	5·17	65·50
Methill Brown Coal . .	65·96	7·78	0·96	0·75	9·23	15·32	...

" All these coals but the first include a considerable amount of bitumen, whose principal constituents are carbon and hydrogen; and this, when exposed to an elevated heat in cast-iron or fireclay tubes, called *retorts*, is resolved into coke, which remains in the retorts, and various volatile products, which are expelled; these latter being tar, an ammoniacal liquid called gas liquor, and a mixture of various gases.

" The tar and gas liquor are by a reduction of their temperature condensed, and the gas, after having undergone a certain process of purification, is conducted into the *gasometer*, from whence, by a regulated pressure, it is transmitted through metal pipes to the various points at which it is to be consumed. The retorts used in the process are usually 7 feet in length and about 1 foot in diameter, the cross section being, not a circle, but an ellipse. Five of them are usually set in the same furnace, and, when sufficiently heated, each is charged with 150 lbs. of coal. It takes from four to five hours to work off this charge, and the products per cent. are the following, being the means of four experiments by Mr Barlow on the Pelton main Newcastle coal :—

		lbs. per ton.	
Gas .	. . .	475·0	21·20
Coke .	. . .	1540·0	68·75
Tar .	. . .	112·5	5·02
Gas Liquor	. .	112·5	5·03
		2240·0	100·00

The specific gravity of the gas was 0·653, and its volume from 1 ton of coal, 9500 cubic feet; or 4·24 cubic feet for every pound of coal.

" The crude aëriform product, as has been already stated, is not a single gas, but a mixture of several, of which the following may be considered as a pretty complete enumeration :—

" Hydrogen	. . . . .	H.
Marsh Gas	. . . . .	C_2H_4.
Olefiant Gas	. . . . .	C_4H_4.
Butylene (Oil Gas), traces	. .	C_8H_8.
Carbonic Oxide	. . . .	CO.
Carbonic Acid	. . . .	CO_2.
Sulphide of Hydrogen	. . .	HS.
Ammonia	. . . . .	NH_3.
Nitrogen	. . . . .	N.
Cyanogen, a trace	. . . .	NC_2.
Vapour of Water	. . . .	HO.
„ of Bisulphide of Carbon	.	CS_2.
„ of Benzole	. . . .	$C_{12}H_6$.

* * * * *

" The determination of the exact composition of a coal gas is a task of considerable difficulty; but if the ammonia, carbonic acid, and sulphide of hydrogen be first removed, its analysis may be accomplished in the following manner :—

" 1. The oxygen is absorbed by a ball of papier-maché soaked with an alkaline solution of pyrogallic acid.

" 2. The heavy hydrocarbons, or luminiferous constituents, are next condensed with absolute sulphuric acid in the manner already explained.

" 3. The residual gas, which is a mixture of hydrogen, marsh gas, carbonic oxide, and nitrogen, is burned in a eudiometer with a known volume of oxygen, the quantity used being more than is sufficient for the conversion of their carbon into carbonic acid, and their hydrogen into water, and the volume left accurately measured. A ball of hydrate of potash is then introduced into the tube, and the diminution of volume noted,—which corresponds to the volume of carbonic acid produced by the combustion of the marsh gas and carbonic oxide.

" 4. There now remain but the nitrogen of the coal gas and any excess of the oxygen used for the combustion. The latter is got by adding at least 2 volumes of pure hydrogen, again exploding in the eudiometer, and noting the condensation. This divided by 3 will be the volume of the oxygen in excess, and of course the nitrogen and the oxygen consumed are had by difference.

" 5. The aggregate of the oxygen in the gas, its nitrogen, and the heavy hydrocarbons, is now subtracted from the volume of gas with which the experiments were begun, and a residue is obtained representing the sum of the volumes of the hydrogen, marsh gas, and carbonic oxide. Let the bulk of this sum in cubic inches be v, the amount of oxygen consumed be v', and the volume of the carbonic acid formed be v''; and designating by x, y, and z, the respective bulks of the hydrogen, marsh gas, and carbonic oxide, we will have the three following equations :—

"(1.) $x + y + z = v$.　(2.) $\frac{1}{2}x + 2y + \frac{1}{2}z = v'$.　(3.) $y + z = v''$.

" Equation (1) does not require any explanation.

" Equation (2) expresses the fact that for its combustion hydrogen requires half its volume, marsh gas twice, and carbonic oxide half its volume of oxygen.

" Equation (3) assumes, what is also the result of experiment, that marsh gas and carbonic oxide, when burned with oxygen, give their own volume of carbonic acid.

" From the solution of this simple equation we deduce the values of x, y, and z.　They are as follows :—

$$x = v - v''$$
$$y = \frac{2v' - v}{3}$$
$$z = v'' - \frac{2v' - v}{3}.\text{''}$$

These extracts are sufficient to show the mode in which the subjects are treated.　We consider the work as an excellent contribution to the teaching department of chemistry; as well worthy of the reputation of its author, and as admirably fitted for the use of students.

On the Popular Names of British Plants ; being an Explanation of the Origin and Meaning of the Names of our Indigenous and most commonly cultivated Species. By R. C. ALEXANDER PRIOR, M.D., F.L.S., Fellow of the Royal College of Physicians of London. 8vo. London : Williams and Norgate. 1863.

This is a curious work, written by a medical man of high literary attainments, and who at the same time possesses extensive and accurate botanical knowledge.　It is a welcome addition to our botanical literature.　The author's acquaintance with the languages of Greece and Rome, and with those of modern Europe, including Scandinavia, and his intimate knowledge of plants, acquired during travels to various parts of the Old and New World, render him peculiarly qualified for the very difficult task which he has undertaken, and give force and authority to his opinions.　We recommend the work as well worthy of perusal, and as containing many interesting remarks on the etymology of the common names of British plants.

The Pines and Firs of Japan; illustrated by upwards of
200 wood-cuts. By ANDREW MURRAY, F.L.S., Assistant
Secretary to the Royal Horticultural Society.   8vo.
Pp. 124. London: Bradbury and Evans. 1863.

Coniferous trees and shrubs have of late years attracted much
attention, and many useful plants of this Order have been introduced into Britain. Expeditions to Oregon, Columbia, and
various parts of North America, as well as to China, Japan, and
other countries, have been successfully carried on; and, while
additions have been made to our botanical knowledge, valuable
plants have been secured for our plantations. Many landed pro-
prietors in this country are now fully alive to the importance of
the subject, and have contributed liberally to expeditions which
had for their object the introduction of hardy conifers likely to
yield valuable timber.

In the present work we have an account of the coniferæ of
Japan from one who has devoted much attention to the subject,
and who was one of the originators of an expedition to Oregon,
the result of which was the acquisition of several valuable conifers
fitted for the climate of Britain. He is also the editor of the
magnificent work which is now being published by Messrs Lawson
and Son. The author states in the preface, that the two British
botanists, who, by their explorations in Japan, have recently
added so much to our knowledge of the vegetable products of that
country,—viz. Mr Fortune and Mr John G. Veitch,—have kindly
placed their stores of specimens and information at his disposal.

The original works on the flora of Japan, the author remarks,
are Kæmpfer's "Amœnitates Exoticæ," Thunberg's "Flora Japonica," and Siebold and Zuccarini's "Flora Japonica." The
first two are works published in 1712 and 1784, and the latter is
a costly illustrated work, of which the portion relating to conifers
was published in 1842. Recent travellers have made many additions, and have corrected errors in former observations. The species
described in this work are eighteen:—*Abies Alcocquiana, A. firma,*
A. Fortuni, A. Jezoensis, A. Kæmpferi, A. microsperma, A. polita,
A. Tsuja, A. Veitchi, Cunninghamia sinensis, Larix japonica, L.
Leptolepis, Pinus Bungeana, P. densiflora, P. Koraiensis, P.
Massoniana, P. parviflora, Sciadopitys verticillata. The work is
executed with great care, and is illustrated by excellent wood-cuts.
As a digest of Japan coniferæ, it is an important contribution to
science.

Flora Australiensis: a Description of the Plants of the Australian Territory. By GEORGE BENTHAM, F.R.S., F.L.S., assisted by FERDINAND MUELLER, M.D., F.R.S. and L.S., Government Botanist, Melbourne, Victoria. Vol. I. Ranunculaceæ to Anacardiaceæ. Published under the authority of the several Governments of the Australian Colonies. 8vo. Pp. 508. London: Lovell Reeve and Co. 1863.

This is the first part of one of those colonial floras which are now in the course of publication under the authority of the Government of this country and of the colonies. Already several important works of a similar kind have been either completed or commenced, such as Bentham's Flora of Hong Kong, Grisebach's Flora of the West Indies, Thwaites' Ceylon Flora, Harvey and Sonder's Flora of the Cape of Good Hope; besides the more expensive illustrated Floras of Tasmania and New Zealand, by Dr Hooker. We may look ere long for a Flora of Africa, which has been to a certain extent explored by Vogel and M'Williams, Barth and Barter in the Niger Valley, Vogel and Petherick in the White Nile and Nubia, Welwitsch in Loanda, Speke and Grant in Eastern tropical Africa, Kirk and Mellor in the Livingstone Expedition, and Gustav Mann on the shores, islands, and mountains of the Bight of Benin. It is to be hoped that the researches of the various travellers in South America, including Humboldt, Bonpland, Kunth, Spix and Martius, St Hilaire, Ruiz and Pavon, Aublet, Endlicher, Cambessedes, Miers, Zuccarini, Gardner, Spruce, and others, may soon be embodied in a work similar to that which is now before us.

The work is to embrace a description of all the plants of the Australian continent and of Tasmania. The Flora of New Zealand will not be included. In the introduction Mr Bentham gives Outlines of Botany, with special reference to Local Floras. This consists of—1. A general view of the Organs of Plants, with definitions and descriptions; 2. The Principles of Classification and Systematic Botany; 3. Vegetable Anatomy and Physiology; 4. Remarks on the Collection, Preservation, and Determination of Plants; 5. An Index of terms and glossary.

The author then proceeds to describe the Australian plants in the class of Dicotyledons, and in this volume he embraces the Polypetalous orders, extending from Ranunculaceæ to Anacardiaceæ, and included in the series Thalamifloræ. The work promises to be one of great value. It is conducted by a botanist *of high reputation,* who is well qualified for the task, and it is carried on with the aid of the unrivalled herbarium of Sir William Hooker, and with the assistance of Dr Mueller, the zealous and able director of the Botanic Garden at Melbourne, who has already done much to illustrate the Australian flora in his "Fragmenta Phytographiæ Australiæ," and other publications. This important flora could not be placed in better hands, and the completion of it will be hailed as a boon to science in general, and more especially to those who are prosecuting botany in the great continent of Australia.

Air-Breathers of the Coal Period: a Descriptive Account of the Remains of Land Animals found in the Coal Formation of Nova Scotia, with Remarks on their bearing on Theories of the Formation of Coal and of the Origin of Species. By J. W. DAWSON, LL.D., F.R.S., F.G.S., Principal of M'Gill University. 8vo. Pp. 81. Montreal: Dawson Brothers, 1863.

In the carboniferous period, the author remarks, though land plants abound, air-breathers are few, and most of these have only been recently recognised. We know, however, with certainty, that the dark and luxuriant forests of the coal period were not destitute of animal life. Reptiles crept under their shade, land-snails and millipedes fed on the rank leaves and decaying vegetable matter, and insects flitted through the air of the sunnier spots. Great interest attaches to these creatures— perhaps the first-born species in some of their respective types, and certainly belonging to one of the oldest land faunas, and presenting prototypes of future forms equally interesting to the geologist and the zoologist.

The author, who is well known as a sound and able geologist, after describing various species of Hylonomus, Baphetes, Dendrerpeton, Hylerpeton, Eosaurus, Xylobius, Pupa, which occur. in the Nova Scotia carboniferous strata, concludes with some valuable remarks on the nature of the animals and on the state of the globe at the time when they lived on it. He states, " In the coal measures of Nova Scotia, while marine conditions are absent, there are ample evidences of fresh-water or brackish-water conditions, and of land surfaces, suitable for the air-breathing animals of the period. Nor do I believe that the coal measures of Nova Scotia were exceptional in this respect. It is true that in Great Britain, evidences of marine life do occur in the coal measures, but not, so far as I am aware, in circumstances which justify the inference that the coal is of marine origin. Alternations of

marine and land remains, and even mixtures of these, are frequent
in modern submarine forests.   When we find, as at Fort Lawrence in Nova Scotia, a modern forest rooted in upland soil forty
feet below high-water mark, and covered with mud containing
living Tellinas and Myas, we are not justified in inferring that
this forest grew in the sea.   We rather infer that subsidence has
occurred.   In modern salt marshes it is not unusual to find every
little runnel or pool full of marine shell-fish, while in the higher
parts of the marsh land plants are growing, and in such places
the deposit formed must contain a mixture of land plants and
marine animals with salt grasses and herbage, the whole *in situ.*"

There are some interesting remarks on the bearings of the facts
on the origin of species, and the author brings forward statements
which convey strong impressions of the permanence of species.
On the subject of transmutation, he gives the following judicious
observations :—

" If we could affirm that the air-breathers of the coal period
were really the first species of their several families, they might
acquire additional interest by their bearing on this question of
origin of species.   We cannot affirm this ; but it may be a harm-
less and not uninstructive play of fancy, to suppose for a moment
that they actually are so, and to inquire on this supposition as to
the mode of their introduction.   Looking at them from this point
of view, we shall first be struck with the fact that they belong to
all of the three great leading types of animals which include our
modern air-breathers—the Vertebrates, the Articulates, and the
Mollusks.   This at once excludes the supposition that they can
all have been derived from each other within the limits of the
coal period.   No transmutationist can have the hardihood to
assert the convertibility, by any direct method, of a snail into a
millipede, or of a millipede into a reptile.   The plan of structure
in these creatures is not only different, but contrasted in its most
essential features.   It would be far more natural to suppose that
these animals sprang from aquatic species of their respective
types.   We should then seek for the ancestors of the snail in
aquatic gasteropods, for those of the millipede in worms or crustaceans, and for those of the reptiles in the fishes of the period.
It would be easy to build up an imaginary series of stages, on the
principle of natural selection, whereby these results might be
effected ; but the hypothesis would be destitute of any support
from fact, and would be beset by more difficulties than it removes.
Why should the result of the transformation of water-snails
breathing by gills be a *Pupa ?*   Would it not much more likely
be an *Auricula* or a *Limnea ?*   It will not solve this difficulty to
say that the intermediate forms became extinct, and so are lost.
On the contrary, they exist to this day, though they were not, in

so far as we know, introduced so early. But negative evidence must not be relied on; the record is very imperfect, and such creatures may·have existed though unknown to us. It may be answered that they could not have existed in any considerable numbers, else some of their shells would have appeared in the coal formation beds, so rich in crustaceans and bivalve mollusks. Further, the little Pupa remained unchanged during a very long time, and shows no tendency to resolve itself into any thing higher or to descend to any thing lower. Here, if anywhere, in what appears to be the first introduction of air-breathing invertebrates, we should be able to find the evidences of transition from the gills of the prosobranchiate and the crustacean to the air-sac of the pulmonate and the tracheæ of the millipede. It is also to be observed that many other structural changes are involved, the aggregate of which makes a pulmonate or a millipede different in every particular from its nearest allies among gill-bearing gasteropods or crustaceans.

It may be said, however, that the links of connection between the coal reptiles and the fishes are better established. All the known coal reptiles have leanings to the fishes in certain characters, and in some, as in *Archegosaurus*, these are very close. Still the interval to be bridged over is wide, and the differences are by no means those which we should expect. Were the problem given to convert a ganoid fish into an *Archegosaurus* or *Dendrerpeton*, we should be disposed to retain unchanged such characters as would be suited to the new habits of the creature, and to change only those directly related to the objects in view. We should probably give little attention to differences in the arrangement of skull bones, in the parts of the vertebræ, in the external clothing, in the microscopic structure of the bone, and other peculiarities for serving similar purposes by organs on a different plan, which are so conspicuous so soon as we pass from the fish to the batrachian. It is not, in short, an improvement of the organs of the fish that we witness so much as the introduction of new organs. The foot of the batrachian bears perhaps as close a relation to the fin of the fish as the screw of one steam-ship to the paddle wheel of another, or as the latter to a carriage wheel, and can be just as rationally supposed to be not a new instrument but the old one changed.

" Again, our reptiles of the coal do not constitute a continuous series, nor is it possible that they can all, except at widely different times, have originated from the same source. To suppose that *Hylonomus* grew out of *Dendrerpeton* or *Baphetes*, and *Eosaurus* out of either, startles us almost as much as to suppose that *Baphetes* grew out of *Rhizodus*, or *Hylonomus* out of *Palæoniscus*. It either happened, for some unknown reason, that many kinds of fishes put on the reptilian guise in the same period, or else the vast lapse of ages required for the production of a reptile from a fish must be indefinitely increased for the production of many dissimilar reptiles from each other; or, on the other hand, we must suppose that the limit between the fish and reptile being once overpassed, a facility for comparatively rapid changes became the property of the latter. Either supposition would, I think, contradict such facts bearing on the subject as are known to us.

" We commenced with supposing that the reptiles of the coal might possibly be the first of their family ; but it is evident from the above considerations, that, on the doctrine of natural selection, the number and variety of reptiles in this period would imply that their predecessors in this form must have existed from a time earlier than any in which even fishes are known to exist, so that if we adopt any hypothesis of derivation, it would probably be necessary to have recourse to that which supposes at particular periods a sudden and as yet unaccountable transmutation of one form into another—a view which, in its remoteness from any thing included under ordinary natural laws, does not materially differ from that currently received idea of creative intervention with which, in so far as our coal reptiles can inform us, we are for the present satisfied."

*   *   *   *   *

" Humble though the subjects of this paper are, we see in them the work of Supreme Intelligence, introducing new types upon the scene, and foreshadowing in them those higher forms afterward to be created. It is this, their Divine origin, and the light which they throw on the plan and order of the creative work, of which we ourselves form a part, that gives them all their interest to us. They are the handiwork of our Father and our God, traces of His presence in primeval ages of the earth, evidences of the unity of His plan and pledges of its progressive nature, adding their feeble voices to the testimony of revelation in respect to the history of creation in its earlier stages, and to the carrying on of that plan which still involves the extinction of many things from the present world, and the elevation of others into new and glorious manifestations. Their place in the system of nature and in the order of the world's progress, their uses in their own time, and their relations to other beings as parts of the great cosmos, are the points that chiefly interest us ; and if any one desires to understand more in detail how they were created, we wish him all success in his inquiries, but warn him not to suppose that this great mystery is to be solved by a reference merely to material agencies, apart from that Spiritual Power who *is the essence of* forces, the origin of laws."

The First Principles of Natural Philosophy. By W. T.
　　Lynn. London: Van Voorst, 1863.

This is in many respects a creditable little work—totally distinct in kind from the peculiar absurdities produced by the
Reddie and Worms school of self-constituted philosophers.

To some of its errors, which, we must say, are attributable to
those of moré pretentious treatises, we shall presently advert.
But we allow that the author expresses himself well and clearly
on many points, and that his treatise is so far what it was intended to be — a *very* simple guide to the first elements of
Mechanics and Optics.

In some cases he has shown considerable acuteness, though
only brought into action by his want of knowledge. Thus his
remark, p. 40, about *intensity* of motion, though quite wrong,
shows that he has an idea of Vis viva as distinguished from
Momentum.

The author has, however, followed too closely authorities
whose value is really much less than is generally believed; such
as the bulk of the Cambridge and Dublin Text-Books. Most, we
may almost say all, of these—are admirable in their mathematical
development : but there is, in general, a fearful amount of loose-
ness, and even inaccuracy, whenever first principles or *experimental*
results are concerned.

Thus Mr Lynn says (p. 20),

"It is at once apparent that if a force acts continuously on a body,
it will generate in that body a continually increasing motion."

He has not, so far as we can see, added the proviso that the force
must *not* act at right angles to the direction of motion.

Again (in pp. 20, 22) he speaks of a *constant*, as distinguished
from an *impulsive*, force—meaning by the erroneous first term a
continuously acting force. He seems also (in p. 23, for instance)
to fancy that an accelerating force must be of constant magnitude.
In the same page, after giving an exceedingly good and complete
investigation of the space described from rest by a body in any
time, when acted on by a uniform force in its line of motion, he
seems to be dissatisfied with his work—for he goes on to say,

"This proposition is so important and fundamental that it will be
proper to give a strict demonstration of it."

Then he goes on with limits, and limiting ratios, as if to make
doubtful to the beginner, in 4 pages, what he has just before
satisfactorily proved (by general reasoning) in 8 lines.

§ 41, as it stands, is simply untrue. It requires most copious and careful limitations, which are not even hinted at in the text.

In § 45 we find another of these cases of rash assertion (about the attraction of spheres)—requiring a great amount of unsupplied qualification.

To mention but *one* more serious blunder—we find that *water is a non-elastic fluid !* (p. 47.)˙ Shades of Canton and Oerstedt, have your discoveries come to this ? Is not elasticity the tendency of a compressed or distorted body to recover its volume or form when the disturbing cause is removed ? Is not water compressible, and *perfectly elastic ?*

The Philosophy of Geology: a brief Review of the Aim, Scope, and Character of Geological Inquiry. By DAVID PAGE, F.R.S.E., F.G.S.   12mo.   Pp. 155.   William Blackwood and Sons, Edinburgh and London, 1863.

The object of this work is to direct attention to some of the higher aims of geological science, to the principles which ought to guide geologists in their generalisations, and to what may be ultimately anticipated of geology in her true and onward progress. "The philosophy of geology is the study of the structural arrangement and composition of this earth, the causes that have produced this structure and arrangement, the laws by which this causation is upheld, and the comprehension of the whole in time as constituting a continuous and intelligible world history." The author first discusses the objects of inquiry, defines what geology is, and shows that speculations as to the origin of the earth are inadmissible. "Geology is but the physical geography of former ages. Every rock system retains some evidence of the conditions of its period ; and the determination of these conditions, the causes that produced them, and the life by which they were accompanied, is the spirit and purport of geology." Certain forces have been called into play, in order to produce geological changes on the globe, and these forces act according to uniform natural laws, so that we are enabled in some measure to argue from the present as to what took place in the past. At the same time we must bear in mind that over limited areas cataclysmal phenomena occur, depending on earthquakes, volcanoes, or floods, which may interfere with our reasonings, and which show the necessity of being cautious in our inductions. We must beware of·the error of those who say that all things continue as they were from the creation, and that nothing has been done *per saltum.* Floods *and convulsions* have occurred, and local phenomena may present

themselves which we cannot account for, and which, in the meantime, we must be content to leave unexplained.

In referring to theories of the earth, the author remarks :—

" The philosophy of our science is thus neither to ignore nor re-
ject hypotheses, but merely to receive them as tentative processes
and provisional aids towards the attainment of veritable theory.
. . . . Granting the value of hypotheses in occasionally directing
the way to a solution of our problems, it cannot be denied that
indulgence in speculation has been the great bane of geology.
. . . . The earlier progress of geological inquiry was encumbered
by its absurdities, which would have been simply ridiculous but
for the discredit they attached to the science, and the obstacles
they threw in the way of its acceptance. There is nothing so
easy as generalisation, where the facts are few; nothing more
difficult than adherence to a course of induction where the field
of observation is wide, and the facts numerous and complicated.
.On this ground the many ' theories of the earth' that prevailed
towards the end of last century and the beginning of the present,
are in some degree excusable ; but now that geology has taken her
proper place among the natural sciences, all such attempts should
meet with the most steadfast discountenance and reprobation."

The subjects of the Direction of Drifts, Chemical Formation,
Metamorphism, Mineral Veins, &c., are discussed. The author
points out that in regard to them there are many difficult problems in geology still to be solved. The question of Time as a
factor in geology is considered; and it is shown that while we
admit a prodigious amount of geological time, we have no means
of approximating to its duration. Calculations founded on the
deposit made by rivers of the present day are very fallacious.
All that can be done at present is to arrange the stratified formations into systems, groups, and series, each section representing a
relative but inadequate amount of time, but occupying its own
proper chronological position. A proper nomenclature and classification of formations is an important object, and in this point of
view palæontology has done great service. At the same time,
there can be no doubt that much remains to be done in these departments. Geologists have often drawn very rash conclusions
from fossil specimens, both as regards their nature and the indications which they afforded of soil and climate. A deficient
knowledge of the zoology and botany of the present epoch is one
great cause of geological error. The following remarks of Mr
Page deserve attention :—

" While we must admit the vast benefits conferred on our
science by palæontology, let us take care that we are not misled by the dicta of its earlier cultivators into beliefs which are at
variance with the known principles of biology and physical geo-

graphy, and which extended observation in geology refuses to confirm. As a general maxim, it is true that the most trustworthy tests of the sequence of the stratified systems are the fossil organisms which these systems contain. It is also true, that over limited tracts like Britain, or even over the area of Europe, contemporaneous deposits are characterised by the same fossil species; but it may not be true that the same species never recur at more stages than one in the same formation, nor may it be true that formations containing the same species in widely separated regions are strictly contemporaneous. The truth is, there has been a tendency of late to carry the argument of fossil evidence beyond its legitimate limits. Under the change of external conditions, a species—say a marine one—may gradually shift its ground many degrees south or north, and so become extinct in its original area, and yet, after hundreds of feet of sediment had been deposited on that area, a reversal of conditions may occur, and establish the species once more on its primal habitat. We would thus have the same species occurring at two different stages of the same great formation, separated, it may be, by some thousands of years in time, though only by a few hundred feet of sediment. Again, under the slow oscillation of sea and land, species terrestrial and marine may have been gradually transferred from the latitudes and longitudes of Europe to the latitudes and longitudes of America; and thousands of years after they had become fossil in the eastern hemisphere, they may have been flourishing in the western."

After considering the appearance of life on the globe, the author calls attention to the theory of Progression, and more especially discusses the Darwinian views as to external condition, embryonic phases, use and disuse of organs and natural selection, and he concludes that all these "are but subordinate factors of one great law, and that we must know much more of the forms that have become extinct, and more of the variations that are now taking place in existing plants and animals before we can hope to approach the solution of the all-important problem."

The origin of man naturally becomes a matter of consideration. On this point the author says :—

"Whatever be the plan of development, it must of necessity embrace the whole scheme of vitality. There can be no severance in the great creational idea of life; and whatever theory be adopted, it must be applicable alike to every constituent member of the system. The highest as well as the lowest, man as well as the monad, forms part and parcel of one continuous evolution; and whatever the ordainings of the past, they exist in the present, and must operate in the future. If by any genetic law the radiata have given birth to the articulata, the articulata to the mollusca, *and the mollusca* to the vertebrata—nay, were it only the great sections of the vertebrata that were so genetically connected—the fishes giving rise to the reptiles, the reptiles to the birds, and the birds to the mammals,—still this would be sufficient to prove man's inseparable association with the same scheme of development. Whatever may be the law that determines the origination of other species, to the same law must we ascribe the origin of man. Philosophy has no alternative. Science has nothing to gain, but everything to lose, by the adoption of any other opinion. We must look, therefore, for man's precursor in the order that stands next beneath him in the zoological scale; and wide as the gap may appear, we may suppose it either bridged over by intermediate forms that became extinct during the tertiary period, or, if the rate of ascent be more rapid in the higher than in the lower orders, to have been passed over *per saltum*, or at most by the intervention of very few such intermediate species. But though one form be descended from another—the higher from the next lower in the scale of creation—such descent differs widely from that of ordinary generation, inasmuch as qualities unknown in the lower form begin to manifest themselves in the higher. Whence then, these newer qualities and higher functions? Clearly, not from the predecessor, who did not possess them; not from the law, which is simply a mode of operation; but from the Lawgiver, who ordained and continues to sustain the method of development. Similar as the framework of the monkey may be to that of man —nay, were it a hundred times more closely resembling—yet every superaddition of reason, gift of speech, moral feeling, and religious sentiment in man, is in reality a new creation—a creation as special as if it had proceeded from the audible 'Let it be' of the Creator. To the devout and philosophic mind the secondary law of causation is the great ' Let it be,' ever ringing through nature as audibly as on the morning of its primal utterance."

Our author, therefore, is opposed to the views of those who think that man is merely an ennobled ape, and that he has a Simian origin. The advocates of this opinion say, that because the lowest ape does not differ from the highest ape in the conformation of its skeleton more than does the highest ape from man, therefore man was originally an ape. This, in our opinion, is a complete *non sequitur*. These statements would merely show that the Creator, in forming animals and man, followed a great type, and that out of a small amount of materials He moulded all the varied forms seen in the animal creation. It shows unity of design and wonderful wisdom. But animals have other characteristics than their bony skeleton affords, and so has man. We must consider their anatomy in a physiological point of view, and then we shall at once see marked and evident distinctions.

On the question of the Antiquity of Man, our author, while he does not agree with Lyell as to the enormous lapse of time since man's appearance on earth, is disposed to believe that man is older than is generally allowed in the received chronology.

"Admitting the same genetic law for the human race as for the rest of animated nature—and, philosophically speaking, there is no other course left for the adoption of science—the question next arises, At what period of the geological scale did man make his appearance ? . . . So far as research has been prosecuted in the different quarters of the globe—and at the outset it must be confessed how insignificant the area that has been examined—no remains of man or of his works have been discovered till we come to the lake-silts, the peat-mosses, the river-gravels, and the cave-earths of the Post-tertiary period. In these have been detected tree canoes and stone hatchets, rude implements of flint and bone, the embers of the fires he kindled, and occasional fragments of his own skeleton. As yet these have been chiefly discovered within the limited area of southern and western Europe, and we have scarcely any information from the corresponding deposits of other regions. Till these other regions shall have been examined—and especially Asia, where man historically flourished long prior to his civilisation in Europe—it were premature to hazard any opinion as to man's *first* appearance on the globe. But taking the facts such as geology finds them—viz., the occurrence of stone implements in conjunction with the remains of Irish-deer, mammoth, hippopotamus, rhinoceros, cave-lion, and other creatures long since extinct in Europe, and this in deposits of considerable geological antiquity—it is evident that man has been an inhabitant of the globe much longer than is popularly believed. It is true that the antiquity of some of the containing deposits, especially the river-drifts, is open to question, and it is also quite possible that the remains of the extinct quadrupeds may in some instances have been reassorted from older accumulations; but even allowing for these, geologists have sufficient in the valley-deposits of France and England, the caves of southern and western Europe, and the lake-silts of the same area, to convince every mind capable of appreciating the evidence, that mankind has existed even there (to say nothing of Asia) long antecedent to the time assigned by the patristic chronologers as the date of his creation.

"But while the nature of the deposits, their situation, and their mode of formation, indicate the lapse of many thousand years (estimating by the usual modes of geological computation), we must be careful not to run into the opposite extreme, and conjure up ages of fabulous duration. Historically we have no means of arriving at the age of these deposits; geologically we *can only* approximate the time by comparison with existing operations; while palæontologically it must be borne in mind that the associated animals are among the most recently extinct or exterminated. It is a sound maxim in palæontology, that the more ancient any specific form is, the more widely it differs from existing species of the same genus. Structural variation is, in fact, the measure of antiquity. . . . . .

" In dealing, then, with the antiquity of man, we have both its lithological and palæontological, and to some extent also its historical, aspects to consider, before we can arrive at any truly philosophical conclusion. Lithologically, we see that great changes have taken place in the physical geography of the districts where human remains have been discovered, and the deposits in which they occur are frequently also of considerable thickness, and of a nature that implies slow and gradual accumulation. As geologists, we may feel convinced that more than six or eight thousand years have elapsed since their formation, but how much more, we have, in the present state of our science, no means of definitely determining. Palæontologically, we perceive that other animals whose remains are associated with those of man do not differ very widely from species still existing, and are therefore constrained to oppose that enormous antiquity which some geologists are disposed to contend for. Historically, all is silent on the subject of these remains ; but while the mammoth and rhinoceros may have died out of Europe within the last five or six thousand years without attracting the notice of the rude inhabitants, the present state of human civilization seems incompatible with such a brief period as five or six thousand years for its development.''

We certainly must differ from our author as to the antiquity of man. Whatever may be said in regard to animals and plants, we can have no doubt as to the creation of man, and we have scriptural data which certainly seem to fix his appearance on earth within tolerably definite limits. We think that geologists are completely at sea on this subject, and want data on which to found their conclusions. The author recommends in other parts of his work cautious induction and a sifting of facts, and we would apply to him in this case the exhortation which he gives to others. No doubt he does not give a decided statement,—his trumpet here gives an uncertain sound, and he feels the necessity of treading warily on ground which may ere long be found to yield under his feet. Those who advocate progressive development, and the transmutation of the ape into man, find great difficulty in setting aside the statements of the Bible as to man's creation, and hence they are glad to lay hold of any speculation as to man's antiquity which they think will throw discredit on the sacred volume, and put it out of the category of a Divine and truthful record. Hence the avidity with which the so-called facts as to the antiquity of man are seized hold of. Already, however, the Darwinian hypothesis has been attacked by able naturalists and geologists such as Flourens and

Dawson, and the foundations on which the Lyellian speculations as to man have been based are tottering. There is great doubt and uncertainty among geologists, and we are satisfied that much remains to be done. At the same time we feel sure, that, as in all former cases, Science and Revelation will be found not to be at variance, and the attempt to throw distrust on the Bible record will fail.

The future of this planet is considered by Mr Page in his work, and he refers to the changes which may be expected to occur ere this state of things comes to an end, and a new race of beings people the earth. As to the future of this world " The Philsosphy of Geology " cannot give us information ; but we know from an un-erring testimony that " the heavens and the earth which are now, by the same word are kept in store, reserved unto fire against the day of judgment ; " and that " the heavens shall pass away with a great noise, and the elements shall melt with fervent heat, the earth also and the works that are therein, shall be burned up ; " and that we are to " look for new heavens and a new earth, wherein dwelleth righteousness." (2 Peter iii.) Here, then, is a prophesied cataclysm, a *per saltum* occurrence totally inconsistent with the progressive-development theory, the supporters of which tell us that " no cataclysm has desolated the whole world, and that we may look with some confidence to a secure future of equally inappreciable length," in which, " judging from the past, we may infer safely that not one living species will transmit its unaltered likeness to a distant futurity."

We have perused Mr Page's clear and interesting volume with much interest. It gives a comprehensive view of the actual state of the science of geology, and contains valuable suggestions as to the future prosecution of it. It brings before the student the difficulties which stand in his way—the danger of rash specula-tion—and the need of accurate data, before drawing conclusions regarding the lithological and palæontological history of our globe. Differing as we do from the author in some particulars, still the cautiousness with which he makes his statements in regard to controverted points, will, we doubt not, show to the young observer that he must not be led away at once by the theories of those geologists who have shown a disposition to dogmatise on these matters. We are all too apt " *jurare in verba magistri,*" and anything which makes us doubt the theories given *ex cathedra,* may be useful in making us sift the points for ourselves and inves-tigate fully the data whence conclusions may have been rashly drawn. We must examine carefully and observe accurately before we can theorize. We think that the book will be valuable to young and local observers, by showing them the work to be done, *the mode in* which it is to be done, and the doubtful speculations *which beset* their path.

PROCEEDINGS OF SOCIETIES.

Botanical Society of Edinburgh.

Thursday, 12th November 1863.—Professor MACLAGAN, President,
in the Chair.

His Royal Highness Prince Alfred was elected by acclamation an
Honorary Fellow of the Society.

The President delivered an opening Address.

The following Communications were read :—

I. *Notes on the Fertilisation of Orchids.* By WM. RUTHERFORD, M.D.

(This paper appears in the present number of this Journal.)

II. *Synopsis of Canadian Ferns and Filicoid Plants.* By Professor
LAWSON.

(This paper appears in the present number of this Journal.)

III. *Notes on some New and Rare British Mosses, and on the Occurrence
of* Trichomanes radicans *in the Island of Arran, Firth of Clyde.*
By Mr JOHN SADLER.

Bryum Duvalii, Voit.

Whilst lately engaged examining the genus *Bryum* in Dr Greville's
collection of *Musci,* recently added to the University Herbarium, I came
upon two barren specimens marked " *Bryum turbinatum,* Ben Voirlich,
1823.'' On close examination, I found them to be the true *Bryum
Duvalii* of Voit. The specimens were collected by Dr Greville himself;
thus he is the discoverer of the plant in Britain, having gathered it fully
thirty years before the late Colonel Madden met with it near Waterford
in Ireland, as recorded in the Society's Transactions, vol. vii. p. 6. The
following are the known British stations for this moss :—

    1. Ben Voirlich, 1823. Dr Greville.
    2. Near Waterford, Ireland, 1852. Colonel Madden.
    3. Hart Fell, near Moffat, 1858. Dr William Nichol.
    4. Helvellyn, 1859. Mr John Nowell.
    5. Ben Lawers, 1860. Mr William Bell.

Mnium stellare, Hedw.

On the Ochils, near Bridge of Earn. August 1863. J. S.

Grimmia leucophœa, Grev.

Rocks between Kinghorn and Pettycur. June 1863. J. S.

Schistidium maritimum, B. *et* S.

Rocks between Kinghorn and Pettycur. June 1863. J. S.

Didymodon recurvifolius, Tayl.

Ben Voirlich. 1863. Mr M'Kinlay.

Dicranum circinatum, Wils.

Ben Voirlich. 1863. Mr M'Kinlay.

Glyphomitrium Daviesii, Schwæg.

Bowling Glen. May 1863. Mr Galt.

In a letter which I lately received from Mr John Robertson, Glasgow, he says:—" I have been to Glen Turret and Ben Chonzie, and the following are some of the few rarities I collected. A solitary tuft of *Tortula tortuosa* in fruit, on Ben Chonzie ; *Bryum roseum,* but not with capsules. In a fruiting state I found *Oligotrichum hercynicum, Tetraplodon mnioides, Splachnum ampullaceum, S. sphæricum, Encalypta ciliata, Climacium dendroides,* &c."

Mr Sadler exhibited specimens of *Trichomanes radicans* which he had received from Mr Walter Galt, Glasgow, accompanied by the following note, dated 26th August 1863 :—" I enclose you fronds of *Trichomanes radicans,* collected by Mr George S. Combe in the Island of Arran, Firth of Clyde. It occurs in two separate patches, one of which is about 3 feet square, seemingly a natural habitat."

Thursday, 10th December 1863.—Professor BALFOUR, President, in the Chair.

A letter was read from Major Cowell, conveying the thanks of H.R.H. Prince Alfred for his election as an Honorary Fellow of the Society.

1. *Notice of the Occurrence of* Polypodium calcareum, *near Aberdeen.* By Mr JAMES ROBERTSON.

Mr Robertson states that he had discovered this plant in August 1862, growing in the debris of a limestone quarry in Scotston Moor, near Aberdeen, along with *P. Dryopteris.* He was disposed to look upon the plant as being wild in that locality. Professor Dickie, however, believes that it has been introduced, and he has learned that a gentleman's gardener in the neighbourhood was in the habit of planting ferns in waste places. Specimens of the plant were exhibited from the Scotston station.

II. *Account of the Vegetation of the Cliffs of Kilkee, County Clare, Ireland.* By N. B. WARD, Esq.

In compliance with Professor Balfour's request, I send a brief account of the vegetation of the Cliffs of Kilkee, and its neighbourhood, which I visited last summer, in company with Professor and Mrs Harvey, and an old friend, Mr Snell, with my daughter. During our stay, we visited Loophead, at the mouth of the Shannon, and an intermediate portion of the cliffs, on which Baltard Castle is situated. Five days were spent at Kilkee, one at Baltard Castle, and one at Loophead. The vegetation at the three places was so perfectly identical, as to lead us to the conclusion, that the same geological structure prevailed throughout, consisting of hard grits, shales, &c., a conclusion which was confirmed by a subsequent visit to Mr Jukes of Dublin.

Kilkee is situated on the west coast of Ireland, exposed to the tide force of 2000 miles of unbroken seas,—the waves of which roll in with such power, as to furnish abundant food to periwinkles, located on rocks 200 feet above high water-mark, and to supply the wants of marine plants which cover the summits of cliffs, varying in height from 150 to 400 feet. That physiological law by which plants, under adverse circumstances, produce their flower and fruit, if they can do nothing else, is here strikingly exemplified. Looking at the stunted character of the vegetation, one might imagine oneself in a high alpine region—many species not attaining more than a tenth or twelfth of their usual size. Thus we find here a number of plants, simulating the appearance of the inhabitants of alpine regions—e. g., *Aster Tripolium*, in full flower, from half an inch to an inch and a half in height ; and equally stunted, not starved, forms of *Samolus Valerandi, Euphrasia officinalis, Jasione montana, Erythræa Centaurium, Ranunculus Flammula*, &c. The higher portions of the cliffs are cushioned by continuous tufts of the common thrift, *Armeria maritima*, which, when in full flower, must be exceedingly beautiful. The lower grounds are carpeted by *Anagallis tenella, Ranunculus Flammula, Hydrocotyle vulgaris*, &c., whilst the hill sides are dotted with generally solitary plants of *Erythræa Centaurium*, the broad-leaved variety ; the little ridges towards the sea are lined with *Glaux maritima*, the drier spots being covered with *Radiola Millegrana*, &c.

The following is a list of all the plants which were seen in flower :—

Ranunculus Flammula, in great abundance in wet places, towards the lower portion of the cliffs, in company with *Anagallis tenella, Juncus bufonius*, &c. ; *Silene maritima, Spergularia rubra*, var. *marina* ; *Radiola Millegrana*, on little elevated ridges ; *Polygala vulgaris, Potentilla Tormentilla, P. anserina, Sedum anglicum*, the great ornament of exposed and bare rocks here and in many other parts of Ireland, from the beauty of its flower and the rich decaying tints of its foliage ; *Aster Tripolium*, many specimens in full flower, not an inch in height ; *Bellis perennis, Achillæa Millefolium, Senecio Jacobæa, Carduus pratensis, Leontodon autumnale, Calluna vulgaris*, very sparingly ; *E. Tetralix*, very sparingly ; *Jasione montana, Campanula rotundifolia, Glaux maritima*, coating the seaward face of the high ridges ; *Anagallis tenella*, everywhere in moist places ; *Samolus Valerandi, Euphrasia officinalis*, rarely exceeding half an inch in height ; *Erythræa Centaurium, Thymus Serpyllum, Armeria maritima* cushioning the whole of the upper surface of the cliffs ; *Plantago maritima, P. Coronopus, Juncus bufonius, Luzula campestris, Carex flava, Aira caryophyllea*, withered ; *Melica cærulea*, very sparingly. The paucity of grasses was remarkable.

III. *Notice of the Discovery of* Fucus distichus, *L., at Duggerna, County Clare, Ireland.* By Professor HARVEY.

In a letter to Dr Greville, Professor Harvey says:—

" In a summer excursion to Kilkee, last July, in company with N. B. Ward, we found what I take to be the true *Fucus distichus* of Linnæus. I have no authentic specimen of the Arctic plant, nor have I seen one, but the specimens exactly agree with the description of authors, as well as with the *figures*, though none of the latter do this plant justice. I enclose specimens for your herbarium. Unfortunately we were rather *late* in the season, and the fruit had mostly dropped off, leaving truncated branches. Some, however, were in fruit. I suppose it is in perfection either in winter or spring, and I mean, if I can manage it, to visit the station next Easter.

" It grows on a remarkable rock facing the sea, near low water *mark*, but rising much above low water *level*—that is to say, the rock is at the outer edge of a long reef, but rises above the reef level. The fucus grows in *patches* on little ledges of the perpendicular side of the rock, along with *Gigartina mamillosa*, &c. It has quite a peculiar *aspect* when growing. The stipes or base of stem is thick and *rigid*, and stands erect, while the fronds are just sufficiently limber to bend over, but not to lie flat, so that the *patch* looks like a miniature grove of weeping willows."

In the University Herbarium (which now contains Dr Greville's algæ) there are specimens of *Fucus distichus* from Faroe, sent by Professor Hornemann, and from Newfoundland, sent by Bory in 1831, which appear to correspond with the Irish specimens.

IV. *Notice of the Occurrence of* Sagina nivalis (Lindblom), *on Ben Lawers.* By Professor BALFOUR.

In October last, Mr J. T. Boswell Syme, wrote to me in the following terms :—" Will you be so kind as to look in your herbarium under *Alsine rubella*, to see if you have got specimens of *Sagina nivalis*, Lindbl. Under this name, I have this plant with the following label from the Edinburgh Botanical Society, '*Alsine rubella*, Ben Lawers, Perthshire, Aug. 25, 1847. Dr Balfour.' I have come to the genus *Sagina* for the new edition of English Botany, and I feel great doubts as to whether or not I should include *S. nivalis*. Babington mentioned it in the third edition of his Manual, but omitted it from the fourth and fifth." On examining the plants in my herbarium and my duplicates, I found that three specimens of *Sagina nivalis* were fastened down on a sheet along with specimens of *Alsine rubella*. The specimens of the former were put together and quite distinct from the latter, implying that I had looked upon them as peculiar. They are marked 'Ben Lawers, 1847,' and they were gathered by me on the 25th August. There were only a few pupils with me, viz., Mr Charles Murchison, Mr F. J. Ivory, Mr Gilby, Mr Hewitson, Mr Hugh Balfour. Among my duplicates I could only detect two specimens remaining, one of which I sent to Professor Babington. I have no doubt that there are others which have been distributed, as a number of duplicates were contributed by me to the Society. In my notes of the excursion, I refer to our getting *Alsine rubella*, and from the indication given, I think that I know the particular locality. Babington refers to the plant as having been gathered on Glassmeal, by Mr Backhouse; and Hooker and Arnott, in the eighth edition of their " British Flora," mention *Sagina nivalis* as found in the Isle of Skye and Clova. They say that the plant is distinguished from *S. subulata*, by being almost quite glabrous. It is possible that their plant may be *S. saxatilis*. The genera *Alsine* and *Sagina* are very nearly allied. The chief characters are derived from the styles and valves of the capsule, which in *Alsine* are usually three, while in *Sagina* they are four or five. In *Alsine rubella* the sepals are distinctly 3-nerved, whereas in *Sagina nivalis* they are obscurely 1-nerved. By this character the plants can be easily separated.

I now show the original specimens gathered by me on Ben Lawers on 25th August 1847.

The following are the characters of the plant as given by Fries :— *Sagina nivalis*, Lindbl. Caulibus cæspitosis, foliis subulatis mucronatis glabris, pedunculis brevibus strictis, sepalis quinque ovatis obtusis membranaceo-marginatis petala integra vix æquantibus. Lindbl. "Bot. Not.," 1845, p. 66 ; Fries, "Summa Veg. Scand.," 156 ; Fries, "Nov. Mant." iii. 31 ; Sagina intermedia, Ledebour " Flora Rossica," i. 339 ; *Spergula saginoides*, β. *nivalis*, Lindbl. in " Phys. Sallsk. Tidskr.," 1838, p. 128 ; Arenaria cæspitosa, "Fl. Dan.," t. 2289.

The plant is found in moist places of the Dovrefeld. Gathered by Lindblom at Sprenbacken in Kundshö, above Kongsvold, and at Drivelven. It is a perennial, and flowers in August and September. The discoverer of this plant states that it bears the same relation to *Sagina saxatilis*, that *Sagina stricta* bears to *S. procumbens*. The species exhibits two forms, one congested and erect, the other lax, with elongated procumbent stalks. The petals are entire, while in *Sagina saxatilis* they are slightly emarginate.

Babington states that he has specimens from Fries in his Herb. Norm. (xii. 51) gathered by Blytt in the Dovrefèld in Norway. He also says: " It is to be remarked that Blytt finds it in Norway, Fenzl has it from the extreme north-east of Siberia, and *Flora Danica* from Greenland."

V. *Remarks on Monœcious Spikes of Maize.* By Mr JOHN SCOTT.

(This paper will appear in next number of the Journal.)

VI. *On the Cultivation of the Quiniferous Cinchona in British Sikkim.* By Dr THOMAS ANDERSON, Superintendent of the Botanic Garden, Calcutta.

The cultivation of Cinchona at Darjeeling, has been carried on successfully. The following is a return of the Cinchona plant in the nurseries at that place, on the 15th June 1863 :—

Cinchona succirubra, ...	1024
C. Calisaya, ...	53
C. officinalis, ...	573
C. micrantha, ...	695
C. pahudiana, ...	2275
Total, .	4620

The cultivation of Cinchona at Darjeeling was attended with very great difficulties at first ; but these have now been overcome, and there is every reason to believe that the plantation will be successful.

In the commencement of June 1863, I supplied Dr Simpson, the European Civil surgeon of Darjeeling, with about two lbs. of fresh leaves of each of the following species : *C. succirubra*, *C. officinalis*, and *C. micrantha*. Decoctions prepared with water slightly acidulated with sulphuric acid, were very bitter, and three patients suffering from well-marked intermittent fever were cured by the administration of these preparations alone. Towards the end of June Dr Simpson and I endeavoured to examine chemically the nature of the leaves of *Cinchona succirubra*, and detected quinine in them.

VII. *On the Cultivation of Tea in India.* By WILLIAM JAMESON, Esq., Surgeon-Major, Superintendent of the Botanic Garden, Saharunpore.

In a former communication I estimated the quantity of waste and other lands fitted for cultivation with Tea, throughout the Kohistan of the North-Western Provinces, and Punjab, and Dhoons, and showed that by them the enormous quantity of 385,000,000 lbs. might be there raised. But in this estimate I excluded the Kohistan of Huzarah and Rawul Pindee, of Cashmere, Jummoo, and the protected Seikh States. The following estimate of the yield of the British territory is nearer the mark, and as a general return when in full bearing, 100 lbs. per acre may be given :—

	Acres.	100 lbs. per Acre.
Kohistan of Rawul Pindee and Huzarah, ...	20,000	2,000,000
Kangra Valley, . . .	35,000	3,500,000
Kazloo, ...	35,000	3,500,000
Munndee, &c. . . .	40,000	4,000,000
Protected Hill States, . .	10,000	1,000,000
Jonsar Bawer, . . .	10,000	1,000,000
Dehra Dhoon, . . .	100,000	10,000,000
Western Gurhwal . . .	180,000	18,000,000
Kumaon, ...	3,500,000	350,000,000
	3,930,000	393,000,000

—a quantity nearly equal to the whole export trade of China, and with high cultivation the figures mig easily be doubled, and thus not only allow an immense quantity for the consumption of the Indian community, but at the same time afford a vast supply for export to other countries.

In February last, at the request of the Lieutenant-Governor of the Punjab, I proceeded to the Kohistan of the Rawul Pindee Districts and Huzarah, there to establish the Tea plant, which has been most successfully done—the plants removed from the Kangra Plantations, and transplanted at Seelah, now growing with vigour.

It is no longer an experimental Tea cultivation in the North-Western Provinces, it having passed from experiment to fact. It has been proved by data which cannot be gainsaid, that the cultivation of the Tea can be profitably conducted ; that the Tea prepared is admirably fitted for the Home and Indian markets ; and that, if properly conducted and backed by capital, the undertaking presents a safe and profitable investment.

VIII. *On some Economic Plants of India.*   By Dr Hugh F.C. Cleghorn.

1. The Box Tree (*Buxus sempervirens*).—This tree, grown at Koclor, has been tested by Dr Alex. Hunter, at the Madras School of Arts, and the wood is found valuable for engraving.

In Mr M'Leod's arboretum at Dhurmsalla the tree grows well. The arboretum contains many introduced Himalayan trees of great interest, as well as many European fruit trees adapted to this hill station. It is, perhaps, the only collection of indigenous alpine trees in the Punjab, the nearest to it being that of Mr Berkeley at Kotghur. I hope the day is not far distant when the Punjab Agri-Horticultural Society will have a hill garden associated with it at one of the sanitaria of the province.

The Himalayan box appears to be identical with the tree common all over South Europe, from Gibraltar to Constantinople, and extending into Persia. It is found chiefly in valleys at an elevation of 3000 to 6000 feet. I have met with it from Mount Tila, near Jhelum, to Wangtu bridge on the Sutlej. It is variable in size, being generally 7 to 8 feet high, and the stem only a few inches thick, but attaining sometimes a height of 15 to 17 feet, as at Mannikarn in Kullu, and a girth of 22 inches as a maximum. The wood of the smaller trees is often the best for the turner and wood-engraver. It is made into little boxes by the villagers for holding *ghee, honey, snuff,* and tinder. At the medical stores in Sealkote it is turned into pill-boxes, and it appears to be adapted for plugs, trenails, and wedges. The wood is very heavy, and does not float; it is liable to split in the hot weather, and should be seasoned, and then stored under cover.

2. The Olive, *Zaitoon*, which has also been tested for wood-engraving at the Madras School of Arts, is another plant of the Mediterranean flora, which ranges from the coast of the Levant to the Himalaya. It varies a good deal in the shape of its leaves and in the amount of ferruginescence, hence the synonyms *cuspidata* and *ferruginea*, but it does not appear to differ specifically from the *Olea europæa* (Mount of Olives), the emblem of peace and plenty. The finest specimens I have seen are in the Kaghan and Peshawur valleys, where the fruit resembles that of rocky sites in Palestine or Gibraltar. The wood is much used for combs and beads—and is found to answer for the teeth of wheels at the Madhopore workshops.

3. *Urtica heterophylla*, a kind of Indian nettle, is plentiful in Simla, having followed man to the summit of Jako, attracted by moisture to an elevation unusual for any member of the family. It is found within the stations of Dalhousie and Dharmsalla, and at many intermediate points. The quantity is surprising wherever the soil has become nitrogenous by the encamping of cattle. The growth at this season (July) also is luxuriant in shady ravines near houses, where there is abundance of black mould; but the sting being virulent, the plants are habitually cut down as a nuisance, both by private persons and municipal committees.

There are other plants of the nettle tribe, particularly the *Boehmeria salicifolia*, "siharu," used for making ropes (to which attention has been directed by Dr Jameson); this plant does not sting, and is abundant at low elevations. The produce of this might be turned to good account, though not yet recognised as merchantable fibre.

4. *Cultivation of Bamboo.*—Mr M'Leod, Financial Commissioner in the Punjab, writes thus to the Commissioners of Umballa and Jullundhur:—

"As it is desired to extend the growth of the bamboo as widely as possible throughout the Punjab, and some of the districts of your division possess them in greater or less abundance, I have to request that you will ascertain whether any of the four following varieties have borne seed during the present year, and inform me of the result of your inquiries.

1. The hollow Bamboo of the plains.

2. Solid Bamboo of the lower hills, of which spear handles and clubs are usually made.

3. The Nirgali or small Bamboo of the hills, growing at elevations from 5 to 8000 feet.

4. The Garoo, or still smaller hill Bamboo, growing at higher elevations, probably up to 12,000 feet.

"It would be interesting also to ascertain, if possi le, from the people, the intervals which lapse between the seasons of flowering of the several varieties. A point on which the more observant might readily furnish information, as, after flowering and yielding seed, the entire tract of bamboo which has seeded, simultaneously dries up and perishes, fresh plantations springing up from the seeds which have been scattered by the old stock.

IX. Dr Alex. Hunter, Secretary of the Agri-Horticultural Society of Madras, transmitted reports as to the cultivation of Peruvian cotton at Chingleput by Dr Shortt, and in the Kistna District by Mr E. B. Foord. Both reports are satisfactory. The following is a statement which he also transmitted :—

Statement showing the Quantity of Cotton carried on the Madras Railway in the Years 1861–62 up to June 1863.

	1861. Indian Maunds, or 82 2-7 lbs.		1862. Indian Maunds, or 82 2-7 lbs.		1863. Indian Maunds, or 82 2-7 lbs.	
January, . . .	6,846	10	694	0	7,886	10
February, . . .	10,519	10	3,535	30	10,305	...
March, . . .	6,134	15	3,346	10	4,471	30
April, . . .	1,390	30	5,466	30	13,597	30
May, . . .	3,046	20	21,795	10	50,500	20
June, . . .	7,238	30	17,457	10	71,193	18
	35,175	35	52,295	10	157,954	28
July, . . .	8,174	20	29,499	20	...	...
August, . . .	6,357	20	20,381	0	...	...
September, . . .	3,721	30	24,979	10	...	...
October, . . .	5,043	30	14,173	0	...	...
November, . . .	6,836	20	16,157	30	...	...
December, . . .	12,607	20	17,151	20	...	...
Total, . .	77,917	15	174,637	10	175,954	28
Average per month in the 1st 6 months, . . .	5,862	25	8,715	35	26,325	31
Average per month in the year, ...	6,493	4	14,355	4	26,325	31

Table showing the Monthly Export of Cotton from Madras, and its Official Value from 1860 to 1863.

	1860.		1861.		1862.		1863.	
	Quantity.	Value.	Quantity.	Value.	Quantity.	Value.	Quantity.	Value.
	Cwt.	Rs.	Cwt.	Rs.	Cwt.	Rs.	Cwt.	Rs.
January,	4,973	82,154	11,162	1,52,904	10,340	2,09,191	7,312	3,35,555
February,	5,605	1,02,290	26,439	3,53,754	9,096	1,87,300	24,485	11,96,878
March, .	7,499	1,09,380	6,035	89,637	12,820	2,52,881	25,780	12,68,855
April, .	10,591	1,66,235	6,748	1,30,242	7,430	1,80,941	12,755	6,44,705
May, . .	5,010	78,105	11,247	1,68,556	17,433	3,89,017	26,260	14,36,706
June, .	1,663	25,132	20,113	3,12,498	36,115	8,80,529	55,875	31,00,255
July, . .	5,829	1,04,696	15,528	2,51,887	35,114	9,46,471	75,269	43,03,488
August, .	21,246	3,22,288	21,061	3,40,859	39,062	15,94,115	...	...
September,	4,661	69,988	5,750	95,312	39,262	18,69,655	...	...
October,	19,003	2,65,371	6,391	1,17,888	5,749	2,85,660	...	...
November,	11,961	1,48,934	7,615	1,44,850	10,264	5,12,605	...	...
December,	6,992	1,01,454	2,315	46,747	18,844	3,28,133	...	...
	105,033	15,76,027	140,404	22,05,134	2,41,529	82,36,498	...	...

Dr Hunter also reports that the American Saw Gin for cleaning cotton *has been* introduced with great success.

X. *Extract of Letter from* WILLIAM JAMESON, Esq., Surgeon-Major, Saharunpore, to Professor BALFOUR, July 9, 1863.

I send two small packets of seeds.

1. Seeds of the Folel or Phulwah (*Bassia lutyracea*) which is now just ripening here. From the seeds of this tree a kind of butter is extracted which is valuable in rheumatism. It is used in lamps, and as it gives a fine inodorous light, it is prized for night-lights. The tree grows to a height of from 30 to 40 feet, flowers in October, and ripens its seeds in July. It was supposed to be confined to Eastern Kumaon and Nepal, but this is a mistake, as it is common at Bhimtul, where I have an extensive tea plantation. Bhimtal is ten miles from the plains and twelve from the Sanatarium at Nynee Tal. The Folel or Phulwah is met with growing at altitudes of from 4000 to 4500 feet. It will do well therefore in your green-houses, but it is not sufficiently hardy to withstand your winters. Where it is met with snow falls annually, but only remains a short time on the ground.

2. Seeds of Bamboo (*Bambusa arundinacea*), which flowered this year in the garden at Saharunpore. Other plants of Bamboo also flowered last year. As the flowering of the Bamboo rarely occurs in our gardens, and as the seeds appear to be good, a small supply may be useful to you.

I also enclose a few seeds of *Eremostachys superba*, which may be a novelty. It flowers in April, and is met with in hot, low localities, as at the Chowki, in the Mohur Pass, in the Sevalik range, and at Jewalah Moki in the Kohistan of the Punjab.

The museum building in the Saharunpore garden is now progressing rapidly, and I trust to see it finished about the end of the season. When filled with specimens, it will, I trust, be one of the most interesting collections in India. I am collecting botanical specimens useful in the arts and sciences from all parts of India, and as soon as the collection is sufficiently extensive a catalogue will be printed.

Two great exhibitions of arts and manufactures are to take place in India,—one at Lahore in November and December 1863, and the other at Calcutta in January 1864. In these we have a move in the right direction, as, under one roof, all the raw products and the articles manufactured in the respective countries will be brought together, and the wants and requirements of each district ascertained; at the same time will be shown what each can give in return, and send respectively into the market.

To open up the country, railways are rapidly extending; but amongst the engineers the cry is—We have no sleepers. Over hundreds of miles in the Himalayas the Cheer (*Pinus longifolia*) is met with in millions, forming trees from 10 to 18 feet in girth four feet from the ground, and in height varying from 80 to 120 feet. These noble trees are every where, I might say, met with in the mountains at altitudes from 3000 to 6000 feet—occurring in two varieties,—one with the wood white and twisted, and easily acted on by the weather, and thus useless in architecture or for railway sleepers; the other, generally met with on the northern slopes of mountains at altitudes of from 5000 to 6000 feet, has reddish-white timber, close-grained and highly resinous. This timber, of which millions occur in the Himalayas, is admirably fitted for architectural purposes, and if kyanised or creozotised would also make first-rate sleepers for railways. Nothing, however, has been done, and the cry of the engineers is,—We cannot get on with our work, because the trees met with in the country yielding timber, fitted for sleepers, are limited. To remove this impression, so far as the North-West Provinces are concerned, I am doing my endeavour, and ere long I trust to see the so-called difficulty to the rapid progress of railways overcome. When once the railways are finished, Government, particularly that of the Punjab, will, through time, find difficulty in feeding the engines, unless every where measures be taken to plant the finest tracts which are now being felled. In the Punjab, only two short lines are open,—one at Moultan, the other between Amritzur and Lahore,—and, with this small drain, firewood has risen 150 per cent. in value. Timber has from time immemorial been felled in the most reckless manner; and only now are the forests beginning to receive the attention that they deserve. Madras and Bombay have for a time been doing something; but as yet no regular plan has been pursued in the North-West Provinces. Numbers of parties were allowed to fell timber, and did so recklessly; so much so that first-class timber of Sal (*Shorea robusta*)—a timber admirably fitted for railway purposes—had, in many of the fine forests at the base of the Himalayas, become scarce, and hence the outcry of the engineers. But there are many other timber trees, admirably fitted for railway purposes, which, through sheer ignorance, have been passed over, such as the Sar (*Pentaptera tomentosa*), Backha (*Anogeissus latifolius*), Dhowlah (*Lagerstrœmia parviflora*), Huldou (*Nauclea cordifolia*), &c. In the Kohistan of the North-West Provinces and Punjab there is no chance of coal being found, the formation being altogether wanting.

I have now established the Cinchona plant in two localities in the Himalayas, in Gurhwal, and the west of Mussouree, at altitudes of from 4800 to 6000 feet. The following species have thus been introduced:— *Cinchona Condaminea, C. succirubra, C. peruviana, C. nitida*, and *C. micrantha.*

XI. Mr M'Nab's *Report on Plants in Flower in the Botanic Garden.*

To give some idea of the mildness of the present season, I beg to lay before the Society dried specimens of 220 species of plants in flower, collected from the open air in the Royal Botanic Garden since the 1st day of December; the largest proportion being the summer and autumn annual and perennial plants, the others chiefly composed of trees, shrubs, and spring flowering plants in the following proportions:—

Annual plants (summer and autumn),	34 species.
Perennial plants. do. do.	118 ...
Trees and shrubs,	38 ...
Ferns,	8 ...
Spring flowering plants,	22 ...
	220

The 220 species are spread over 50 natural orders in the following proportions:—

Natural Orders.	No. of Species.	Natural Orders.	No. of Species.
Ranunculaceæ,	9	Caryophyllaceæ,	11
Berberidaceæ	2	Hypericaceæ,	1
Fumariaceæ,	1	Geraniaceæ,	3
Cruciferæ,	17	Rutaceæ,	1
Resedaceæ,	2	Rhamnaceæ,	1
Violaceæ,	4	Leguminosæ,	7
Polygalaceæ,	1	Rosaceæ,	17

Natural Orders.				No. of Species.	Natural Orders.				No. of Species.
Myrtaceæ,	.	.	.	2	Labiatæ,	.	.	.	13
Onagraceæ	.	.	.	1	Verbenaceæ,	.	.	.	1
Portulacaceæ,	.	.	.	1	Primulaceæ,	.	.	.	4
Umbelliferæ,	.	.	.	4	Plantaginaceæ,	.	.	.	2
Araliaceæ,	.	.	.	1	Polygonaceæ,	.	.	.	2
Loranthaceæ	.	.	.	1	Thymelæaceæ,	.	.	.	1
Caprifoliaceæ,	.	.	.	2	Euphorbiaceæ,	.	.	.	2
Valerianaceæ,	.	.	.	1	Urticaceæ,	.	.	.	2
Dipsacaceæ,	.	.	.	3	Corylaceæ,	.	.	.	2
Compositæ,	.	.	.	26	Garryaceæ,	.	.	.	1
Campanulaceæ,	.	.	.	3	Coniferæ,	.	.	.	4
Ericaceæ,	.	.	.	13	Iridaceæ,	.	.	.	2
Aquifoliaceæ,	.	.	.	1	Liliaceæ,	.	.	.	2
Jasminaceæ,	.	.	.	1	Juncaginaceæ,	.	.	.	1
Apocynaceæ,	.	.	.	3	Cyperaceæ,	.	.	.	1
Gentianaceæ,	.	.	.	1	Gramineæ,	.	.	.	8
Polemoniaceæ,	.	.	.	3	Filices,	.	.	.	9
Boraginaceæ,	.	.	.	4					
Scrophulariaceæ,	.	.	16				Total,		220

SCIENTIFIC INTELLIGENCE.

BOTANY.

The Progress of Tea Cultivation in Northern India.—As the Russian war gave an immense impetus to the growth of fibrous and oil-giving plants by the natives on the plains of Northern India, so the mutiny has been followed by a still more remarkable extension of the cultivation of tea, by English settlers and native landowners, along the belt of the Himalayas, between the altitudes of 2000 and 5000 feet for 1500 miles from Suddya to Peshawur. Official reports enable us to learn exactly the extent of that development up to so recent a period as May 1863. To ascertain the number of planters, extent of grants, and amount of produce at the close of the present year, we may add one-half to all the figures we are about to give. A glance at the "Calcutta Gazette" will show the enormous extent of tea land advertised as applied for by capitalists in Assam. Our share-list, which does not represent private owners, almost every week contains the name of a new tea company. There are several young plantations, which annually double their produce; and Dr Jameson's reports of the Western Himalayan Gardens abound in remarks, to the effect that the out-turn of tea a short time hence will be immensely increased. To the capitalist, the recommendation of a tea plantation is the annually increasing returns it gives from the third to the seventh year, when the plant attains perfection. Nine-tenths of the gardens now in existence are not four years old.

We shall begin our survey at the border line which separates China from Assam, and proceed westward. Chinese tradition points to India as the original home of the tea plant; and the connection between the two countries was so intimate, as proved by Buddhism, that we accept the fact on which the tradition is based. We had hardly obtained possession of Assam, when in 1825 Mr Bruce, still an uncovenanted officer at Tezpore, discovered the indigenous tea-plant. For some time Government nursed the experiment of cultivation, till in 1839 their gardens became the property of the long mismanaged but now most prosperous Assam Tea Company. Two years after this a few plants, grown from China seed, were introduced from Kumaon into Darjeeling, but no tea was made there till 1846, when an Assam planter visited the bright spot. No plantation was formed till 1856 at Kursiong. In 1855 a common labourer discovered the indigenous plant in the Cachar valley, and thus gradually was begun that cultivation in the three great tea districts of Bengal proper, to which Lord Canning's land policy gave such an impetus and stability in 1861. While in the plains, the sepoys strove to wrest from us our empire; on the hills white settlers were laying the foundation of a trade which will yet enrich the land; missionaries were building new stations; engineers were surveying Central Asia; and political officers were upholding our honour in the midst of Mussulman fanatics, who clamoured to be led to victory over the infidel.

In the entire province of Assam there were in May last 246 tea estates, of which 76 belonged to companies, and 170 to private owners. Of these 96 had been acquired during the year. The area of the whole was 122,770 acres, of which 20,144 were under cultivation, or an increase of 4144 during the year. These acres yielded 2,150,068 lbs., or 358,979 lbs. more than in the previous year; taking each pound at 1s. 9d., the whole produce may be estimated at L.190,000. Allowing for long indifference to the plant, this may be considered the result of ten years' labour since the Assam Company revived. But in the six years since 1856 no less than 177 grants of land, covering 558,078 acres, had been applied for in Cachar, or almost every available foot of tea land in that rich valley. On 78 of these grants containing 146,218 acres, 17,594 acres were cultivated, and of these 9426 had been cleared during last year. The tea manufactured, with seed sold, is estimated at L.47,614, and in the current year the value will be double. In six years planters here drew from the treasury of a district previously uninhabited no less than L.173,058. Where there was hardly a human being before, there are now 150 English planters, employing 15,317 coolies, and the number is increasing every month. At Darjeeling there was last year 12,366 acres cleared, of which 9102 were cultivated by 7447 coolies. The out-turn was 40,446 lbs. of tea and 3280 of coffee; and the official estimate for this year is three times this amount of tea, which is likely to drive coffee out of cultivation altogether, as it has done at Hazareebaugh in South Bengal, of which we can give no statistics. Tea cultivation is said to have been successfully attempted on the Kymore hills of Shahabad.

As Bhotan intervenes between the tea districts of Assam and Darjeeling, so Nepaul absorbs a large extent of the tea-bearing area. Starting from the Kali river on its western border, we are at once in the great tea tract of the North-Western and Punjab Provinces, which covers 35,000 square miles away to the Huzara hills, into which, near Peshawur, the plant has just been introduced. Dr Jameson estimates the produce of this tract, when in full bearing, on the moderate scale of 100 lbs. per acre, at 93 millions of lbs, or the whole quantity now exported by China. Going still westward, in Kumaon there are 11 plantations, of which two belong to Government; in East Gurwhal 5, and in Debra Doon 21, of which one belongs to Government and eight to Hindoos. Thus, in the North-Western Provinces there are 38,556 acres of tea grants, of which 4596 were under cultivation by 3080 labourers, and produced 33,960 lbs. last year. Passing on westward by the hill road from Dehra and Mussoorie, we enter the Punjab, where there are 9518 acres of tea plantations held by 23 owners or companies, of whom five are Sikhs and one is

Government: 5 are in the Simla district, 2 in the Kooloo and 2 in the Mundee territories, and 14 in the Kangra valley. Government is introducing the cultivation into the Huzara hills, and has given notice that it will sell in fee-simple its Holta plantation, as well as, probably, the four North-Western gardens, to the highest bidder. The experiment in the Punjab dates from 1851, the year of the Great Exhibition. We shall combine all these figures into one comparative table :—

PROVINCE.	Date of beginning.	Acres granted up to 1862–63.	Acres planted in 1862–62.	Number of Estates.	Number of Planters.	Number of Labourers.	Out-turn in 1862–63.
BENGAL.							
Assam, . . .	1826	122,770	20,144	246	[250]	[20,000]	2,150,068
Cachar, . . .	1856	558,078	17,594	177	155	15,317	327,670
Darjeeling, .	1856	12,366	9,102	[40]	[50]	7,447	40,446
Hazareebagh, .	1859	......	...	...	...	......	......
N. W. PROVINCES.							
Kumaon, . .	1848	9,900	1,500	11	[20]	11,260	30,850
Gurwhal, . .	...	9,900	544	5	[8]	696	15,500
Dehra Doon, .	...	18,787	2,572	21	[40]	1,254	56,540
PUNJAB.							
Simla, Koolo, and Mundee,	1860	2,400	[500]	9	[12]	[1,000]	[500]
Kangra, . · .	1847	7,118	[1500]	14	[25]	[3,000]	[2,500]
Huzara, , . .	1863	......	...	...	...	......	......
		74,319	53,456	523	560	49,974	2,623,074

The three great tea districts vary in several particulars. Labour has to be imported into Bengal; it abounds in the North-Western and Punjab Provinces. On the other hand, Bengal enjoys cheap and easy transit, while in the last two this is the chief obstruction, which the railways will do much to remove. The tea of Assam and Cachar is from indigenous seed, and is stronger than, if not so fine as, that of the other provinces which use China seed. It is preferred by the home dealers to mix with inferior China tea. It will be observed, that in Bengal about two and a-half millions of lbs. were produced last year. From all these circumstances the Bengal tea is chiefly exported, while that of the Western Himalayas is consumed on the spot. The tea districts of the Eastern and Central Himalayas are long likely to be in the hands of English producers alone, while in the Punjab Himalayas, where the hill peasants are so enterprising, the cultivation promises to be conducted on the China method, in small patches by every village, and sold to brokers for manufacture. The eagerness of all castes and classes of our native subjects for tea is well known, and the coarser the flavour the better they consider it. At present tea is sold at a rupee a pound. When it becomes a shilling, and even sixpence for the coarsest kinds, what a trade will arise! All India with its hundred and fifty, and Central Asia with its fifty, millions, will then be the market. And if the Western Himalayas alone, from the Kali river to Peshawur, can yield 393 millions of lbs., what will not the Eastern half, with all Assam, Sylhet, Tipperah, and Munipore, produce? Nay, without extravagance, we may assert that the whole of the hills between Suddya and the Yangtse-kiang, the Chittagong and Burmese hills, and the Yoma range to the valley of the Irrawaddy on the east, and Negrais on the south-west, are tea-bearing tracts, within easy reach of the Bay of Bengal.

In 1862 the import of tea into Great Britain amounted to 114¾ millions of lbs., or an increase of more than 25 millions over 1860. Of the 96½ millions imported in 1861, China sent 92,145,365, Japan 1,348,911, and India, Singapore, and Ceylon 1,983,785. But while the average price of the China and Japan tea was 1s. 5d. per lb., that of India was 1s. 8¾d, or the highest on the list. Since then Mr Gladstone has reduced the duty by one-third ; and who can estimate the increase of consumption this will cause? If, though new to the manufacture, Indian tea-planters can obtain nearly fourpence a pound more for their tea than China, they ought, by careful preparation, and by strict honesty on their own part and that of their London brokers, to make their tea still more eagerly sought after every year. The most cautious will admit that there is practically no limit to the future of the tea trade of Northern India.—*Friend of India*, Sept. 17, 1863.

Tinder used in the Punjab.—Dr Cleghorn states that the tinder of the Hill shepherds, "Kuphi," is furnished by the woolly *tomentum* on the surface of a composite plant, *Onosseris tomentosa*, figured in Royle's "Illustrations" as *Chaptalia gossypina*. The plant is found everywhere at 7000 or 8000 feet, and the Kuphi taken to the plains. It is mentioned by Royle and Jameson that an inferior cloth is manufactured from the woolly down of the leaves. Other composite plants at high elevations are furnished with a somewhat similar downy substance.

Sissoo Tree (Dalbergia Sissoo).—Dr Hugh Cleghorn writes from Simla, 5th August 1863, to the Agri-Horticultural Society of the Punjab as follows:—"I enclose a photograph of a Sissoo avenue at Mozuffurgurh, executed by William Coldstream, Esq., C.S. The picture shows the remarkable growth which the Sissoo attains under favourable circumstances of soil and situation, and gives confidence in extending the culture of a timber tree which is so much valued in the Punjab. The measurements of three of the largest specimens are as follow :—

	Ft.	In.
1. Girth 4 feet above ground, ...	11	0
Girth 11 feet above ground, ...	9	5
Height between 50 and 60 feet.		
Cubic contents from base to 11 feet above ground (approximate), ...	91	0
2. Girth 4 feet above ground, ...	11	7
Girth 11 feet above ground, ...	11	4
Cubic contents from base to 11 feet above ground (approximate), ...	114	0

Both trees have an unbranched trunk for 11 feet, and then throw out two branches more than half the diameter of the trunk. On the road to Shereabah, 1½ mile from Mozuffurgurh, is another Sissoo, girth four feet above ground 12 feet 3 inches. These trees were planted by Mozuffer Khan, who built the town and made the garden in which Nos. 1 and 2 are standing. The age is about seventy years." I send the above extract from Mr Coldstream's letter accompanying the photograph, as the systematic collection of observations showing the rate of growth of different kinds of trees is needed in India.

Botanic Garden, Calcutta.—Dr Anderson's Report of the Royal Botanic Gardens Calcutta, during the past official year, appears in the *Calcutta Gazette.* The planting out of the groups of natural orders was commenced last rainy season. Two specimens of all the species of a considerable number of orders were placed in their proper sites. The botanical name and native country of the plant, painted on a large zinc label screwed down on an iron rod of about two feet high, were placed near the best developed specimen of each species. 2500 seedlings of mahogany were reared for planting along the portion of the Ganges and Darjeeling Road between Titalyah and the foot of the Himalayas. Three hundred seedlings of *Polyalthia longifolia* were prepared for forming an avenue along the Circular Road in Calcutta. Six hundred and eighty plants of *Diospyros embryopteris* were supplied for planting along the banks of the Circular Canal. Of the fifteen Wardian cases, each with 150 Cinchona plants obtained from Ootacamund, four were sent to Dr Jameson at Saharunpore, and eleven were obtained by private individuals for experimental cultivation on tea estates Four hundred and twenty plants were added to the stock in the Darjeeling Nursery. Planters should remember that the only time the introduction can be attempted with any hope of success, is from 15th November to 15th February. The tedious duty of arranging the Herbarium, commenced by Dr Thomson in 1856, is about half completed; and, as two European botanists have been appointed for the work, it will be finished in four years. Dr Thomson's botanical library was purchased for the gardens for L.300.

MISCELLANEOUS.

Letter from Mr Robert Brown, *Botanist and Collector to the British Columbia Association of Edinburgh, dated Victoria, Vancouver's Island, July* 24, 1863.—When I had last the honour of communicating with you, I was about to sail for Alberni. After being delayed for several days both in Victoria and Barclay Sound, I arrived there on the 28th May, and received a very warm reception from the little community, and excellent accommodation in a large house belonging to Messrs Anderson and Company. On the 3d June I started in a little schooner, the Codfish, manned by two men, bound on a trading expedition with the Indians along the coast, and which I considered a good opportunity of visiting various places which otherwise I would have no opportunity of botanising over. We cruised cautiously along the coast, visiting and trading with the savages; while I pursued my researches on shore for nearly a fortnight. In this way we visited or traded with members of the Opischesats, Shesats, Ouchleclous-ets, Oh-i-ats, Uclul-u-ets, Toquats, Clay-o-quots, Ah-ous-ats, Mam-ous-ats, Kel-simats, Ash-quots, Nuchlachlets, Eh-ut-usets, Kaioquots—all little tribes, mutually hating, and frequently at war with each other. But the most famous tribe and locality which I visited was the Nootkas, in Nootka Sound—a tribe which had been traded with by no white man since the murder of Captain Sta last January. I was particularly anxious to visit Nootka, as this was the locality visited by Cook, La Perouse, Meares, Vancouver, and also Menzies, who collected here the few plants he had from this part. Though it is not my intention to describe particularly the botany of the localities I visited, as the absence of books and specimens for reference must make an account most imperfect and uncertain,

I may mention that I obtained *Pinus ponderosa, Rubus nutkanus, Cupressus nutkatensis*, &c., all species interesting from the locality, and worthy of the seed being sent home. As this account of my travels must, for various reasons, be merely a very short outline, I cannot enlarge on the various sights and adventures I saw and mingled in on a journey which people tried to persuade me was dangerous in the extreme, and which only a sense of my duty to the expedition led me to undertake. In every case the Indians treated us well, though this might perhaps be due to the close watch which was kept on them, and the precaution used to avoid treachery. Circumstances have since shown that this was not unnecessary. We returned on the 20th June with all the *éclat* of a successful expedition. The weather, however, during the whole of our trip, was very wet, the rain pouring from morning to night, for sometimes a week at one stretch. I landed, however, on every opportunity, going as far as the dense bush would allow me, and returning regularly every afternoon drenched to the skin (for the Indians had stolen my waterproof coat), with no place to dry my clothes, for a temporary fire was only lit in a grate on deck to cook our food (twice a day), and the rain soon extinguished it. In the little " pigeon-house," called a cabin, we had barely room to squat about, surrounded by trading articles—blankets, beads, tobacco, brass wire, paint, calico, &c.; and often, when turning into our " bunks" at night, we found that the rain had penetrated and soaked our beds. As you may well suppose, our paper got but an imperfect drying. On one occasion, in despair I had to take it under my blanket at night. Of course, all these inconveniences were counterbalanced by many other pleasant occurrences—such as seeing strange scenes, wild men, new places. On the whole, with all the anxieties and dangers we were surrounded by, I do not know whether I would not have preferred it to an excursion in Switzerland or Clova—at all events, I do not regret it now, as it taught me, what no amount of book-learning or teaching could ever train me to—namely, a familiarity with danger, and with the savage tribes through whose country I must travel solitary for a long time yet to come. I now understand the method of treating Indians, their customs, and the Chinook jargon—the ordinary language of communication with the whites—tolerably well. I can now strike bargains, engage men, and travel alone without any assistance from the whites. I need scarcely tell you how valuable this proves to me in such a very rough roadless country, only peopled by a civilised population in a few places in the vicinity of the mines, &c., and where nothing can be procured except at enormous prices.

On the 24th June I started on another excursion to explore the botany of the interior lakes, hitherto almost unknown. As the sawmills at Alberni were stopped at that time to undergo some repairs, four men accompanied me in my excursion. We proceeded up the Somass River to a lake, on the lower arm of which Anderson and Co. have established a " logging camp." After exploring a small arm six miles long, and sleeping for a night among the moss, and having the dubious pleasure of hearing " the wolf's long howl from Unalaska's shore" (by the way, we were only a few days' sail from Unalaska), owing to a storm arising on the lake we were prevented from returning, as we had expected, to the backwoods camp. We returned at daybreak perfectly ravenous. That same morning we started in a canoe, with an Indian guide (Quasson, third chief of the Opischesats), up a long arm of the lake, which no one knew the termination of. We paddled along all day, landing here and there along the wooded banks to cook our dinner, or to " prospect" on an *island.* I found a *Thuja* allied to *Thuja gigantea* (Craigana, Balf.), but, I think, distinct ; and once I picked the cones of *Abies grandis*, and one very like *Pinus contorta*, Dougl. I "blazed" the tree, and will return in the autumn to it, if I do not find it elsewhere. Just as the sun was setting, we came to the termination of the lake, which I had named "Sproat's Lake," after the resident partner in the firm of Messrs Anderson and Co., my friend and countryman Gilbert Malcolm Sproat. It is about 23 miles in length in its longest axis. Here we found a river flowing in. We ascended this in our canoe for a mile, when the navigation becoming difficult we camped, and enjoyed a sound sleep by a blazing fire, notwithstanding the mosquito and various other pests, of which, to those who have camped "*sub Jove frigido*" in North-West America, there requires no explanation. Next morning, after discussing our pork and biscuit, we took to the woods, ascending the banks of the river through a fine open valley, comparatively thinly covered with fine timber, and free of undergrowth—a great rarity for Vancouver's woods. If the wood was clear of undergrowth, we found a pest which incommoded us considerably—*Panax horridum*, well deserving its name, with its long waving stem, crowned with a pecten-like head of leaves, and a raceme of white flowers. In many places the ground was covered with snow a foot and a-half deep, through which some bulbous plants were protruding their flowers. In the course of the day I made a very agreeable discovery in the examination of what I thought the *Abies Canadensis* of the Atlantic slope ; but which I found represented by a species which may be little known in England. It has only been recently described by my friend Dr Albert Kellogg, secretary of the Californian Academy, in their Transactions, vol ii. p. 8, under the name of *Abies Bridgei*. Though I will take care to send you plenty of it later in the season, I subjoin a short description of it :—

Leaves—Evergreen, solitary, linear cuneiform obtuse, somewhat flattened, fleshy slightly, grooved above, ridged beneath, very minutely scabrous, serrate petiolate, somewhat two-ranked. *Cones*—Numerous, solitary, terminal, pendent, elliptic-ovoid, about twice the length of the leaves. *Scales*—About thirty or more oblong, roundish, concave, margin entire, thin, translucent, finely corrugate striate on the back ; base abrupt, subauricled, stoutly attached to the ligneous axis. *Bracts*—Three-lobed ciliate, villous ($\frac{1}{4}$ in. long.) *Seeds* (including wing) scarcely less than the scale, wings oblong oblique, broader at the base, somewhat suddenly narrowed above, obtuse, laterally warped or carinated, seeds proper, ovate, light-brown or drab colour, uniformly marked by three minute ovate glands on the side looking towards the base of the cone. A tree 80 to 100 feet in height, of dark verdure and graceful appearance. The branchlets are very villous, slender, and drooping. The timber is said to be firmer, finer, and straighter grained than the Canadian hemlock spruce, which it represents on the Pacific coast. It is certainly closely allied to *A. Canadensis*, but I believe, with Kellog, that it is distinct. Even the Canadian woodsmen, who are very apt to forget that similarity is not identity, and apply "old country" names to anything at all resembling what they are familiar with in Canada or Maine, recognise this.

About three o'clock P.M. we called a halt to take counsel, after having tracked along the banks for about twelve miles, the river dashing over rocks or flowing calmly over gravelly spit, through wooded meadows or high trap banks, backed by snow-peaked hills, from which the melting snow was leaping down in cataracts, now hid among the dark pines, and again bounding over some rock, until it fell in a sheet over the bank of the river. On starting, we had only taken a few biscuits, expecting to find the river merely a freshet from the mountain, and that we could proceed only a short way; but where the river was as broad as when we first ascended, it became prudent, as there was no appearance of game, to hold a survey upon our provisions, when it was discovered, to our dismay, that we only had two biscuits; so I ordered a retreat.

We reached our camp just before dark, very tired and very hungry, and sweetly did we that night enjoy our bed of pine branches, prefaced by a most substantial supper of salt pork, hard biscuit, and tea—our provisions on all such journeys in this country. While cutting the branches for my bed, I found *Taxus brevifolia* (? Lindleyana Murr.), but without fruit. Our fire was of *Abies Douglasii*. *Abies Menziesii* was not uncommon, random cut, but the commonest was the former; also *Abies Bridgei* and *Pinus Strobus*, with *Rubus spectabilis*, *R. nutkanus* and *R. leucodermis*, with *Ribes* (two species). I did not see *Ribes sanguineum*, so common in our English shrubberies, but it is excessively common on the southern part of the island, particularly the Indian Reserve at Victoria. I observed also *Vaccinium ovatum*, and occasionally the Indian Salsul (*Gualtheria Shallon*), a very pretty shrub. Next morning, looking about the neighbourhood, we re-entered our canoe, hollowed out of *Cupressus nutkatensis*—the mats we sat upon being made of the *liber* of the same tree, ropes of the same material, and occasionally of *Thuja plicata*. This river I named after one of our party, Mr Taylor, an Edinburgh man, who had formerly accompanied me on my first journey on the island, when we were lost in the Nauaimo Mountains, wet and half starved, for two days. Some of the mountains round I dedicated to some of our Edinburgh friends, and which will afterwards appear, when I send home to the Geographical Society, when I have leisure for it in the winter, the chart of my journey. The Indians told us that it flowed out of a lake near Clay-o-quots Sound, and that the Clay-o-quots sometimes descend it. We reached the camp (logging) for supper, and slept among some hay that night sorely harassed by aphaniptera. Next day we started again in our canoe down the south arm of "Sproat's Lake," and landed three miles off in the wood. We now started off, each man with his blankets on his back, over an Indian trail. After travelling for about two miles through dark shady woods, over a tolerably good trail marked by knotted twigs, we were startled by the appearance of a man. It was Kan-ash, second chief of the Opischesats. On learning from "Quasson" that I was "Hyass-ty-hee" (great chief), he was exceedingly friendly, but parted "uyon silex" (very sulky) because I would not give him an "as-pop" (gift) for passing over his land, and for the future use of some supposed "chapatz" (canoe) which was on the "Uyass-aw-ak" (the Great Lake), to which Quasson was conducting us.

About two o'clock, the Great Central Lake, forty-five miles long, burst upon our view, the landing-places scattered round with bones of bears and deer, the remains of Indians who had formerly camped here. We found a canoe hid in the bush, but of such a miserable character as to be quite unfit for our purposes, only holding three at a time, and then rapidly filling with water; so we were very unwillingly forced to abandon our project of ascending the lake to its head. We camped that night near the lake, and returned next day to the mouth of the Somass River, where, after a good deal of trouble at the Opischesat village, at one time threatening to be serious, I obtained a canoe, and swept down the river, and thence into the Alberni Canal, all safe and in sound health and appetite, after our novel journey. Such is a short account of one out of many similar journeys. I have gone more into detail concerning it than some others equally interesting. But as I cannot enter into anything like particulars *regarding* all *my* travels, you may accept this as a specimen. Some were better; many a great deal worse. Of course, on such excursions a botanist can only collect one, or at most two specimens of each plant he meets with; and frequently, so limited are his means of conveyance in an almost trackless country, he can only take such as he has not seen in other places, noting, however, the occurrence of all others, in addition to all such observations as ordinary travellers make. So many mishaps come under one's precious load before it gets to a place of safety, that sometimes I have been mortified to find, after a laborious excursion, the tangible results, notwithstanding all my care, consisted only of a few blackened, indifferently dried specimens. I was preparing to cross the island by the Indian trail to Quallicom, and thence by Nauaimo and the valley of Corvitchen to Victoria, a distance of about 100 miles, when the rain began to fall in torrents, and so fell for ten days in a continuous pour, until, losing all hope of being able to effect this journey for the present, I came down to Victoria in the Thames, where I arrived after twenty hours' steaming.

I here take the opportunity of recording my thanks to Messrs Anderson and Co. for the great kindness and assistance they rendered me through Mr Gilbert M. Sproat, the resident partner, and without which it would have been utterly impossible to have undertaken anything at all approaching to such a lengthened exploration. On my arrival in Victoria, I found that my visit to the Nootkas had got noised abroad, and all the newspapers were anxious to have the first account. By request of his Excellency the Governor, I called upon him next day, when he was pleased to express great satisfaction at my safe return, and at the mode in which I had conducted the exploration. Among other favours, he offered me letters to the magistrates and district officers in British Columbia. Governor Douglas, some people say, is about to resign; and if so, his loss, I am sure, will be a deep blow to the success of the colony, for, notwithstanding many differences of opinion from interested or disappointed people, no man will ever be found with such a thorough knowledge of the wants of the country and all that concerns it.

Finding many plants in flower round Victoria which I had not obtained elsewhere, I remained here until the above date, wandering far and near, and frequently passing the night in some surveyor's tent away in some wild part of the district. During this period I was occupied part of the time in making arrangements for my more important future journey, and had the pleasure of gaining the friendship of many most pleasant people —so pleasant, indeed, that one feels almost sorry to have to leave them, and take to a vagabond life again in the wilds, where he may pass long days without looking on a human face.

I propose to start on Thursday the 30th on another and longer journey, and one which will be most profitable, not only because undertaken during the seed season, but also over a more varied and extensive tract of country. From Victoria, I go to Port Angelos, W.T., thence to Port Townsend, and *viá* Olympia, Fort Nisqualley, Seattle, and Whatcom, over the Cascade Mountains to the Fraser River, along the Fraser to Fort Alexandria, and down either the Bentinck arm trail, or that by Bute Inlet. This will, I expect, occupy me until the end of October, and will, I hope, yield something to reward the outlay of the Association, though I know I will be devoured by mosquitoes. You will perceive that part of it lies through American territory, but I do not suppose that the Association is so fastidious as to object to seeds gathered over what our Government was compliant enough to assign as the boundary line of 1846; and, indeed, the country is not so rich as to allow of picking and choosing. I have obtained the Government credentials necessary for the journey.

Spontaneous Generation.—This subject has occupied the attention of the French Academy for some time, and various communications have been read on the subject. The experiment of M. Pasteur on this subject are conclusive. He has shown by a series of admirable and carefully executed experiments, that no organisms appeared in solutions, unless the germs were introduced from without. M. Flourens states, that he considers M. Pasteur's experiment as decisive; and that there is no such thing as spontaneous generation. Germs and sporules are so abundant in the atmosphere everywhere, that there is no excluding them from fluids exposed to the air. M. Pasteur shows that even the mercury trough was the receptacle of a multitude of germs, and that these entered the tubes during manipulation. The experiments, therefore, required the utmost caution in order to be free from all sources of fallacy.

Call to participate in presenting a Testimonial to the distinguished Botanist, Dr Carl Friedrich Philipp von Martius of Munich.—In March of the coming year 1864, Dr von Martius will reach the fiftieth anniversary of his graduation in medicine. During his labours as a teacher for so many years, he has gained the highest esteem of his numerous colleagues, friends, and pupils, and has rendered the greatest services to science. All who have been privileged to live with him, or who at times only may have been inspired by his presence, will seize with pleasure an opportunity like the present, to give expression to their feelings of esteem, friendship, and gratitude. The recognition, from his contemporaries, which may fall to the lot of the man of action is in truth his highest reward, and there is much bitterness in life which only thus is sweetened and sunk in forgetfulness. Here, arriving at an epoch in his life, we see a man who is honourably known not only in his own country, but far beyond its limits; who, indeed, enjoys in foreign countries a reputation such as falls to the lot of few in his department of science. He has earned this reputation by his unexampled activity; and his works, which may serve as models alike of profound investigation and of beautiful delineation, are so numerous, and several are so magnificent in their plan, that they cannot but excite the greatest wonder. It is true that any mark of recognition in such a case is only an imperfect expression of the feelings which suggest it. To express these, and, as far as possible, to embody them in an appropriate form, the undersigned have deemed it most suitable to commemorate this jubilee occasion by presenting Dr von Martius on the day of honour (the 30th of March) with a medal struck in gold. In order to procure for this end the necessary and not inconsiderable means, co-operation from all quarters is necessary. Considering the great number of friends, and the multitude of his former pupils, now for the most part in the position of medical practitioners or apothecaries, we hope we do not deceive ourselves in reckoning upon numerous contributors. We beg that every contributor should give at least two thalers cons., and hope that such a contribution may make it possible to prepare for every contributor a copy of the medal in bronze which will be accompanied by a list of the contributors.—*Erlangen, November* 1863. (Signed) Dr Ludw. Radlkofer, U.-Professor, Munich; Dr Aug. Schenk, U.-Professor, Würzburg; Dr Adalb. Schnizlein, U.-Professor, Erlangen.

Geographical Discovery in New Zealand.—Dr James Hector, the Government geologist now engaged in the survey of New Zealand, has discovered a navigable river flowing into Martin's Bay, on the west coast, near Milford Haven. This river descended and entered a lake ten or twelve miles long, and from one to two miles broad. Passing through this lake, he found another river falling into it at the eastern end. This river he also entered and ascended for a considerable distance. He then left his vessel, and proceeded through a country fit for settlement. After only forty-six hours' march, he found himself at Queenstown, on Lake Wakitepu. Martin's Bay will now be the nearest port to Melbourne and Sydney. The future town in Martin's Bay may entirely eclipse Dunedin.—*Times' Correspondent.*

OBITUARIES.

The late Rev. Stephen Hislop of Nagpore.[*]—The newspapers of the day have recently recorded the death, in melancholy circumstances, of this gentleman, whose geological researches in Central India will doubtless render an obituary notice of him interesting to many.

Stephen Hislop was born at Dunse, in Berwickshire, on the 8th of September 1817. He received his education first in the schools of his native village, and subsequently at the universities of Glasgow and Edinburgh. On the "Disruption" of the Scottish Establishment in 1843, Mr Hislop, then a student, cast in his lot with the seceding party, and was sent out next year to found a mission at Nagpore, in Central India. Schools were in process of time established at the three stations of Nagpore, Seetabuldee, and Kamptee. These ultimately acquired great influence, being attended by no fewer than 700 pupils. During extensive tours, undertaken for missionary purposes, Mr Hislop paid keen attention to the physical character of the districts traversed, and various geological discoveries of a remarkable character were the result. Several officers, after a time, joined in the inquiry, and rendered effective assistance. A brief notice of the Nagpore discoveries was sent by Mr Hislop, in April 1853, to the Bombay Branch of the Royal Asiatic Society; and a more detailed paper, in our joint names, was read before the Geological Society of London on the 21st July 1854, and printed in their "Quarterly Journal" for August 1855. Various supplemental papers were subsequently drawn out by Mr Hislop, which also appeared in the journal. When afterwards at home on sick leave, he obtained aid from the British Museum and other sources, and described the fossil shells from Central India, a large number of which were new to science. Professor Owen had already named and pointed out the characters of one remarkable fossil, a new labyrinthodont reptile. Professor T. Rupert Jones, from whose kind assistance and sympathy much advantage had all along been derived, added a memoir on the Cyprides. Mr A. Murray, F.R.S.E., took up the subject of the insect remains. Sir Charles Bunbury appended a paper on the more antique series of the fossil plants. The more modern series of plant-relics, chiefly beautiful fruits, believed to be of Eocene age, has not yet been figured.

Though geology was the chief, yet it was not the only subject of inquiry at Nagpore; and it was in connection with another department of research that the lamented missionary met his death. In December 1847, as Mr Hislop, with his colleague, was passing the village of Takulghat, twenty miles south of Nagpore, he observed a circle of large unhewn stones. Further examination revealed, that there were no fewer than ninety such circles, some single, others double—all close together, and spreading over an area of about four square miles. Permission was subsequently sought and obtained from the late Rajah of Nagpore to make excavations among the circles; and in the centre of one of them, at the depth of three feet from the surface, was found an iron vessel like a frying-pan, with a handle on either side, which had rusted off and was now lying detached. The bot-

[*] This obituary is from the pen of the Rev. Robert Hunter, late of Nagpore, and is inserted in "The Reader."

tom of the vessel was covered with little pieces of earthenware, neatly fitted to each other like mosaic work, possibly designed to protect human ashes, of which, however, there were only doubtful traces beneath. The hostility of one of the petty native officials at Takulghat prevented the excavations from being as complete as had been intended, and it was all along felt that they should be resumed when a favourable opportunity presented itself. A couple of months ago, Mr Temple, the chief commissioner at Nagpore, who had succeeded to the authority of the deceased Rajah, was encamped four miles from Takulghat, and, feeling that the time had come for paying renewed attention to the stone circles, wrote requesting Mr Hislop to join him and superintend a new series of excavations. The missionary in consequence set off for Takulghat on the 3d September, and aided Mr Temple during that day and the next. As the night of the 4th approached, Mr Temple left first for his camp, and Mr Hislop was to follow when he had gathered up the trophies and examined a school in the neighbouring village. It was the depth of the monsoon, and, though no rain had fallen at Takulghat, it must have done so very heavily somewhere in the vicinity. In consequence, Mr Temple, on reaching the backwater of a river, which had been quite shallow in the morning, found it 10 feet deep. He left behind him a native to warn Mr Hislop of his danger, and conduct him to a ford some distance higher up the stream. When night fell the native left his post, and went back nominally to meet the missionary. Two horsemen left at Takulghat, as an escort were also missing when the hour of departure came. Mr Hislop, thus unwarned, rode up to the fatal post, unattended, at a canter, and was immersed before he suspected danger was near. There were indications to show that he must have been thrown from his horse, as the animal was plunging in the water. Alone, and under the cloud of night, he struggled hard with his destiny. But all his efforts were unavailing to avert the fatal result. When, two hours later, his horse appeared in Mr Temple's camp without a rider, alarm was excited, and parties having been sent out to scour the jungle, the body of the missing missionary was found in comparatively shallow water, still holding with a death-grasp handfuls of grass, showing that he had succeeded at one period in reaching the bank, but been too feeble to raise himself from the water.

Mr Hislop was tall, wiry, and able to endure astonishing fatigue. His natural and moral courage were heroic. None could long associate with him without being constrained to admire his fortitude and decision of character. His powers of observation were of the first order. His mind was unimaginative, but logical and painstaking in no ordinary degree. His general knowledge was extensive. He could address a native audience in Mahratta with much effect, and had preached in that tongue through a great part of Central India, enlivening the tedium of the journeys (mostly performed on foot) from village to village, by geologising as he went along. His personal piety was deep and sincere, and he possessed much influence over the Europeans, as well as the natives of Central India. He leaves a wife and four children. An obituary notice of him, from the pen of Professor T. Rupert Jones, in the November number of "The Geologist," concludes with these remarkable words :— " Taken away suddenly from his family, his friends, and his native church and schools, he will live in our memory as a beloved man, just and good, and as an acute observer, cautious and conscientious, not courting praise, nor even notice, but delighting in work and truth, as a loving student of nature, and a faithful servant of God."

Mr P. A. Munch, the Historian of Norway.—The following notice has been transmitted by Mr Christopher Holst, Secretary of the Royal University of Norway, Christiania:—The Royal University of Norway has the honour of informing you of the loss which it has just sustained in the person of Mr P. A. Munch. Mr Munch, whom the voice of the people has proclaimed the national historian of Norway, was born at Christiania on the 15th December 1810. His father, Edward Munch, a protestant minister, made him take his first classical studies at the school of Skien, and sent him at a later period to study law at the University of Christiania. Young Munch soon drew public attention towards him by a remarkable intelligence, a lively imagination, and a wonderful memory. He abandoned law for historical studies, and was appointed Professor of History at the age of thirty-one, 16th October 1841. He devoted himself from that time exclusively to the history of his country, whose monuments are found almost as numerous out of Norway as within its actual limits. His researches led him successively into Sweden, England, Ireland, and Normandy, where he sojourned at different times. Little by little his name acquired a European celebrity, and he was elected member of several learned foreign societies. In 1857, the Storthing having given him a grant, in order to enable him to make researches at Rome into the ancient history of the Scandinavians, he devoted several years to the fulfilment of this mission in the archives of the Vatican. He gave himself up to these studies, which would have deterred a less persevering energy, with a sagacity which is only equalled by the strange boldness of his conclusions. Centuries did not succeed in hiding from him any of their secrets, and his penetration into the obscurity of past times enabled him to illuminate with glowing hypotheses the chaos of our early history. Is it not hypothesis alone which could open to history the tracks which the critic will clear at a later time? Mr Munch is the author of a number of historical, geographical, philological, and political writings. But his great work, his special title to glory, and at the same time to the eternal gratitude of his country, is his History of Norway (Det norske Folks Historie), a remarkable scientific and critical work, by which he has made known to the Norwegians their national origin. Unhappily this monument—raised by a choice spirit to the honour of a people who were formerly powerful, and whose annals were associated in the Middle Ages with those of the most of the great European nations—remains uncompleted. Death has stopped this work at the date of the union of Calmar (1397). Mr Munch was suddenly carried off at Rome on the 25th of March last, in the fifty-second year of his age. He leaves a sorrowing widow, one son, and four daughters.

PUBLICATIONS RECEIVED.

1. Air-Breathers of the Coal Period; a Descriptive Account of the Remains of Land Animals found in the Coal Formation of Nova Scotia. By J. W. Dawson, LL.D., Principal of M'Gill University —*From the Author.*

2. An Inquiry into the Nature of Heat, &c. By Zerah Colburn. London, 1863.—*From the Author.*

3. Canadian Naturalist and Geologist, for February, April, June, and August, 1863.—*From the Editors.*

4. Bulletin de l'Académie Royale des Sciences, des Lettres, et des Beaux Arts de Belgique, Nos. 6–10, 1863.—*From the Academy.*

5. Journal of the Chemical Society, for September, October, November, and December 1863.—*From the Society.*

6. Journal of the Asiatic Society of Bengal, No. 2, for 1863.—*From the Society.*

7. Transactions of the Tyneside Naturalists' Field Club, Vol. VI., Part 1.—*From the Club.*

8. Catalogue of the Army Medical Museum, Washington, 1863.—*From the Surgeon-General of the United States Army.*

9. Report of Lieut.-Colonel J. D. Graham, United States Topographical Engineers, on Mason and Dixon's Line. Chicago, 1862.—*From the Author.*

10. Transactions of the Academy of Science of St Louis. Vol. II. No. 1, 1863.—*From the Academy.*

11. Annual Report of the Board of Regents of the Smithsonian Institution for 1861.—*From the Institution.*

12. Report of the Superintendent of the United States Coast Survey, for 1859 and 1860.—*From the United States Government.*

13. Dr Daubeny on Climate. An Inquiry into the Causes of its Difference, and into its Influence on Vegetable Life. 1863.—*From the Author.*

14. American Journal of Science and Arts for September 1863.—*From the Editors.*

15. Quadrature du Cercle. Par un Membre de l'Association Britannique pour l'avancement de la Science (James Smith); traduit par Armand Granges, Bordeaux, 1863.—*From the Translator.*

16. Essays on Digestion. By the late James Carson, M.D.—*From the Publisher.*

17. First Principles of Natural Philosophy. By William Thynne Lynn, of the Royal Observatory, Greenwich.—*From the Author.*

18. Proceedings of the Literary and Philosophical Society of Manchester. No. 2, Session 1863–64.—*From E. W. Binney, Esq.*

19. Victoria Toto Cœlo; or, Modern Astronomy Recast. From a Paper on the Theoretical Motions of the Earth, Sun, Moon, and Planets, &c. 1863. By James Reddie, F.A.S.L.—*From the Author.*

20. Manual of the Metalloids. By James Apjohn, M.D., Professor of Chemistry in the University of Dublin.—*From the Author.*

21. A Familiar Epistle to Robert J. Walker, from an Old Acquaintance. Saunders, Otley, and Co., London.—*From the Publishers.*

22. The Velocity of Light; its Astronomical Data and Experimental Proof. Letter from G. F. Chambers, Esq. to J. Reddie, Esq.—*From Mr Reddie.*

23. The Philosophy of Geology; a Brief Review of the Aim, Scope, and Character of Geological Inquiry. By David Page, F.R.S.E., F.G.S., &c.—*From the Publisher.*

24. Tables of Heights in Sind, the Punjab, N.W. Provinces, and Central India, determined by the great Trigonometrical Survey of India. Calcutta, 1863.—*From the Office of the Survey.*

THE

EDINBURGH NEW

PHILOSOPHICAL JOURNAL.

Notes on the Mummied Bodies of the Ibis and other Birds, found in Egypt. By A. Leith Adams, A.M., &c., Surgeon 22d Regiment.

I procured many mummied specimens of the *Ibis religiosa* (Cuvier), from Thebes and Lower Egypt, and found the following discrepancies in the measurements. It must be understood that the specimens were procured from different localities, and not two from the same tomb.

From the occipital protuberance to the tip of the bill the following were the dimensions in seven—viz., (1.) $8\frac{6}{10}$ in.; (2.) $8\frac{8}{10}$ in.; (3.) 9 in.; (4.) $7\frac{9}{10}$ in.; (5.) $7\frac{9}{10}$ in.; (6.) $9\frac{4}{10}$ in.; (7.) $7\frac{4}{10}$ in. The bills of these measured from the gape to tip—(1.) $6\frac{4}{10}$ in.; (2.) 7 in.; (3.) $8\frac{2}{10}$ in.; (4.) $5\frac{4}{10}$ in.; (5.) $5\frac{4}{10}$ in.; (6.) $7\frac{6}{10}$ in.; (7.) $5\frac{4}{10}$ in. Thus the total length of the skull varied from $7\frac{4}{10}$ inches to 9 inches, and the bills from $5\frac{4}{10}$ to $8\frac{2}{10}$ inches.

The length of the *humeri* in the above varied from $4\frac{4}{10}$ to $5\frac{4}{10}$ inches.

The *ulnæ* could only be measured in four of the number, and ranged from $5\frac{4}{10}$ to $6\frac{4}{10}$ inches.

The *tibiæ* in five of these specimens varied from $5\frac{6}{10}$ to 7 inches.

The *tarsi* in the seven varied from $4\frac{4}{10}$ to $5\frac{4}{10}$ inches.

The middle toe in six of the specimens varied from $2\frac{6}{10}$ to $3\frac{4}{10}$ inches.

NEW SERIES.—VOL. XIX. NO. II.—APRIL 1864.

The largest claw ou one specimen measured $\frac{7}{10}$ inch.

There can be no question that all I examined belonged to one species ; and of many more noticed in various parts of Egypt, I did not discover one that, in the dimensions of its bill, would agree with either the black Ibis[*] or the glossy Ibis. I think that the discrepancies just shown may be fairly attributed to the effects of long domestication as well as sex. The manner the sacred Ibis was permitted to range over the country, especially in towns and villages, but perhaps more in and about the temples, would, in the vast ages embraced by Egyptian history, be sufficient to produce not only a considerable difference in the size of the bird, but probably also in its colouring ; add to this, the filthy narrow streets among the mire and refuse of large mud-built cities, like Thebes and Memphis, where, under conditions akin to what is noticed with poultry, it is by no means likely that the pure white of the plumage was at all preserved. Hence, Herodotus[†] may have fastened on a variety and considered it another species. His description of the white Ibis is clear and distinctive, but that of the other " all black, with legs like a crane," is not so evident ; certainly it is not applicable to the glossy Ibis, and, except in the colouring, the black Ibis (*Ibis sacra* of Temminck[‡]) will not agree ; moreover, as far as I can discover, the last named species is not a native of any portion of North Africa. Cuvier[§] states, that his specimens from Lower Egypt had larger bills than those found at Thebes. The largest billed specimen in my collection was from the Ibis pits near Sacarah. It is likely, however, that the bird did not thrive so well in the north country ; indeed we know from Roman historians, that when introduced into the temple of Isis at Rome, after the conquest of Egypt, it soon pined away and died. Savigny states, that the *Ibis religiosa* is to be met with at the present day on Lake Menzaleh, near the Damietta mouth of the Nile ; but all inquiries I have made during and since my excursion to Egypt failed to substantiate his assertion. With Vierthaller,[||] I am inclined to agree that the sacred

---

[*] Falcinellus igneus, Gould, B. E., plate 47.   [†] Euterpi, I. xxvi.
[‡] Temm. Man. d'Ornith., 2d Ed., 2, 596.   [§] Ossemens fossiles.
[||] Naumannia, 1852, p. 58.

Ibis is not now a native of Egypt or Nubia. He fixes its northern limit on the Nile at 14° or 15° of north latitude; and states, that it migrates to Chartum in July, and breeds there on the shores of the White Nile. Heuglin records it in his list of birds collected on the Red Sea.* Bruce gives a faithful description of the bird.† Between the *Sacred Ibis* and the *Ibis bengalis* there seems a very close alliance. The earlier naturalists, such as Buffon, Belon, Perrault, &c., either never saw the bird, or were constantly confounding it with the *Tantalus ibis* and the storks. Belon, in his description of an ibis in the menagerie at Versailles, is evidently noting the characters of the black stork;‡ moreover, the Egyptian vulture has been frequently mistaken by travellers for our bird, and the common buff-backed heron§ at the present day pays the penalty of death from many Nile voyagers, in order to be preserved as the sacred Ibis, on account of its white colour, for in no other respect is there any resemblance. A difficulty to be accounted for with reference to the presence of the Ibis in Egypt during the existence of the ancient race is,—How did they obtain the bird? and that we can only conjecture by its having been gradually introduced, and having there propagated itself. Eggs have been found along with mummied birds; and in the Antiquarian Society of Edinburgh there is a collection of eggs found at Thebes by my late lamented companion Mr Rhind,‖ which it would be well to compare with those of the existent *Ibis sacra.* From the enormous numbers of mummied bodies found both at Thebes and between the pyramids of Sacarah and Gizah, there cannot be a doubt but that the birds were very numerous. To have regularly imported young or old from the upper country would scarcely have been possible, and we have no proof whatever of the bird having ever been indigenous in Egypt or Nubia; besides, there is no reason why it should not have bred freely in a domesticated state, especially as long as it held its position among the sacred birds, which Herodotus tells us were

* Ibis, vol. i. p. 34. † Ibis, vol. vii., App. p. 271.
‡ Obs. de Belon, Paris, 1555. § Ardea bubulcus (Savigny).
‖ See that author's late work, "Thebes, its Tombs and their Tenants," p. 52.

preserved by a penalty of death to whomsoever killed one, either by "accident or design."* Its extirpation may have been gradual, but the Romans at first do not appear to have been the cause, whatever may have taken place subsequently. The temple of Isis was honoured by its presence, as the drawings in Pompeii and ancient writers fully prove. It is likely, however, as the religion of the ancient race began to suffer from the inroads of the early Christians, that as the latter increased so did the Ibis decline; and to a bird which had received so much attention, neglect was likely to have soon brought about a complete extinction of the species.

In comparing the past with the existent race, there are two points to be considered : *first*, the age of the mummied birds. Sir Gardner Wilkinson writes me—" It is difficult to ascertain the date of an ibis, or other bird mummy, because they have seldom hieroglyphics with king's names." He believes, however, that many at Thebes are of the time of the nineteenth dynasty, *i.e.* B. C.

In the *second* place, the modifications in the dimensions of the skeleton are without doubt owing to its having been domesticated and subjected to artificial influences, such as affect our tame animals.

There is a fact in favour of there having been only one species. Among all the coloured drawings on the tombs of Beni Hassan, Thebes, &c., one species is only represented, and that is clearly the white Ibis. The perfection, nay I may say beauty and brilliancy, of many of the paintings and delineations of the Ibis, are unrivalled by anything of the sort now-a-days. Moreover, it is to be expected that as its figure had to be used in the hieroglyphic writings more than that of any other bird, by dint of constant practice, and in spite of the conventional style of the Egyptians, it is likely they would, after so long practice, arrive at a high point of excellence in tracing its outline. From the countries now frequented by the Ibis, there is every reason to believe that at least it did not migrate to Egypt or Nubia in winter ; moreover, it is demonstrably a tropical species, and

---

* Euterpi, i. 15.

does not venture beyond the confines of the tropics. I doubt, moreover, if it has any well authenticated claim whatever to a place in European lists.

The following analysis of the stomachs of several opened by me testify to its varied diet. With reference to reptiles forming its chief food, or that the species shows any particular predilection for these animals, is, I think, at best doubtful. In common with its allies, such as the glossy and black Ibis, &c., it no doubt devoured frogs, lizards, and small snakes, along with fish, shells, and coleopterous insects, and even carrion ; in fact, the species was a sort of public scavenger, just as the Egyptian vulture and carrion crow are to the present race, and the former, the adjutant, and ground kite, are to the natives of Hindustan.

Examination of the Gizzards of the Mummied Ibis, from Tombs at Thebes, and near Memphis. *

Specimen a. Contained numerous portions of small beetles, with the legs of larger species, also small objects like the naked seeds of wheat or barley, and several angular fragments of stone. The legs of this specimen were in a perfect state of preservation, owing to a solution of bitumen having been painted on them, which has the appearance of japanning.

Spec. b. The lower extremities, from not having been preserved in the way just described, and merely covered with a bituminous bandage, were very friable. The claws were much elongated, overgrown and twisted ; the outer claw and toe so much deformed as in all probability to have greatly impeded the bird's movements. The skin on the forehead was much shrivelled, and the bill and tarsi were of a reddish-brown colour. The gizzard contained the back-bone and ribs of a fish of small size. The plumes on the back were distinctly traced in this specimen ; but excepting the centre portions of the wing quills, the original colouring of the feathers was completely destroyed by the bitumen.

Spec. c. Wherever the bitumen had been sparingly applied to the naked parts of the lower extremities or the

* See note at conclusion of this paper.

bill, both appear of a reddish colour. The bones of this specimen were much smaller and more friable than many of the others—probably a young bird. The gizzard contained a mass of black peat-like substance, with portions of the elytra of small beetles.

Spec. d and e. The gizzards of these two were empty. One was the largest specimen of all I examined.

Spec. f. The gizzard contained several large quartz pebbles, interspersed with numerous univalve spiral shells, evidently of the family Paludinæ.* Along with the aforementioned pebbles, a small oblong bead was found, with a hole through it, the same as now found on the wrists and forming necklaces on the human mummies.

Spec. g. A mass of small beetles of different species; a few with brilliant green elytra;† also the cast of what may have been the shell of an Helix.

Spec. h, i, j, contained the usual amount of undeterminable peat-like substance, interspersed with abundance of elytra and parts of beetles of divers sorts.

Examination of a Mummied Hawk from the Necropolis of Thebes.

From the small portions capable of removal, and the size and measurements of the bones, I take this to be a female of *Circus pallidus*, which is at present one of the common rapacious birds of Egypt and Nubia.

Skull much fractured. The upper mandible, $1\frac{7}{10}$ in.; humerus, $3\frac{8}{10}$ in.; ulna, $4\frac{1}{10}$ in.; femur, $2\frac{8}{10}$ in.; tibia, $3\frac{8}{10}$ in.; claw of middle toe, $\frac{4}{10}$ in.

The right radius and ulna appear to have been fractured at the time the bird was buried, showing that the bird may have been killed by accident or design.

Examination of the Body of a Mummied Eagle, found at Thebes.

This is in all probability the common spotted eagle (*Aquila nœvia*), which affects the river valley for a long way, even above the first cataract. The following mea-

---

* See note at conclusion.

† Several identical with the species common everywhere on the banks of the river.

surements correspond with a ♀ of *A. nævia* in my possession, killed at Thebes.

Head, as taken in the Ibis's, $4\frac{9}{10}$ in. ; cere to tip, $1\frac{6}{10}$ in. ; humerus, $6\frac{4}{10}$ in. ; ulna, $7\frac{7}{10}$ in. ; femur, 4 in. ; tibia, $5\frac{7}{10}$ in. ; tarsus, $4\frac{6}{10}$ in. ; great toe, $2\frac{6}{10}$ in. ; hind toe and claw, $2\frac{8}{10}$ in. ; hind claw, $1\frac{6}{10}$ in.

Sternum, antero-posterior measurement, $3\frac{6}{10}$ inches. Gizzard contained the feathers and tarsi of a small bird, with the usual amount of peat-like carbonaceous substance.

Examination of the Body of a Mummied Hawk, evidently a Kestrel (Falco tinnunculus, *L.*)

Head, $2\frac{6}{10}$ in. ; bill from cere to tip, $\frac{7}{10}$ in. ; humerus, $2\frac{8}{10}$ in. ; ulna, $2\frac{8}{10}$ in. ; femur, $1\frac{6}{10}$ in. ; tibia, $2\frac{6}{10}$ in. ; tarsus, $1\frac{6}{10}$ in. ; great toe, $1\frac{7}{10}$ in. ; claw of do., $\frac{6}{10}$ in.

The above dimensions are somewhat short for the ♀, but agree with ♂ of *F. tinnunculus.* Stomach was empty.

The birds found in a mummy state were evidently subjected to the process by injecting the bituminous substance into the trunk by a wound in the abdomen. In none of those examined had the brain been removed or disturbed. After freely bedaubing the outer surface, the tips of the wings and tail were more or less twisted together, and the legs either bent at the tibia-tarsal joint, and placed on the front of the breast by the sides of the wings, or stretched out at full length, as was usually the case with short-legged birds, as the Kestrel, Eagle, &c. Long-necked birds had the head brought down and placed on the belly, whilst hawks, &c. were preserved in the natural position. There seems, however, to have been no rule as to position of the head and extremities, the object being to so form the mummy that it might be easily placed in the jar, after which the mouth was sealed up, and the whole deposited in tombs and pits among others of the same description. It appears that the latter was the case, more especially in Lower Egypt, whereas at Thebes the Ibis and other birds have been found with merely the usual thick walls of bandage around them.

I have often unrolled a large mass of bandage, and

found only a leg or wing of an Ibis, from which I conjecture that these may have been portions of mutilated carcases, possibly half-destroyed by dogs, &c.; and as the bird was so highly venerated, every part of it, wherever found, was preserved with the greatest care.

From the evidence of historians, and what can be inferred from a study of the mummied Ibis, I think we may fairly conclude, that the bird represented on the monuments, and preserved in pits, was identical with the *Ibis religiosa* of Cuvier, *Ibis æthiopica* of Bonaparte, the *Tantalus æthiopicus* of Latham, &c. We can also show that it was domesticated, and, in all probability, bred freely in Egypt, roaming over the cultivated tracts in and about certain towns, villages and temples, at least as late as the first and second centuries of the Christian era. I believe it disappeared with the religion in which it figured so conspicuously, and as the Christians increased, so the Ibis decreased. One may contemplate a few survivors among the ruins of Karnak, or on the battered walls of Thebes and Memphis for a few years after their overthrow, just as if the Hindu religion was to be overturned and a few sacred bulls were to linger on the scenes of their former majesty.

The vast numbers of the mummied Ibis met with, especially about ancient Memphis and Thebes, and the scarcity in other places, lead me to suppose that the bird was not universally distributed over Egypt; indeed, like the other sacred animals, it had its patron cities, Hermopolis being the chief, as is stated by historians; the site, however, of this city has not been clearly defined, and by some it is conjectured to have been one of the many names for Memphis; at all events, the bird was excessively common in and about the Pyramids.

With reference to other birds, it appears that many of the more common species were mummied,* the Kestrel, in particular, which, however, does not seem to have been at all so plentiful in comparison with the last. There can be no doubt, however, that hawks were often kept in cages in and about the temples. The Kestrel, the bird of Re, Horus,

* See author's " Notes and Observations on the Birds of Egypt and Nubia," 'n the " Ibis" for 1863.

and a host of other deities, must have enjoyed unbounded freedom and protection ; and it is a curious circumstance now-a-days, with reference to this species, that as it is one of the most common rapacious birds of Egypt, so is it far tamer in that country than anywhere else I have noticed. Can the feeling of security which pervaded the old race be still lingering on? It is not evident why other than sacred birds should have been preserved ; but as the process of embalming was almost exclusively confined to the priest-hood, who seem to have followed out whatever practices their own fancies suggested, they most probably gave direc-tions that all dead animals should be brought to their temples, without reference to individual species, which, among the hawks especially, is not always very easily determined.

The circumstance that, even on ceasing to occupy its position as a sacred bird, not one Ibis remained in the country—which ought not to have been the case had the climate, &c., been suited to its habits and constitu-tion—surely goes some length to show that the bird was a foreigner, and, when left to its own resources, soon pined away and died ; possibly the cold of winter tried it most, when it had been accustomed to withdraw more from under the ample shelter of the temples and among the dwellings of the natives. Besides, the artificial habits ac-quired by a long domestic condition had rendered the spe-cies in many respects almost akin to poultry ; although, as far as the mummied specimens go, there is every appear-ance in the development of the bones and muscles of the wings to lead to the belief that the bird could make good use of these organs.

Note.—The following notes on the beetles found in the gizzards of the specimens of Ibis unrolled by Dr Adams have been supplied by the kindness of Andrew Murray, Esq., and will be read with considerable interest:—

" The contents of the gizzards submitted to me consisted of small lumps of bituminous matter containing numerous fragments of insects. The numbers of individuals which these represented must have been very considerable. There were ten heads of a large species of *Calosoma*, besides a corresponding quantity of fragments of other parts of the body. There was a tibia of *Ateuchus sacer* (a large insect), several legs of large Pimelias and Blaps, and a multitude of debris of smaller insects.

"In none have I been able to see any material difference between them and the individuals of the present day. The species which I have been able to identify are the following, and I mention the portions I have found in order to indicate the probable value of my opinion, viz.:—

"Calosoma rugosa, Sch., Dej. Spec. des Coleopt. ii. 202.

"Heads, thoraces, elytra, abdomen, legs, and tarsi—almost every part of the insect, but all separate.

"This species is not (as far as I am aware) now to be found in Egypt. It is found in the Cape of Good Hope, and at least as far north as Natal. Dejean quotes the Cape of Good Hope as the locality of his specimens. Bcheman gives Natal. But it is not quoted in any of the Mediterranean lists (which of course include the records of Egyptian species), and I am not acquainted with any instance of specimens having been found anywhere else than the Cape. It is represented on the west coast by a somewhat smaller but very similar species, *C. imbricatum.* Its representative in Europe is *C. inquisitor.* Although I do not see anything to warrant the fragments not being referred to *C. rugosum,* they are not absolutely the same. They are rather smoother, perhaps smaller. The thorax has the edging of its margins not so much raised, and the rugose punctuation finer.

"*Sphodrus,* sp.

"The termination of an elytron. There is an Egyptian species named *picicornis* by Klug, to which this may perhaps belong. I have not seen it.

"*Hyphedrus senegalensis,* Aube, Dej. Spec. des Coleopt. vi. 453.

"The entire body, except the head and legs.

"This specimen corresponds with Aube's description of *H. senegalensis,* but I have not seen an authentic type of Aube's species. As the name implies, the species comes from Senegal. There is no species recorded as being found in Egypt which at all comes near the mummied fragment. I have therefore the less hesitation in referring it to *senegalensis.*

"*Ateuchus sacer,* Linn. Syst. Nat., t. 1, part 2, f. 545, 18.

"A single broken tibia.

"Although the fragment is small it is well marked, and shows that it belonged to the true Egyptian *sacer,* and not to the other Mediterranean variety *picis,* or any other of the varieties of that species.

"*Scaurus tritis,* Fab., Sol. Ann. Soc. Ent. Fr., vii. 165.

"Fragment of elytra.

"*Scaurus striatus,* Fab., Sol. Ann. Soc. Ent. Fr., vii. 165.

"A head.

"*Adesmia* (perhaps) *microcephala,* Sol. Ann. Soc. Ent. Fr., iii.

"Fragments of elytra.

"*Ocnera* (Fisch.), (*Trachyderma*) *hispida,* Fab. Klug Symb. Phys. ii. pl. 12, f. 8.

"Two fragments of the thorax and part of elytra.

"*Pimelia,* sp., Dej.

"A number of legs belonging to one or other of the large Egyptian species of Pimelia, such as *P. coriacea,* Dej., *P. barbara,* Sol., *P. cribripennis,* Sol., &c.

" *Opatrum subsulcatum*, Dej.

" Numerous elytra, and parts of the body and legs.

" *Sclerum*, sp.

" Several elytra and abdomen. Query *S. lucasii* near *S. foveolatum*, but larger and more distinctly marked.

Some of the gizzards contained remains of land or fresh-water shells. The only one sufficiently perfect to ascertain was submitted to Dr Baird of the British Museum, who writes to me—' Your shell appears to be identical in shape and size with the *Paludina bulimoides* of the Nile. This species varies much in colour and markings, but not in form.' The colours in this case were destroyed by the process the birds had undergone.—W. J.

On the Circulation of the Atmospheres of the Earth and the Sun. By Joseph John Murphy, Esq. (Plate I.)

Were the atmosphere not acted on by heat, it would be everywhere at rest, and every level surface, at whatever height, would be an isobarometric surface, or surface of equal barometric pressure. The earth's rotation cannot produce currents, but it modifies them when they are produced by the action of heat.

The greater heat of the equatorial regions expands the air, and thus causes the upper surface of the atmosphere to stand at a higher level there than in the polar regions. This difference of level produces an outflow in the upper strata of the air towards the poles; this outflow causes a partial vacuum in the lower strata, and an inflow of air towards the equator at the earth's surface. This inflow constitutes the trade-winds. The east component of the motion of those winds is due to the fact that they come from a higher latitude, where the earth's rotation is less rapid; they carry their less velocity with them, and thus have a relative motion from the east, or against the earth's rotation.

The upper currents, on the contrary, coming from a lower latitude, where the earth's rotatory velocity is greater, carry their greater velocity with them; they consequently move more rapidly than the earth itself in the latitudes to which they are impelled, and become south-west winds in the northern hemisphere, and north-west in the southern.

The upper and lower currents exercise friction on each other, and so tend to destroy each other's momentum, and the eastward momentum lost by the one must exactly equal the westward momentum lost by the other. But in addition to this the lower current must lose momentum by friction against the earth's surface. Consequently, the west component of the momentum of the upper current is much greater than the east component of the momentum of the lower one, and this preponderance of force causes the upper currents to communicate their own westerly motion to the lower ones. At the equator the easterly motion of the trade-winds must still prevail in a slight degree at all heights in the atmosphere. At a very little way towards the poles, the westerly motion begins in the upper stratum, thence the upper stratum of westerly motion deepens, and the lower one of easterly motion thins out, until about lat. 28° (taking the mean of both hemispheres) the former appears at the earth's surface. From thence to the poles, the air, in both its upper and its lower strata, constantly circulates round the globe from west to east, constituting what Maury calls the counter-trades. Every east wind in higher latitudes is either merely a local phenomenon, or a polar extension of the trade-winds.

Professor Coffin, in one of the earlier volumes of the Smithsonian Transactions, maintains, on the authority of certain registers, that the prevalent direction of the wind in very high latitudes is from the east. I do not understand the reasoning by which he endeavours to account for this, and I suspect it is a merely local or perhaps temporary phenomenon. Sir James Ross met nothing like it in high southern latitudes, and, as we shall see further on, observations bearing on the great atmospheric currents are of more importance when made in the southern than in the northern hemisphere. We have every reason to believe that in the northern hemisphere, during the summer half of the year at least, the pole of greatest cold does not coincide with that of rotation, and this would produce very complex and quite incalculable motions.

The principle of reaction makes it it impossible that the winds can have any effect in either accelerating or retarding the earth's rotation. A west wind moves round the earth's axis more rapidly than the earth, and tends, by its friction to accelerate the earth's rotation. An east wind, for the opposite reason, tends to retard it; and the two sets of forces exactly neutralise each other. The friction of a wind is approximately as the square of its velocity, and the unbalanced effect of any wind on the earth's rotation = the the east or west component of its force × the area it covers × the radius of the parallel of latitude. The last factor gives *leverage*.

Were the whole equatorial region occupied by the trade-winds, and the whole of both circumpolar regions by the counter-trades, and were the east and west components of the force everywhere the same, the dividing lines, in order to produce the above-mentioned compensation, would be at 20° 19′ 20″ north and south nearly,* but they actually are at about 28°, showing that, in order to produce the compensation, the force of the west winds must be greater than that of the east ones. The greater force of the west winds is a necessary consequence of the law of the conservation of areas, in virtue of which, if friction were absent, the air at any latitude would be moving round the earth's axis with an absolute velocity inversely as the radius of the circle of latitude. The excess or deficiency of the absolute velocity of a mass of air, as compared with the earth's velocity of rotation at the same latitude, is the velocity, west or east, of the wind. A mass of air in moving towards the pole will consequently gain absolute velocity, and increase in relative velocity as a west wind; in moving towards the equator, on the contrary, it will lose absolute velocity, and increase in relative velocity as an east wind. But the utmost increase towards the equator will be finite; towards the pole, on the contrary, friction apart, it would be infinite; the velocity at the pole would be infinite, in consequence of the radius of the circle of latitude there being nothing, which is physically interpreted by saying, that in the absence of friction no air would reach the pole—being kept away by centrifugal force.

Friction prevents the centrifugal force of these aerial

* My friend, Mr Harlin, Fellow of St Peter's, Cambridge, has calculated this for me. It is identical with the parallel that bisects the solid hemisphere.

vortexes from having so great an effect as this ; but their centrifugal force does produce a sensible effect in keeping the air away from their centres, and heaping it up at their margins.   The barometer stands at a maximum at about lat. 28°, from which it falls towards each pole.   This depression is much greater in the southern hemisphere than in the northern ; in the highest explored latitudes of the south, the barometer stands at least an inch below its mean level elsewhere.   The reason I assign for this difference is, that the vortex is much more perfectly formed in the southern hemisphere than in the northern, owing to the unequal heating of the continents and oceans in the latter, which produces cross currents.   Of course, the centrifugal force is chiefly due to the velocity of the upper strata, as that of the lower is reduced by friction against the earth's surface.

The excess of barometric pressure at lat. 28° over that at the poles, and the comparative absence of centrifugal force at the earth's surface, determine a motion of the air from lat. 30° towards each pole, and thus are produced the south-west winds of the middle latitudes of the northern hemisphere, and the north-west winds of the southern.*   But these can occupy only a comparatively thin stratum.   In the highest strata of every latitude, polar as well as equatorial, there must be a flow of air from the hotter to the colder regions, from the equator to the poles, and a return current underneath it, in the contrary direction.   In the circumpolar vortex, consequently, there is a motion from the equator above and below, and a motion from the pole

* The cause of the polar depression of the barometer was first, I believe, pointed out by me in a paper read at the Belfast Natural History Society in the winter of 1855–6.   The whole theory of atmospheric circulation in extra-tropical latitudes was first cleared up by Professor James Thomson, in a paper on the "Grand Currents of Atmospheric Circulation," read at the British Association in 1857, of which an abstract is published in the Transactions for that year.   His discussion of the subject is also published with some fuller particulars in the "Proceedings of the Belfast Natural History Society" for 6th April 1859.   I never published my paper, as I afterwards became convinced that it contained serious errors, but Professor James Thomson, in the last-mentioned paper, has referred to me as having first explained the cause of the polar depression of the barometer.

at an intermediate level. The motion from the poles is in the direction of the centrifugal force, that from the equator is against it.

There are two regions of barometric maxima, lat. about 28° or 30°, and three of minima, at the equator and at the poles. From the maxima, air flows at the surface of the earth to the minima, appearing in the tropical regions as the trade-winds, in the circumpolar as the counter-trades. The polar minima are produced by centrifugal force, and a barometer placed at any height above the sea-level, in the region of the polar minimum, will consequently stand below the normal level for that height; for centrifugal force acts at all depths in a vortex. But the equatorial minimum is produced in a totally different way, namely, by the ascent of rarefied air and outflow above. In that region, consequently, a barometer placed in the lower strata, where the currents are flowing inwards towards the barometric minimum, or comparative vacuum at the equator, stands below its normal level; but if placed in the upper strata, where there is an outflow of air towards the poles, it will stand above its normal level for the height. This is because an outflow can only be the effect of increased pressure and an inflow of diminished pressure.

In Plate I. the inner circle represents the earth, and the circles concentric with it represent level surfaces in the atmosphere. The dotted lines represent isobarometric surfaces. It will be seen that at all elevations they fall towards the earth's surface in nearing the poles. At the lower heights, they fall towards the earth's surface in nearing the equator, but at the greater heights they rise higher in nearing the equator.

Within the triangular spaces enclosed by the lines drawn from west and east to the earth's surface, the wind is from the east; outside of these, it is from the west.

The arrows marked *a*, indicate the trade-winds; *b*, the upper return trade-winds, which extend to the poles; *c*, the winds at middle height in the higher latitudes, which blow towards the equator; *d*, the winds at the earth's surface in the higher latitudes, which blow towards the poles.

It is obvious that in any other planet which rotates on

its axis, and is hotter at the equator than at the poles, the system of atmospheric circulation must be essentially the same as that of the earth.

The sun is such a planet. Its rotation has long been known, and Secchi of Rome has ascertained that its equatorial regions are sensibly hotter than the polar. No cause, I believe, has hitherto been assigned for this difference.

Mayer, Mr Waterston, and Professor William Thomson, have brought forward very strong reasons for believing that the sun is receiving a constant supply of heat by the fall of meteors from external space into his atmosphere ; and Mr Carrington and another observer have simultaneously observed two meteor-like bodies of intense brightness suddenly appear on the sun's disc, and rapidly move across it from west to east. If, as is all but certain, meteors are small planet-like bodies, it can scarcely be doubted that the meteors, which supply the sun with heat, move round the sun from west to east like the entire solar system, and, like it, exist in a space of the form of a very oblate spheroid, having its greatest diameter nearly in the plane of the sun's equator. Consequently, the largest proportion of meteors must fall on the sun's equatorial regions, making them hotter than the poles.

It can scarcely be doubted that the meteors must enter the sun's atmosphere with a tangential velocity not much short of that of a planet revolving at that distance. We know that the sun's rotatory motion is incomparably less than this, and, consequently, the meteors, moving from west to east, ought to make the sun's atmosphere move round his body in the same direction, and with greatest velocity in the equatorial regions, as most meteors will 'fall in there. At the same time, the difference of temperature between the sun's equator and his poles, combined with his rotation on his axis, will tend to produce a system of circulation similar to that of the earth's atmosphere, and the actual circulation will be the resultant of this and of the motion from west to east, produced by the infalling meteors. Mr Carrington's comparison of the motions of the solar spots at different latitudes,* affords proof that such a circulation

* Proceedings of the Royal Astronomical Society, 13th April 1860. Mr

is what really exists. Assuming the sun's period of rotation to be 25·38 days, he has computed the mean daily drift
of the spots, in longitude and latitude, to be as follows, the
+ sign indicating pole-ward motion in latitude and eastward in longitude. †

At 50° north — 64′ in longitude + 11′ in latitude.
   30 ,,   — 25   ,,   + 5   ,,
   18 ,,   — 14   ,,   + 1   ,,
    8 ,,   + 8   ,,   — 5   ..
   11 south + 10   ,,   — 3   ..
   19 ,,   — 10   ,,   + 1   ,,
   29 ,,   — 21   ,,   + 4   ,,
   45 ,,   — 85   ,,   — 2 (uncertain).

We thus see a regular decrease in eastward motion from
the lowest to the highest latitudes in which spots are observed, being what I have inferred from the meteoric theory;
but the **exact opposite** of that which is observed in the
earth's atmosphere, and which exist in any atmosphere
which is acted on, like the earth's, only by greater heat in
lower than higher latitudes, combined with the planet's
rotation; for in any such planet the motion of the whole
atmosphere must be westward in the equatorial, and eastward in the middle and higher latitudes.

In order to explain the motion of the spots in latitude,
it is necessary for me to assume that they are formed, and
float in the lowest stratum of the sun's atmosphere.

Were the sun's atmosphere acted on only by the mechanical force of the in-falling meteors, the centrifugal force would
heap up the air at the equator, and barometric pressure
would be greatest there; and this excess of pressure would
produce currents from the equator to the poles at the sur-

Carrington's facts, which I quote, are most valuable; but I confess I do not
understand the reasoning by which he tries to account for them.

  † It is true that the absolute motions in longitude assigned by Mr Carrington are quite untrustworthy, as the true period of the sun's rotation is
not yet determined. But what we have to do with is the *differences* in the
motions at different latitudes. If it is true that the sun's atmosphere is impelled round his body, it follows that the rotation of his body must be slower
than has been inferred from observations of the spots. Mr Carrington thinks,
on the contrary, that it is more rapid than he has assumed, in order to construct the table.

face of the sun, where the centrifugal force would be diminished by friction ; just as we have seen that a similar
cause produces currents in the earth's atmosphere from Lat.
30° to the poles.

But this is not what we observe. Mr Carrington's table
shows that if lines are drawn round the sun at about Lat.
15° north and south, the currents in which the spots drift,
flow on the polar side of those lines towards the poles, and
on the equatorial side towards the equator. We may infer
that a parallel of latitude from which currents flow on both
sides must be a place of barometric maximum. In the
earth's atmosphere, as we have seen, there are two such
barometric maxima, but they are at Lat. 28°, about 13°
nearer the pole than those of the sun; and in the earth's
atmosphere they nearly coincide with the boundaries between the westward trade-winds and the eastward counter-
trades. It is, I think, safe to assume that such coincidence
must take place in the sun's atmosphere as well as in the
earth's ; for there must be an outflow of air at the surface
of the earth from both sides of a zone of barometric maximum ; and the effect of the planet's rotation on a wind flowing to a different latitude, will be to give an eastward direction to one towards the pole, and a westward direction towards the equator. I have shown that were the east and
west component of the velocity everywhere the same, the
boundary of the east and west wind regions would be at Lat.
20° 19′ 20″, in order to have no effect on the planet's rotation. But I have further shown, that the force of the eastward winds of the higher latitudes, must of necessity be
greater than the force of the westward winds of the lower
latitudes ; so that in order to effect the above-mentioned
compensation, the boundary must be on the polar side of
the parallel of 20° 19′ 20″, and with it, if I am right, the
zone of barometric maximum.* But in the sun, as we see,
that zone is on the equatorial side of 20° 19′ 28″.

If my reasonings are correct, were the sun's atmosphere
acted on only by the meteors, the barometric maximum

---

* The frictional force of a wind is a function of its velocity and the nature
of the surface it passes over; but we have reason to believe that the sun's
surface is everywhere alike, being everywhere liquid from the intense heat.

would be at the equator; were it acted upon only by the forces that act on the earth's atmosphere, the barometric maxima would be on the polar side of Lat. 20° 19′ 20″ (in the earth's atmosphere they are about 28°); but they are intermediate between the two, about Lat. 15°, I infer from this, in addition to other facts, that the sun's atmosphere is acted on by both sets of forces, and that the observations tabulated by Mr Carrington show a resultant effect from the two.

The solar spots are most numerous in the zones north and south of the equator, and never appear near the poles; they are seldom seen on the equator itself. It is very probable that they are cyclones, and we know that cyclones cannot be formed on a planet's equator, though they may drift on to it. But this will not account for their absence near the poles; on the contrary, were all other things equal (which, however, is not the case in the earth's atmosphere), the tendency to the formation of cyclones would be greatest at the poles; as it is, there the rotation of any planet is most rapid in relation to an axis drawn perpendicular to its surface. The production of spots in the lower latitudes is probably due to the greater number of meteors that fall in there, causing greater mechanical disturbance, as well as by the greater heat of those latitudes, which must give rise to a more energetic vertical circulation of the atmosphere. Such vertical circulation is certainly proved to exist by the phenomena of the sun's atmosphere, especially by the "rose-coloured protuberances" seen during solar eclipses, which are in all probability cumulus clouds.

The following short *resumé* of the most novel and important points of this paper was read as a communication from me in Section A of the British Association at Newcastle, 1863.

"Secchi of Rome has ascertained that the sun's equator is sensibly hotter than his poles. That this should be the case follows from the meteoric theory of solar heat. The asteroids which revolve round the sun and fall into its atmosphere as meteors, probably occupy, like the entire solar system, a lenticular space having its greatest diameter nearly coincident with the sun's equator, and if so, a greater number of meteors must fall on the equatorial than on the polar regions of the sun, making the former the hottest. The meteoric theory will also account for the currents in the sun's atmosphere observed by Mr Carrington. He finds that the spots in the lowest latitudes drift most rapidly from W. to E. Were the sun's atmosphere, like the earth's, acted on by no other motive-power than the unequal heating at different latitudes, the relative direction of the currents would be the reverse of this, in virtue of the well-known principles of the trade-winds and " counter-trades," and this would be true at all depths in the sun's atmosphere. But if meteors are constantly falling into the sun's atmosphere, moving from west to east with a velocity scarcely less than that of a planet at the sun's surface, and in greatest number in its equatorial regions, there is a motive power which is adequate to drive its atmosphere round it from west to east, and with greatest velocity at the equator. The intensely bright meteor-like bodies, which Mr Carrington and another observer simultaneously saw traverse the sun's disc, moved from west to east, and they were almost certainly asteroids falling into the sun."

Remarks on the Sexuality of the Higher Cryptogams, with a Notice of a Hybrid Selaginella. By JOHN SCOTT, Royal Botanic Garden, Edinburgh.[*]

Modern researches, on the reproductive phenomena of Cryptogams, have induced a number of botanists to accept the doctrine of their sexuality, this function being attributed to the organs known as the Antheridia and Pistillidia. Amongst those botanists who deny the sexual hypothesis, as applied to Cryptogams, a difference of opinion exists ; one class attributing a sexual function to the above organs as occurring in the genera Pilularia, Marsilea, Salvinia, and Isoetes, but strangely arguing, that such an import cannot possibly be attributed to these organs in the other orders ; while another class,—with a more consistent scepticism,—

[*] Read before the Botanical Society of Edinburgh, 10th March 1864.

refuse to attribute a sexual import to these organs in any order of the class, and regard all as strictly agamic.

It would be mere surplusage, on my part, to give to the Society even the briefest *resumé* of the nature of the evidence on which the sexuality of Cryptogams is based, inasmuch as the writings of Henfrey, Berkeley, Suminski, Hofmeister, &c., have rendered it sufficiently familiar to all, and must satisfy all who have accepted the doctrine that nothing short of hybrids, artificially produced between distinct species of Cryptogams, will induce a universal acceptance of the hypothesis of sexuality as applied to these plants.

Several supposed instances of hybridity have been recorded by authors, but these not being results of direct experimentation, do not by any means place the question beyond the reach of doubt. For example, Hofmeister, in his work " On the Higher Cryptogams," p. 181, states that Bayrhoffer " suggested certain mosses, found by him growing wild, were hybrids between *Gymnostomum pyriforme* and *G. fasciculare* on the one side, and *Funaria hygrometrica* on the other side." Hofmeister, however, remarks that he " has not yet succeeded in producing such hybrids experimentally, although he brought together antheridial plants of *Gymnostomum pyriforme* and plants of *Funaria hygrometrica*, with their antheridial shoots cut off. The mutilated plants of *F. hygrometrica* always perished."

In the case of ferns, it has been asserted that true hybrids exist in the genus Gymnogramma. Braun, in his "Plantarum novarum et minus cognitarum adumbrationes," notices several supposed hybrids belonging to the above genus which have appeared in gardens; and similar notices have from time to time appeared in the " Gardeners' Chronicle." That now well known segregative individualising power of the fern-spore—if I may term that subordination of the specific formative tendencies in that organ to those casual variations of the segments or pinnæ upon which it originates—ought to make us extremely cautious in ascribing a hybrid origin to any forms that may appear amongst these plants. Furthermore, the hermaphrodite nature of the prothalli, and the juxtaposition of the antheridial and archegonial cells, render the occurrence of hybrids, in the *true* ferns, much less probable, I believe, than in any other order of Cryptogams. The Botrychiums and Ophioglossums, as shown by Hofmeister and Mettenius, afford much higher facilities for successful casual hybridization than occurs in the true ferns. Inasmuch as in the former the antheridial and archegonial cells occur on opposite sides of the prothalli, so that an equal, or even higher facility, is thus afforded for the conjunction of distinct individuals than the pure hermaphrodite conjunctions. In the latter—or true ferns—on the other hand, where the antheridial and archegonial cells are produced upon the same side of the prothallus, and this being the under, an examination of the individual relations of the prothalli in a single pot will, I think, suffice to show that the crossing of distinct individuals must here be a most exceptional occurrence ; unless, indeed—as so generally occurs in the higher plants—nature has provided certain external agents.

In the Selaginellas, the only genus of the Lycopodiaceæ whose reproductive phenomena are known,* the greatest possible facilities are afforded for hybridization by the unisexual characteristics of their spores, and their production in distinct organs ; one kind of spore—microspore—producing spermatozoa; the other—macrospore—producing the archegonial cells. From these relations of the reproductive organs, it might be supposed that hybrids would be easily raised experimentally between different species. The only points to be studied being a slight regard to systematic affinities, and the relative time required for the development of the

---

* Hofmeister has the following remarks on the above point :—" The reproduction of those Lycopodiaceæ which bear powdery spores of one kind only, is still a mystery. Repeated sowings of the spores of *Lycopodium clavatum, inundatum,* and *Selago,* have yielded me no results ; but I have lately often observed. that in spores of *Lycopodium Selago,* which had been sown for from three to five months, numerous small spherical cells had been formed, similar to the mothercells of the spermatozoa of *Selaginella helvetica.* I have not yet found spermatozoa inside these vesicles. De Bary has lately discovered that the spores of *Lycopodium inundatum* produce a body composed of a few cells, whose structure is not unlike that of the archegonium of a fern. It is probable, from these observations, that the similarly formed spores of Lycopodium, Psilotum, &c., are of different sexes, and, as in *Equisetum arvense,* produce partly *archegonia and partly spermatozoa.*"—" On the Higher Cryptogams," p. 398.

spermatozoa and archegonial cells.* I have found, however, that this is far from being the case; for, after numerous experiments, the subject of the following remarks is
the only one to which I can with certainty assign a cross
origin. The history of this plant may be thus briefly
told :—I placed *thirty macrospores* of *Selaginella Danielsiana* on the surface of a pot of moist sand; over these I
strewed thickly the *microspores* of *Selaginella Martensii*,
and then closely covered all with a small bell-glass. In
case of differences in the time required for the perfect development of the male and female organs of the respective
species, for some time after the first sowing, I frequently
added fresh microspores of the latter species, *S. Martensii.*
Ultimately *one* of the macrospores produced a *germ-plant,
all the others proving abortive.* The gradual development
of this germ-plant I have watched with interest, and I have
now the pleasure, through the kindness of Mr M'Nab, of
placing it and its parent forms upon the table for the examination of this Society.

Previous to my noticing the individual and relative characteristics of hybrid and parents, there are one or two other
points on which I beg to make a few remarks, by way of
obviating certain objections which may be advanced against
the hybrid nature of my seedling; they are as follows :—
A. Braun (" Plantarum novarum et minus cognitarum adumbrationes," 1857, Appendix, p. 16) considers that *Selaginella
Martensii* and *S. Danielsiana* are conspecific; and taking the
former for the normal or typical form of the species, calls it
S. Martensii normale; the latter *S. M. compacta.* Three
other forms, considered by some as distinct species, have
also been referred by Braun to *S. Martensii* under the
following names :—*S. M. flaccida, divaricata,* and *congesta.*

---

* A single illustration will show the necessity for attending to the period
required for the development of the spermatozoa and archegonia in the species
tried. Thus, in the closely allied *Selaginella denticulata* and *S. helvetica,* the
spermatozoa and archegonial cells are developed in the former species about
six weeks after sowing; whereas, in the latter species, according to Hofmeister, the *microspores* lie *five months,* and the *macrospores* between *six and seven
months,* before they produce their respective spermatozoa and archegonia. We
thus see that here, as elsewhere, in the vegetable kingdom, other points than
recognised systematic affinities must be attended to in hybridizing.

Mr Moore informs me by letter that he is inclined to agree with Braun in uniting under *S. Martensii* the forms he has so placed; and furthermore, considers that the seedling form which I have raised goes to confirm this view, by showing that varying forms are capable of being produced from the spores; that, in fact, so far as he could judge from the pressed specimen (which I sent him for examination), I had merely produced *S. M. normale* from the *S. M. compacta.* Mr Moore continues, however, that in plants so peculiar as these Lycopods, a good deal of their natural appearance is lost under pressure. In the present instance, the Society will observe, by a comparison of the hybrid and parent plants with the pressed specimens upon the table, the truth of Mr Moore's remarks, as respects the affinities in judging from the dried specimens alone, and, moreover, the need for that express reservation added to the above view, inasmuch as it is at once obvious by a comparison of the living plants, that though nearer *S. Martensii* in the characters of the leaves, it—the hybrid—has much more affinity with *S. Danielsiana* in its general habit.

In consequence, then, of this view of Braun and Moore, respecting the conspecificness of the parent forms, I can only give a provisional significance to the hybridity of my seedling; satisfied, however, that even by an ultimate agreement amongst systematists as to the genetic affinities of the parent forms, it will simply cause a substitution of the term " mongrel," for that of " hybrid," at present given. And thus, that in either case, it will afford a stronger argument in support of the sexuality of the higher cryptogams than any, so far as I am aware, which has yet been recorded.*

On the supposition, however, that Braun has rightly regarded the parent forms of my seedling as conspecific, it may

* Mr Moore, in answer to a query as to the occurrence of undoubted hybrids amongst the above plants, writes me as follows :—" I am not aware of any well authenticated instances of hybridization among Cryptogams. I have always regarded the varieties of Gymnogrammas (which do sometimes present an appearance intermediate between two known sorts) as sports—chiefly, however, from the want of any direct evidence of hybridity."

be argued that as the *S. Danielsiana—S. M. compacta—*is the more incipient form of the two experimented upon, it may yet have a *tendency* to produce, by a *truly parthenogenetic process,* a varying offspring from its spores. To this I can answer only as follows, but the answer, I think is satisfactory. *First,* When the *macrospores* of the *S..Danielsiana* and *S. Martensii* are sown *alone, neither—*and I speak from an extensive series of experiments—*will produce a* SINGLE *plant ;* clearly demonstrating, as I think, a sexual reproduction dependent on the mutual action of both kinds of spores. That consequently *parthenogenesis,* in so far as my experience goes, *does not occur in either of these forms ;* nor indeed *in any of the species of Selaginella which I have tried, if sufficient care be taken to exclude the microspores.* Again, *secondly,* When the *microspores* and *macrospores* of the *S. Danielsiana* and *Martensii* are each purely commixed and respectively sown in distinct pots, they reproduce themselves perfectly, as I have in several instances proved by experiments. That the Society may be enabled to judge as to the truth of this statement, I have placed upon the table seedling plants of both forms, all of which betray at once their respective parents. Conjoining, then, the latter with the foregoing evidence, *i.e.,* the non-development of the macrospores when sown alone, and the facility with which both forms reproduce themselves when *the two kinds of spores* are mixed ; and comparing them with the previously given history of the presumed hybrid, we are thus, as I am inclined to think, afforded, *firstly,* most conclusive evidence of the existence of true sexual organs in these plants ; and, *secondly,* indubitable proofs of the mixed origin of the seedling plant.

Let us now see in how far this view of the mixed origin of the seedling plant is supported by an individual and comparative examination of the morphological characteristics of the latter and its parent forms. First, for the individual characteristics :—

1. *Selaginella Martensii.—Spike* sessile, linear, somewhat attenuated, from 8 to 10 millimetres long. *Bracteas* ovate, acuminate, denticulate. *Microsporangia* ovate, subtruncate, tumid, ⅓ of a millimetre. *Microspores* reddish-orange, ₁/₁₀ of a millimetre, somewhat wrinkled and granulated.

Macrospores, greyish-white, $\frac{1}{5}$ of a millimetre, reticulated.
Stem ascending, flexuose at the extremity, branches spreading. *Leaves* oblong-ovate, oblique, falcate, somewhat blunt,
3 to 4 millimetres long; anterior base sub-dilated, margin
ciliated, posterior base rounded, margin denticulated. *Stipuliform leaves* oblong or oblong-ovate, acuminate, denticulate, carinate, recurved, 2 millimetres long; exterior
base auricled, and bordered with a few long hairs; interior
rounded, and nearly entire.

2. *Selaginella Danielsiana.*—*Spike* sessile, short and thick,
6 to 7 millimetres long. *Bracteas* ovate, acute, denticulate.
Microsporangia oblong, tumid, $\frac{3}{4}$ of a millimetre. *Microspores* brownish-grey, wrinkled and granulated as in *S. Martensii*, $\frac{1}{35}$ of a millimetre. *Macrospores* white $\frac{2}{7}$ of a millimetre, reticulated. *Stem* ascending, branches short, rigid,
erect, sub-fastigiate. *Leaves* ovately-oblong, 4 to 5 millimetres long; anterior base dilated, and sparingly fringed
with long hairs; posterior base sub-truncate, margin entire.
Stipuliform leaves ovate, acuminated, carinate, recurved,
3 millimetres long; exterior base auriculate, margin sparingly
ciliated; interior base rounded, margin entire.

3. *Selaginella Danielsiana-Martensii.*—*Spike* sessile, short
and thick, 3 to 4 millimetres long. *Bracteas* ovate-triangular, shortly mucronate, denticulate. *Microsporangia* ovate-oblong, half a millimetre long, tumid. *Microspores* brownish-grey, finely granulated, *size very variable*, from the 38th
to 30th of a millimetre; *a high percentage apparently imperfectly developed*. *Macrospores* white, $\frac{1}{4}$ of a millimetre,
obscurely reticulated. *Stem* ascending, branches numerous,
short, rigid, erect, and sub-fastigiate. *Leaves* oblong-ovate,
bluntish, slightly oblique, 2 to 3 millimetres long; posterior
margin denticulated; anterior margin entire. *Stipuliform
leaves* lanceolate-ovate, shortly mucronate, carinate, 1 to 2
millimetres long.

Again, *secondly*, by a relative comparison of the hybrid
and parent forms, we have something like the following
results:—*First*, in the short, erect, rigid, and somewhat
fastigiate branches (destitute of any principal or leading
shoots) of the hybrid plant we have a marked characteristic
of the female parent, the *S. Danielsiana*. As in the latter
species, the right and left forks of the terminal bud are in

general imbued with an equal degree of the vegetative force, so that both forks being developed alike, the plants thereby assume a dwarf, compact, bushy habit. In the male parent—*S. Martensii*—on the other hand, the right and left forks of the terminal bud are alternately more vigorously developed, so as to give rise to an apparently principal axis, or leading shoot, with a right and left series of branches, and a·lax, somewhat spreading habit to the plants. *Secondly*, In the form of the leaves, and their somewhat lax rachidal disposition, the hybrid exhibits more affinity with the male than the female parent, the only difference being a decreased size. In the denser cellular structure of these organs, however, and likewise in the deep lustrous green, with the brownish-tinted stems, the hybrid again approaches the female parent. In the form of the stipuliform leaves and bracteas, it differs from either parent, and here approaches another of the forms which Braun has referred to the *S. Martensii*, viz., *S. M. congesta.* *Thirdly*, In respect to the characteristics of the· organs of fructification, there is a great similarity in the three forms, those of the hybrid being the smallest. There is one point, however, in connection with them, worthy of a passing notice, namely, the relatively great variability in the sizes of the microspores of the hybrid—a high percentage of which are badly developed—as compared with those of the parents; while the macrospores, though smaller than those of the latter, present in general very trifling relative differences, and so far as I can judge, until I have time to test their germinative capabilities, perfectly developed. We have here a curious and interesting—real or apparent—analogy, with that which occurs in the phenomena of sterilisation in the hybridisation of the higher plants. Hybridists have shown, that in the latter class of plants, the pollen is more susceptible to the sterilising action than the ovules, and that in general, perhaps invariably, as has been maintained, we find that if the anther-cases contain a few grains of perfectly developed pollen, the ovaries also will contain a higher percentage of ovules capable of fertilisation.*

* I believe an exception, of which I will satisfy myself at the approaching

On the Chemical and Natural History of Lupuline. By M. J. PERSONNE. Translated by GEORGE LAWSON, LL.D.,[*] Professor of Chemistry in Dalhousie College, Halifax, Nova Scotia. (Plate II.)

Note by Translator.—Considering the great importance of the hop in an economical point of view, we might expect our scientific and manufacturing works to contain a somewhat satisfactory statement of the chemical products of the hop, and of the nature and development of the remarkable organ by which these products are secreted. This, however, is far from being the case; and intelligent brewers in Canada, puzzled by the contradictory statements that have been put forth, have frequently applied to me for information on this as on other scientific points connected with their art. I have therefore thought that a translation of M. Personne's Memoir, published some years ago in the " Annales des Sciences Naturelles," might not be without its use. In some of its bearings, the subject is of much interest in a strictly scientific point of view. It is obvious, likewise, that an acquaintance with the chemical properties of Lupuline is important, not only to the brewer, but to the hop-grower, the exporter, the manufacturer of hop-extract, and, indeed, to every one who has to handle an article so prone to change its character, and, consequently, its commercial value, from apparently trifling causes. The Canadian brewers having a favourable grain-market, and an unlimited supply of excellent water in the great lakes, almost entirely devoid of organic matter, have the means of manufacturing excellent beer. But much of the hops used requires to be imported from England. Canadian hops are grown to some slight extent at Kingston, more abundantly about Pictou, and Belleville, C. W., and especially farther to the westward; but the best qualities of hops are always imported. The Canadian hop gives greater

flowering period, to the above law, occurs in the bigeneric hybrid of the *Rhododendron Chamæcistus*, and the *Menziesia empetrifolia*—the *Bryanthus erectus* (Graham), inasmuch as I have found apparently well-developed pollen grains in the anther-cases, yet I have repeatedly failed in fertilising this plant with its own pollen, or that of either parent.

[*] *Read before* the Botanical Society of Edinburgh, 10th March 1864.

bitterness, but is deficient in delicacy of aroma. Were pains taken (and I have reason to believe that hitherto they have not been taken) to select suitable varieties from the Kentish hop-gardens, and to ascertain, more precisely than we as yet know, what are the special influences of certain soils and climates, no one can doubt but that a great improvement would result in the character of Canadian hops. All attempts in this direction must proceed upon a correct knowledge of the nature of the substances which give the hop its economical value ; and although M. Personne's memoir is more complete and satisfactory than any other that has been published, yet it is to be hoped that by again calling attention to the subject, additional information may be obtained on points that are still imperfectly made out.

The cones of the hop (*Humulus Lupulus*) employed in therapeutics, and especially in the manufacture of beer, owe their properties to a multitude of yellow corpuscles, resinous and odorous, which are separated very freely in bruising the ripe and dry cones. These small bodies have been successively called by the names of Lupulin, Lupuline, and Lupulite. It is to these that the hop owes its bitter and aromatic flavour ; for if the scales and the fruit are deprived of this yellow powder, the cones lose those properties on account of which they are sought after.

The importance of this substance has been known for a sufficiently long time. In 1821, Dr Ives of New York attempted to determine its principal constituents, and endeavoured to introduce it into therapeutics under the name of Lupulin. In France, almost about the same time, Planche likewise concluded that it was a proximate substance, and named it Lupuline, because, said he, "This substance is to the hop what quinine is to cinchona or strychnine to nux-vomica."

In 1822, MM. Payen and Chevallier made the most complete chemical analysis which we have of this substance. They thereby demonstrated the complex nature of lupuline, and, consequently, the error of Planche ; but the small quantity of substance upon which these chemists worked, did not permit them to study sufficiently well the bodies which they had obtained from it.

Lastly, in 1827, M. Raspail published, on the organisation of lupuline, the unique work which exists on this subject. That author sought to demonstrate the analogy of this body with the pollen, as much by the investigation of its structure as by that of the action which the various solvents and chemical reagents exercised upon it. He designated it under the name of *pollen of the foliaceous organs,* "because its office," said he, "is to fecundate the bud, just as that of the pollen of flowers is to fecundate the ovary." I review farther on the observations of M. Raspail.

Structure and Development of Lupuline.

The lupuline obtained from cones that have arrived at maturity presents itself in the form of a yellow powder, whose tint varies according to the length of time which has elapsed since it was gathered. In the fresh state, it has a greenish-yellow colour, which afterwards passes into a golden yellow, deepening more and more the longer it is kept, especially when exposed to contact with air. The form of the lupuline, when it has arrived at its complete development, may be compared to that of an acorn with its cup. Just as some acorns are more or less lengthened at the base, so also some of the grains of lupuline are more or less elongated. The length of these grains varies between $\frac{12}{100}$ths and $\frac{70}{100}$ths of a millimetre, and their thickness between $\frac{44}{100}$ths and $\frac{55}{100}$ths; but in general the two parts of the lupuline, the superior and inferior, are strictly proportional. We shall later see the reason.

In comparing the lupuline with an acorn, I do not mean to say that it is, like it, composed of two solid parts, one of which encloses the base of the other. The comparison can only be applied to the external form, for they differ in all other respects. In fact, the surface of the two parts, superior and inferior, of the lupuline, is perfectly continuous, only the superior, at its point of insertion on the inferior, is bent a little inwards towards the centre, and it is the slight curve which it makes that gives it the acorn form.

These two parts present on the exterior, even under a magnifying power of from 200 to 300 diameters, a structure apparently similar. Both appear to be composed of cellules more or less irregular, which, however, are frequently disposed with a certain regularity from the centre to the circumference ; they are sometimes ranged in radiating series from the summit of the superior part, and from the base of the inferior to the circumference or median line, which unites them. The cells, therefore, increase in size from the two extreme points to the (median line) point of junction. But as I said just now, this structure is only *apparent* in the upper half ; because if we succeed in making a longitudinal section in the direction of the axis of the grain of lupuline, and adjust the same, when placed under the microscope, in such a manner that the plane parallel with its axis shall be in the focus of the instrument, it will be seen that the lower half of the grain is a sort of cupule, composed of a single layer of cells. It is by the base of this cupule that the grain is attached to the epidermis of the bracts, calycine leaves, &c.· It is observed, besides, that the upper half consists only of a very thin continuous membrane, and that the cells, which are depicted upon its surface, are nothing more than the imprints of utricles, the origin of which we give further on in describing the formation of this organ, this singular gland. The space embraced between this membrane and the interior of the cupule is occupied by a yellow liquid, the nature of which we shall examine fully farther on. The cellules which compose the cupule are also filled ; it is these that secrete it, as we shall presently see.

One sees already that this description of lupuline differs essentially from that given by M. Raspail in his " New System of Organic Chemistry," 1833, page 175. Here, in effect, is what he says :—" Examined by the microscope, this yellow powder (the lupuline) is seen to be composed of vesicular organs, rich in cellules, varying in size about the $\frac{1}{4}$th of a millimetre, and of about the form of that represented in figure 6 of plate v. (of his work.) Each of these grains, when dried, is of a beautiful golden yellow, somewhat diaphanous, flattened, presenting on some part of one of its two surfaces the mark of its point of attachment, by which the grain has been originally attached to the organ which produced it, which mark I usually designate by the name of hilum. . . . When these grains are examined, as recently obtained from the still living female cone, they are found to be pyriform, with a peduncle terminated by a hilum," &c.

And farther on, § 387, pp. 176, 177, M. Raspail attempts to prove that the grains of lupuline emit pollen tubes, and that these are produced in contact with water. The conclusion of this paper will show the cause of the error of this observer.

Let us now study the origin of lupuline.

It commences like a hair, by one cellule l (fig. 3, Plate II.), which is developed between cells of the epidermis e. This cellule, projecting to the exterior, is divided by a transverse partition at the level of the external surface of the epidermis. The utricle a, ovoid or elliptical, which results from this division, is in its turn divided transversely (fig. 4, a). The two new utricles enlarge; the superior a (fig. 5) is more dilated than the other, and is filled with somewhat granular matter; the inferior p forms a short pedicel, which unites the former to the epidermis e, by means of the primitive cell l. Thus far the multiplication goes on by transverse division; it now proceeds vertically. The terminal cellule a divides longitudinally into two, as shown by figure 6 at a. Each of the two utricles which thus originate produces in its turn, either one after the other (figs. 7 and 9), or simultaneously (figs. 8 and 10), two cellules, so that by this time the pedicel p is terminated by three cells (fig. 7), or by four, as in figure 8. The figures 11 and 12 show more advanced stages of this subdivision. There now appear some new utricular elements in the interior of the terminal cells. Figure 13 presents a degree of multiplication still more advanced; in it may be clearly observed, in a a a a, the four terminal cells of figure 8, and that they have divided in a radial manner and parallel to the circumference. In figure 14, which indicates a later phase, may' also be observed the four original divisions; but the cells of each of these are still more numerous than in the preceding figure. It not unfrequently happens that the utricular multiplication parallel to the rays is more marked than that which occurs in the other direction, in which case the section appears as in figure 15. It is at this stage of development of the lupuline that its edges become raised; then from the discoid state it becomes cup-shaped. Figure 16 represents some of these cupules which are almost arrived at the perfect state. They have longitudinal striæ, interiorly and exteriorly; that is to say, in the direction of the utricular multiplication parallel to the rays. These elegant cupules appear sessile in consequence of the pedicel not being elongated.

When the enlargement of the cups has ceased, other phenomena take place in the interior of their tissues. Each cupule consists, at this time, of a layer of cells, which is covered with a cuticle on its two faces, the interior and exterior; then commences the secretion of the yellow liquid before mentioned. It is poured out on the whole internal surface of the cupule between the secreting cells and the cuticle which covers them. The latter, detached from the cells by this flowing, is gradually raised completely from the whole extent of the internal surface (fig. 17 *d*), and finally pushed up like the finger of a glove; it is now that the lupuline takes the form of an acorn (figs. 18 and 19), to which I have compared it; it is then arrived at its most perfect stage of development.

It is curious to observe under the microscope the rising of the cuticle. It may be caused artificially, by placing the cupules in water slightly alkalised, which penetrates their walls better than pure water. They may be seen successively passing through all the intermediate stages between the form *l* of figure 16 and that of figure 18.

If we examine the fresh but perfectly developed lupuline in water, it is seen to swell gradually, becoming turgid by endosmose, then all the cells of the cupule appear as a perfect network, and it is then evident that the imprints marked upon the cuticle disappear almost completely. The enlargement increases to the point of bursting of the grain, and it then emits a perfect cloud, formed by a multitude of

small globules of essential oil ; it frequently happens that
these globules, by uniting, form a globule somewhat large,
which is very well seen on the summit of the grain in front
of the rent.

This rent is generally made at the junction of the cuticle
with the edge of the cupule. The cuticle is raised, as a
cover, and, as the cupules open, the cuticle is detached, and
swims away in the surrounding liquid. Occasionally during
this action it occurs as much in the wall of the cuticle as
in that of the cupule, according to their greater or less resistance.

An alkaline solution and alcohol act more rapidly than
water, because, by dissolving more readily the resinous
matter which impregnates the walls of the grains, they
render the penetration more easy.

It has never been possible for me to observe the pretended pollen tubes seen by M. Raspail, in examining the
fresh lupuline. But if we examine lupuline that has been
kept for some time, we observe a very few grains which are
with difficulty impregnated with liquid in this or that place,
and which, breaking a long time after most of the others,
permit the exudation of a viscid matter. This matter,
moulding itself in the aperture which gives it passage,
slightly resembles, to a certain extent, a pollen tube, and
it was this most probably that was seen by M. Raspail ;
but it requires only a slight examination to account for the
appearance, which is most certainly due to the interior
matter of the grains having been dried dissolving with
difficulty.

The lupuline is produced on the ovaries, on the inferior
surface of the bracts and on that of the leaves. It is equally
met with on the stem and on the stipules ; but it is only
on the ovary and on the scales of the cone that the lupuline arrives at its complete development. On the leaves,
on the stipules, and on the stem, it is never met with except in the state of cupules more or less advanced, or all
simply of discs, which readily wither up and are shed.

The lupuline is then a gland, which contains a complex
liquid, of which we now proceed to investigate the nature.

Chemical History of Lupuline.

The matter contained in the gland, which I designate by the name of *Lupuline*, has a very complex composition ; its constituent principles may be classed into two groups : the one embracing those that are volatile, and are obtained by distillation with water ; the other those that are fixed, or at least not volatile with steam.

Examination of the Volatile Principles.

The product of distillation consisted of a solution decidedly acid, which reddened tournesol paper, and upon which floated an essential oil, coloured occasionally of a most beautiful green.

The proportion between the quantity of essential oil and of acid of the liquor distilled varied according to the quality of lupuline employed in the operation. Besides, the lupuline when as fresh as possible, furnished at once a less acid liquor and a greater quantity of essential oil than the older lupuline, which gave, on the contrary, more acid and less essential oil ; the latter is likewise drier and more resinous than that obtained with the freshest lupuline.

The quantities of essential oil which I obtained with the lupulines of different ages, have given me the following proportions :—With recent lupuline I obtained as much as 1 from 100 of the essence, while with older lupuline I have had not more than 0·61 from 100, that is near my proportion.

Volatile Acid of Lupuline.

If we next separate the essential oil from the acid liquid obtained, as I have described, by distillation of lupuline with water, and saturate the liquid with some carbonate of soda, and then evaporate to dryness, it yields as residue a mass of a soapy nature, which liquefies by heat and becomes very solid on cooling, it is with difficulty permeable by water, which, however, ultimately dissolves it completely ; it, in short, comports itself like the compounds of fatty acids with alkalies.

This mass, dissolved in a small quantity of water and

then treated with sulphuric acid diluted with its weight of water or with gelatinous phosphoric acid, yields some sulphate or phosphate of soda which remains in solution in the aqueous liquid, to the surface of which is seen to float a brown oily liquid, diffusing a strong and disagreeable odour of butyric and valerianic acids.

Subjected to distillation this liquid furnished, by many successive rectifications, a product which boiled at + 175 degrees (= 347° Fahr.), and distilled without alteration at this temperature; the first portions carried over water in excess, which was thus separated with sufficient ease.

This acid, obtained in a state of purity, is a liquid, slightly oleaginous, very fluid, colourless, with a strong and persistent odour of valerianic acid; its flavour is acid and piquant; it produces a white stain on the tongue in the manner of energetic fatty acids; it is not solidified by a cold of −16 degrees (= +3°. 2 Fahr.), and remains perfectly limpid; it burns readily with a smoky flame. The specific gravity of this acid is found to be 0·9403 at +15 degrees (=59° Fahr.). It corresponds to that of valerianic acid, which has been found to be 0·937 at +16·5 (=61°·7 Fahr.)

I omit here the description of all the analyses which I have made for ascertaining the composition of this acid. All lead to the formula of valerianic acid. I have purposely multiplied its combinations with oxide of copper, oxide of silver and baryta, in order to be well satisfied of its true constitution. But the odour alone of lupuline, especially of that which has been kept for some time, does not admit of doubt of the existence of this acid among the bodies which this substance contains.

Volatile Oil of Lupuline.

This crude essential oil—that is to say, such as has been given by distillation of lupuline with water—is an oleaginous liquid, more or less fluid according to the state of the lupuline which furnished it, and of a specific gravity less than that of water. It has at the same time a somewhat intense colour of yellowish green, more frequently of a beautiful green; its odour recalls slightly that of the hop;

but this odour does not resemble that of valerianic acid when the oil has not undergone oxidation or contact with air.

Subjected to distillation, it enters into ebullition at +140 degrees (=284° Fahr.), and distils for some time at +150° (=302° Fahr.) to 160 degrees (=320° Fahr.), but the temperature rises gradually, and when the process is finished, is +300 degrees (=572° Fahr.)

The portion of this essence obtained between 150° (=302° Fahr.) and 160° (=320° Fahr.) is a sufficiently thin liquid, slightly amber-coloured, of an odour which does not resemble that of the hop, and of a specific gravity of 0·8887. It has not an acid reaction, but, on exposure to air, it acidifies and becomes resinous; it is slightly soluble in water, to which it communicates its odour, and the solution exposed to the air acidifies rather readily; it is soluble in alcohol and in ether. With a cold of −17 degrees (+1°·4 Fahr.) it lost a little of its fluidity, but its transparency was not altered, even after four or five hours' exposure to that temperature. It deviates to the right the rays of polarized light. Its rotatory power (Dextrogyrate) has been found by the red glass to be +2·7⤳ for the length of 0^m·080; it is then of

$$\frac{+^6 2·7}{80 \times D}$$

Nitric acid gives at first a beautiful purple colour; afterwards, if heated a little, the reaction becomes more lively, and the products furnished are a resinous matter and valerianic acid.

Potash in solution fails to attack it at a boiling temperature; but if we form an emulsion with a concentrated solution of potash, and expose the mixture for some time to contact with air, we find that there are produced valerianate of potash and a resinous matter.

Fused potassa transforms it into carbonate and valerianate of potassa, with disengagement of hydrogen and of a hydrocarbon liquid.

This reaction of potassa is very important, because, after some useless trials, and a great number of analyses, it rendered clear the true nature of this essence, placing it by the side of the essential oil of valerian.

In fact, the composition obtained by analysis of the crude essence, may be represented by the formula $C_{56}H_{46}O_6$; that of the essence distilled between $+150$ and 160 degrees ($302°$ and $320°$ Fahr.) by the formula $C_{22}H_{18}O_2$.

In submitting these essences to the action of fused potassa, there were obtained products in which the quantity of carbon and of hydrogen increased each time that they were submitted to a renewed action of potassa, while the proportion of oxygen decreased. Finally, after many successive treatments, we finished by having a perfectly pure hydro-carbon.

This hydro-carbon is a colourless liquid, which boils at $+160$ degrees ($320°$ Fahr.) It does not acidify by contact with air; it is as difficult to be altered by contact, for a score of days, with pure oxygen. Its composition, deduced from analysis, is represented by the formula $C_{10}H_8$; it is consequently the same as that of the oil of turpentine and of *bornéène*, which M. Gerhardt has found in the essential oil of Valerian. But this body, although possessing the composition of oil of turpentine and of bornéène, is not the same, but isomerous with these last; for I have not transformed them into solid camphor of Borneo, neither by the action of nitric acid nor by that of potassa. Kept for some time on a solution of potassa, it acquired the odour of thyme, sufficiently to show an approach to thymol.

We see that the action of fused potassa upon the essential oil of hop, consists in setting free a hydro-carbon liquid $C_{10}H_8$, and in retaining an oxygenated body, which it transforms into valerianic acid and carbonic acid; results absolutely similar to what M. Gerhardt has obtained with the essence of valerian.

It is not easy to separate the oxygenised principle of this essential oil, because it is found to be retained in the thickish resinous matter, which does not permit the separation without great difficulty.

The essential oil of lupuline is clearly, then, to be considered as a complex oil, constituted by a hydro-carbon $C_{10}H_8$, and a body containing oxygen of the formula $C_{12}H_{10}O_2$ analogous to *valerol* of the essential oil of valerian. The formula of the crude oil $C_{56}H_{46}O_6$, may be represented by

3 $(C_{12}H_{10}O_2) + 2 (C_{10}H_8)$; that of the oil rectified between $+150$ and 160 degrees (302° and 320° Fahr.) by $C_{12}H_{10}O_2 + C_{10}H_8 = C_{22}H_{18}O_2$.

The process by which it may be obtained as free as possible from extraneous matter, consists in preparing a tincture of lupuline with alcohol, of 36 degrees; to treat this liquid with an alcoholic solution of tartaric acid, which forms a precipitate somewhat abundant, of bitartrate of ammonia. The liquid separated from the precipitate, is added to a little water, and submitted to gentle heat in a capsule exposed to the air; the alcohol, in evaporating, leaves separate, at the end of two or three days, the resinous matter of the solution, acid and bitter. This bitter liquid is then deprived of the excess of tartaric acid which it contains, and then made to digest with some carbonate of lead recently precipitated; the mass, evaporated at the lowest possible temperature, is treated by boiling alcohol, which dissolves only the bitter matter.

Resinous Matter.

The resinous matter is very abundant in lupuline; it forms itself alone nearly two-thirds of its weight; it retains always a certain quantity of the volatile oily products, which gives to it a variable consistence, and preserves at the same time the peculiar odour of lupuline. It is oxidized by contact with air, especially in presence of water, and its colour then passes from a golden yellow to a deep brown tint, at the same time that it hardens. It is largely soluble in water, to which it communicates the property of lather by agitation. This solution presents an acid reaction, and is completely altered by evaporation in contact with air.

The alkalies dissolve it in the cold, and separate an insoluble part. This resin, insoluble in the alkalies and in water, is soluble in alcohol; it is dry, friable and inodorous. The alkaline solution, saturated by an acid, sets free the resinous matter with its original properties, and retaining some valerianic acid which is got by distillation. Lastly, nitric acid with heat attacks this resin with energy, but without producing special reaction which would serve to characterise it.

To obtain this resin as pure as possible, the lupuline must be exhausted by long boiling in water, which drives off the volatile products, and dissolves the bitter matter. The insoluble residue, composed of resin and of disintegrated tissue of the lupuline grains, well washed and dried, is then treated by boiling alcohol, which sets free, when cooled, a certain quantity of *waxy matter ;* the alcoholic liquor, filtered after cooling, furnishes the resin by evaporation of the alcohol.

The wax is contained in the cells which compose the cupule of the lupuline grain ; it exists also in the scales which constitute the cone of the hop, and by treating these scales with boiling alcohol, it is procured in sufficient quantity. It is dry and pulverulent, inodorous and tasteless ; it begins to soften at +80 degrees (176° Fahr.), and is fully melted at +100 degrees (212° Fahr.) Strongly heated, it gives two volatile products, which diffuse an odour of wax ; it burns without residue, producing a white shining flame ; this matter resembles, as we see by its properties, the wax of the sugar cane.

Explanation of Plate II.

Fig. 1. Cone of Hop.

Fig. 2. Terminal bud enveloped by the stipules, *s s*, on which are marked the granulations, which represent the cupules and the discs indicated by the figures 14, 15, 16, &c.

Fig. 3. Lupuline originating ; *e e*, epidermis ; *l*, primordial cellule of lupuline, by which it is attached to the epidermis ; *a*, cellule produced by the preceding, and which gives rise to the following modifications :—

Fig. 4. *e*, epidermis ; *l*, primordial cellule ; *a*, cellule divided transversely into two ; the inferior division constitutes the pedicel of the lupuline, the superior forms the gland of the same.

Fig. 5. *e e*, epidermis ; *l*, primordial cellule ; *p*, pedicel ; *a*, cellule containing grey matter with granules.

Fig. 6. *p*, pedicel : *a*, cellule divided into two longitudinally.

Fig. 7. *p*, pedicel ; *a*, represents one of the two cells of the preceding figure, subdivided longitudinally into two ; *a'*, is another cellule, not so parted.

Fig. 8. *e*, epidermis ; *p*, pedicel ; *a*, gland formed of four cellules.

Fig. 9. Gland, represented in figure 7, front view ; *a*, is the cell not divided ; *a'*, the cell which is parted into two longitudinally.

Fig. 10. Gland *a* of the figure 8, front view.

Fig. 11. The same gland more advanced, in which are seen many cells originating by the intra-utricular mode of multiplication.

Fig. 12. The same gland, seen on the face, and a little farther advanced.

Fig. 13. Gland more advanced, in which the four cellules of figures 10, 11, and 12, are subdivided parallelly to the ray, and parallelly to the circumference; each of the cells is indicated by *a a a a.*

Fig. 14. Gland in which the utricular multiplication is still more advanced. The four mother cells of fig. 10 are still visible, and indicated by *a a a a.*

Fig. 15. Shows the aspect which the glands present when they have acquired a somewhat considerable size; *e,* epidermis; *l,* the gland.

Fig. 16. Glands more advanced. The edges of the discoid glands, as seen in preceding figures, are here raised, forming cupules, *l, l; e,* epidermis.

Fig. 17. Cupule from the internal (or upper) surface of which the cuticle *d* is detached, and elevated by the secretory products.

Fig. 18. Lupuline, which has acquired its complete development; *c i,* secreting cupule or proper gland, surmounted by the cuticle *c s,* raised up by the products of secretion.

Fig. 19. Grain of lupuline enlarged; *c i,* cupule or gland proper; *i,* point of attachment; *c s,* elevated cuticle. There is seen on this last the impression or trace of the cellules of the cupule, on the cavity of which this cuticle was applied.

Fig. 20. Longitudinal section of a grain of lupuline; *c i,* cupule composed of a single layer, which secretes the contained liquid; *c s,* cuticle detached from the internal surface of the cupule by the secreted liquid.

The figures are from the pencil of M. Trecul.

Remarks on the Sexual Changes in the Infloresence of Zea Mays. By Mr JOHN SCOTT.*

The florets of the Indian corn, *Zea Mays,* as is well known, are unisexual, and so placed that the male florets form a terminal panicle, or raceme, and the females inferior lateral spikes. In the male panicle the spikelets are two-flowered; both florets perfect and characterised by two glumes, two squamulæ, and three stamens. In the female spike, the spikelets are also two-flowered, but in this case the inferior floret is neuter; two paleæ alone being developed, the superior fertile, and possessing two or three paleæ, an oblique, sessile ovary, and a long compressed style, bifid, and pubescent at the apex.

In the abnormal specimens which I now submit to the Society, the male and female florets, in place of being

* Read before the Botanical Society of Edinburgh, 10th December 1863.

arranged as above, on distinct axes of the plant, occur more or less irregularly on a single axis. Thus, in specimen No. 1, we have a female monoicous spike; *i. e.* a female spike, producing both male and female florets. In this case the basal portion of the spike is normal, presenting several circles of the perfect grain; the upper and major portion of the spike, on the other hand, has every floret converted into the male form; each spikelet, be it observed, producing *two perfect male florets*. In other cases, however, the upper portion of the spike retains its feminine character, while the basal portion assumes the male; or again, we may have an irregular intermixture of male and female florets over the whole spike.

The metamorphosis of the female into the male floret is not, however, always complete, and this is more especially so in such cases as the latter, where there is no definite arrangement of the male and female floret. From the special theoretical interest now attached to these imperfectly metamorphosed florets, in their association with others perfectly metamorphosed, I will here describe one or two of the most instructive which have come under my observation. *First*, In the *superior* floret of a female spikelet, the style was abortive, ovary rudimentary, squamulæ developed (though smaller than those of a normal *male* floret), glumes lanceolate-acuminate; in the *inferior* floret the stamens, squamulæ, and glumes were perfectly developed; so that the normally *neuter* floret of the female spikelet was in this case converted into a *perfect* male; whereas a very imperfect metamorphosis has been effected in the case of the fertile female floret. Again, *second*, in another spikelet, from a female spike, I found the *superior* floret presenting all the characteristics of the *normal male* floret; while the *inferior* (though still retaining its neutrality of function) —stamens and pistils being alike abortive—presented by a pair of lanceolate acuminate glumes, and two minute cuneate squamulæ, an evident tendency to assume the male form also.

If we now turn to an examination of the male panicles, we are at once struck with the rare occurrence of the monoicous structure in them, as compared with the occurrence of such a structure in the female spikes. Somehow—and it is difficult to understand why it should be so—the female organs in this instance, as indeed in most other unisexual plants, are much less prone to become developed in the male flowers than are the male organs in the female flowers.* On this account, then, I trust the Society will bear with me while I briefly attempt to describe the few male monoicous panicles, which I have been fortunate enough to obtain; they are as follows :—

First, In specimen No. 2,—a terminal panicle,—the primary axis bears male and female florets; the florets of the former are perfect in the upper portion of the axis, which they exclusively cover, but in the lower portion, where they approach the female florets, the superior floret in the majority of the spikelets is alone perfect; while in the inferior floret the stamens occur in a more or less rudimentary condition. The female florets of the primary axis are *all imperfect*, the ovary existing in a rudimentary form, and the stamens utterly aborted in the *superior* florets; whereas, in the inferior florets of the spike-

---

* May we not regard this as probably indicative of those homological distinctions between the male and female organs of plants, insisted upon by Schleiden and Endlicher, at least as modified by Dr Dickson, in his interesting paper "On the Nature of the Cormophyte." (*Vide* Society's Transactions, vol. vi. p. 95.) Dr Dickson there states, that he is "inclined to believe that there exist in reality two modes of placentation, the one where the ovules are produced by a process of gemmation from the carpellary leaves (*parietal*) ; the other, where the ovules spring from the prolonged floral axis (*central*). In this modified sense, then, a strong argument against the Schleidonian theory of placentation is completely neutralised. I refer to the *inverse* convertibility of male and female organs in certain plants. For example, in the willows, we have some excellent illustrations in the Society's Transactions. Thus, in vol. i. page 118, the Rev. J. E. Leefe has illustrated the gradual modifications of the pistillary into staminal organs in the *Salix Caprea ;* while Mr Lowe, vol. v. p. 113, has given us, *vice versa*, all the conceivable intermediate stages in the transformations of the staminal into the pistillary organs in *Salix Andersoniana.* Now, as these cases of the willows naturally come under the division assumed to possess a *parietal placentation*, their evidently disproving tendencies are utterly invalidated. And thus, even in view of such anomalous occurrences, we may justifiably reiterate the above suggestion, as to the difference in the *inverse* convertibility of the male and female organs, in at least the case of the maize, where the floral axis, as terminal shoot, undistinguishable in the cavity of the germen as a special organ, bears a single seed-bud (Schleiden's " Principles of Botany," p. 386), and is thus referrible to the division characterised by a central placentation.

lets, staminal and pistillary organs are similarly aborted. Again, in the arrangement of the male and female florets in the secondary axes, a similar plan is observed to that of the primary axis; the female florets occupying their basal portions, but in several instances *perfectly* developed, the upper portions being covered with *perfect male* spikelets.

Secondly, In specimen No. 3,—a terminal panicle,—we have a very irregular intermixture of perfect and imperfect male and female florets, along with several structurally *hermaphrodite florets*. Generally speaking, however, in this specimen, as in No. 2, the upper portions of both primary and secondary axes still retain their normal-male-sexual characteristics, and the basal portions assuming the female characters. By a careful examination, however, I have detected several structural peculiarities in certain florets of the latter part, to which I am inclined to attribute a highly important theoretical signification, as will be seen subsequently. The following are the most instructive :—*First*, In the *superior floret* of a spikelet, presenting the broad glumes and paleæ of the female florets, I found a rudimentary *hypogynous stamen ;* while in the *inferior floret* the glumes were *ovately*-lanceolate, squamulæ as usual in normal *male* florets, stamens developed, but *destitute* of *pollen*. *Second* spikelet ; glumes of *superior floret broadly*-lanceolate, squamulæ cuneate, obliquely truncate, larger than those characteristic of the male florets, ovary and style incipient, as in the above; but in this case I found *two* rudimentary *stamens*, one consisting of the filament alone, the other of the filament and a pellucid rudimentary anther, presenting the appearance of a glandular hair; the only modification the *inferior* floret of this spikelet had undergone from its normal male condition, was the non-development of the pollen, the anther-cases being quite empty. *Third* spikelet; glumes of *superior* floret *ovately*-lanceolate, squamulæ cuneate, minute, anthers destitute of pollen ; *inferior* floret functionally a perfect male, glumes *broadly*-lanceolate, squamulæ as usual in male florets, anthers containing pollen.

These, then, are the more interesting peculiarities which I have observed in the structure of the florets in the above *panicle ;* and there is just one other point in connection with it to which I will here specially direct attention,—namely, the remarkable irregularity observed in the relative arrangement of the male and female florets. The most striking case is presented by one of the secondary axes; the florets in its basal portion are nearly all converted into more or less perfect *females,* whilst those above retain in like manner the male characteristics. Associated with the latter, however, and near the upper extremity of the axis, *two solitary female* florets are at once observable by their prominently developed grains. On examination of the spikelets bearing these, I find that the female morphogenesis is complete, the *superior* floret alone fertile, the *inferior* neuter. This individual isolation of the florets, occurring as they do in distinct parts of the axis, and surrounded by normal male florets, and perfect metamorphosis, excellently illustrates the occasional independence of such phenomena on mere physical conditions.*

Thirdly, In specimen No. 4—a terminal panicle—a somewhat different arrangement is observed to that which we have seen followed in specimens Nos. 2 and 3. In the latter two, the basal portions of the axes produce female florets, and the upper male; whereas in the former, No. 4 specimen, the opposite of this occurs, namely, the upper portions of primary and secondary axes converted into *compact spikes* of *female* florets, while the lower portions,

---

* Dr Lindley in treating on the changes of sex under the influence of external causes (*Introduction to Botany,* vol. ii. 4th ed. p. 80), states "that Mr Knight long ago showed that a high temperature favoured the development of male flowers, and a low one that of female;" furthermore, that this eminent horticulturist "entertained little doubt that the same fruit-stalks might be made to support either male or female flowers in obedience to external causes." Dr Lindley illustrates these conclusions by experiments on water-melons and cucumbers. From personal observations, however, on several monoicous plants, I cannot think that these influences are at all definite as to their influence on sex produced. The above laws are still less applicable to dioicous plants; and certainly, upon any theory of special creation; on a subjective consideration of the vegetable individual, I fail to see why they should not be equally potent in the one case as in the other. Such cases as those above noticed in the maize, in which collateral florets assume distinct sexual characteristics, induce me to believe that in general the influence of physical condition on the change of sex is subordinate to certain innate, specific, formative qualities; in short, an inherited tendency to produce the characters in question.

retaining their normal characters, continue to produce the racemose male spikelets. In the majority of the secondary axes, however, the basal spikelets are nearly all aborted ; whereas this portion of the primary axis is covered with perfect male spikelets, which, as they extend upwards, are *abruptly* metamorphosed into a short terminal spike of female florets. These, like those on the secondary axes, are all very imperfect, the ovary and style existing in a more or less rudimentary condition, and occasionally presenting the rudiments of one or two hypogynous stamens.

Hitherto our remarks have been chiefly confined to a mere description of the sexual metamorphoses in the florets of maize, though I have more than once alluded to their possible connection with, and elucidation of, certain highly important points in theoretical natural science. For the sake of clearness in the exposition of the theoretical bearings of these metamorphoses, I will now give a brief *resumé* of the foregoing illustrations. First, then, we have stated that the inflorescence is normally unisexual—the female florets borne on inferior lateral spikes, the male on terminal racemes or panicles. Our illustrations, however, show that these structural arrangements undergo important modifications. Thus, we have first the female spikes assuming a monoicous structure, and this without any regard whatever to the relative axial arrangement of the male and female florets ; the same part on distinct axes indifferently producing perfect and imperfect male or female florets, as well as collateral mixtures of both ; showing us most conclusively, their morphogenetic independence of the mere external conditions of life. Again, individual florets of these female spikes present themselves with a structure intermediate between that of the perfect male and female ; and then manifest a most interesting and instructive co-related order in the development of their organs. Thus the *superior* and normally *fertile* floret of a spikelet with abortive style and rudimentary ovary, had assumed the characteristic squamulæ and glumes of the *male* floret ; while the normally *neuter* floret had assumed *in toto* the male characteristic.

In the terminal male panicles, with a similar series of

changes to those which we have noticed in the female spikes, there are also the important additional illustrations of *structurally hermaphrodite* florets.*   Thus, in one of the spikelets we had a *superior* floret with incipient ovary and style, a *rudimentary stamen*, and the characteristic paleæ and glumes of a *normal* female floret; while the *inferior* floret differed from a normal male only in its *broader* and *shorter* glumes, and *barren* stamens.   Again, in the *superior* floret of another male spikelet, with glumes somewhat intermediate between those of the normal male and female florets, we noticed an incipient ovary and style, and *two* rudimentary stamens; while the inferior floret of the same spikelet retained its *male* characteristics.

What now are we to say as to the cause of these changes? We see the unisexual florets of the maize not only undergoing inverse metamorphoses, *i.e.*, the male converted into female florets, and the female into male florets, but also assuming every conceivable intermediate stage between these and a structural hermaphroditism.   Now, it is well known that similar sexual changes occur in—at least the female florets—many other monoicous and dioicous plants; *e. g.*, in the *Melandryum prœtense*, and the *Lychnis dioica*, the female flowers occasionally become bisexual by the development of the stamens.   I may also state that I have observed bisexual (female) flowers on the *Littorella lacustris*, *Bryonia dioica*, and *Ricinus communis.*†

Seeing, then, that unisexual flowers undergo such serial transformations in their sexual characteristics, we, on the ordinary theory of creation—*i. e.*, assuming species as the original units—might justly expect a similar series of changes in the characteristics of bisexual flowers.   This, however, as is well known, is not the case; no instance can be

---

* I may state, that although I have failed in illustrating structural hermaphroditism in the female spikes, cases are already recorded. C. F. Gärtner, in his "Beiträge zur Kenntniss der Befruchtung," notices the occurrence of solitary stamens in the female florets of *Zea Mays;* he also states that he has observed in the conversions of male into female florets, a solitary stamen associated with the pistillary organ of the latter.

† I will not here enter on details as to the occurrence of the above, as I hope at some future time to lay them before the Society in a notice of my observations and experiments on the subject of *Vegetable Parthenogenesis.*

adduced of a bisexual species undergoing sexual metamorphoses similar to those above described in the unisexual maize.   Moreover, supposing that the sexual characters of hermaphrodite plants had exhibited masquerading tendencies similar to those of unisexual plants, it is at once evident, that, upon any theory of special creation, the *cause* of such changes in either case is equally unintelligible.   On the other hand, if, with Mr Darwin, we believe that species are the modified descendants of previously existing species, these phenomena are no longer enigmatical, but clearly the results of definite and well-known laws.   I need only refer to Mr Darwin's interesting papers on the distinct sexual forms of the dimorphic species of Primulas and Linums, "Jour. Linn. Soc.," vol. vi. p. 77, and vol. vii. p. 69, by way of illustrating, as has been remarked, "the possibility of a plant becoming dioicous by slow degrees."   Now, if we reflect on this dimorphism of the Primulas and Linums, those differences in the variability of the unisexual, relatively to the bisexual flowers, are, I believe, readily explicable on the supposition that the latter—*i.e.*, the hermaphrodite structure, as Professor A. Gray has maintained—*vide* "Sill. Amer. Jour.," vol. xxxiv.—"is the *normal* or *primary* condition of flowers."   In fine, then, in accordance with the theory of modification with descent, I, inferentially guided by that principle of reversion to type so much insisted upon by those opposed to derivative hypotheses, look confidently at such sexual changes as those above described, as retrogressive tracings of the graduated modifications by which an original hermaphrodite progenitor gave rise to a monoicous offspring.

New Researches on Hybridity in Plants.   By M. Ch. Naudin.   Translated from the Annales des Sciences Naturelles, by George May Lowe, Esq.[*]

(1.) *On the Sterility and Fecundity of Hybrids.*

A century ago, Kœlreuter demonstrated by proofs which no other observer has ever surpassed in exactitude, and which

[*] Read before the Botanical Society of Edinburgh January 14, and March 10, 1864.

still retain all their value, the fact of the sterility of hybrids being absolute in some cases, but only partial in others. These two facts, since so frequently confirmed, cannot now be disputed. In a former paper I gave some examples which serve to illustrate them.

We have seen *Nicotiana-californico-rustica, N. glutinoso-macrophylla, N. glutinoso-angustifolio-macrophylla, Digitalis luteo-purpurea* and *Ribes Gordonianum,* sterile both by the stamens and ovary—the former being totally destitute of pollen well formed, and the latter incapable of impregnation by the pollen of the parent plants. But as the pistil does not in every case present any appreciable deformity, it is natural to seek in the ovule itself the true cause of this inaptitude to receive impregnation.

It has been fully proved by many cases of hybridity, in which, in the same ovary one portion of the ovules resists impregnation, whilst the other becomes converted into embryonic seeds capable of germinating—that this defectiveness exists in the ovule, and not in the more exterior parts of the pistil.

We have seen this in the three hybrid generations of *Luffa acutangulo-cylindrica,* also in *Luffa amaro-cylindrica, Cucumis Meloni-trigonus, Nicotiana rustico-paniculata,* and *paniculato-rustica,* &c. *Cucumis myriocarpo-Figarei* is a not less convincing proof, since among 100 fruits which were developed and ripened under the influence of pollen derived from the maternal species, 19 at least were destitute of seeds, and each fruit, among the small number which contained any, only yielded one seed. I might mention, in support of this fact, the example of *Mirabilis longifloro-Jalapa,* though in this case the ovary is uniovular. The stigmas of this hybrid were all equally developed, and in this respect not inferior to those of the parent species ; yet eleven attempts to impregnate it with the pollen of *Mirabilis longiflora* were made without effect, and even ten were necessary with that of *M. Jalapa* to determine the increase of a single ovule. In the *Luffa* hybrids just mentioned, and also in the case of *Cucumis Meloni-trigonus,* however poor the pollen might have been which was employed to fertilise their ovaries, it is beyond doubt that the number of good grains deposited on their stigmas far exceeded that of the ovules which were developed into seeds.

This, it is true, is only hypothetical, but it is extremely probable. It remains to be confirmed by the anatomical examination of the ovule, and it would be very interesting to discover in what part the defectiveness exists; but this is a peculiar kind of research, very difficult, very minute, often uncertain in its results, and which one cannot enter upon without being well accustomed to it, and provided with excellent instruments, two things in which I am deficient.

I therefore contented myself with verifying experimentally the fecundity or the sterility of the ovaries, which was more expeditious, and probably more conclusive; but it is not less a subject to be recommended to professed micrographers.

That the sterilising action of hybridisation exerts much more force on the pollen than on the ovules is a most indubitable fact, and one well known to all hybridologists. This need not surprise us, since the pollen is, of all parts of the plant, the most elaborated, the most animalised, if such an expression can be used. Frequent chemical analyses prove that it is in these granules that the phosphorised and azotised materials are more accumulated than elsewhere, and thus it may be conjectured that it is this high organisation which is injured in hybrids, where the whole vegetation suffers from the disturbance which results from the intermixture of two specific essences created to live separately. The hybrids of which I have given an account present several examples. We have seen that *Mirabilis longifloro-Jalapa* yields pollen unfit for fertilisation, whether it be applied upon the stigmas of the hybrid, or upon those of its two parents, whilst in twenty-one attempts to impregnate it with the pollen of these last (*M. longiflora* and *M. Jalapa*), there was only one which took effect, and enlarged the ovary. This result is quite in accordance with those which M. Lecoq (" Revue Horticole," 1853, pp. 185 et 207) announced that he obtained from the same hybrid, the pollen of which he always found useless, but he was able to fertilise it by that of *M. Jalapa*. The difference in *the strength* of the pollen and the ovules becomes still more

manifest in *Nicotiana glauco-angustifolia* (and it would undoubtedly have been the case with *N. glauco-macrophylla* if
the experiment had been made on it), where the whole
pollen mass is defective and inert, whilst the ovary becomes filled with seeds, when it is fertilised with the pollen
of *N. Tabacum* and *N. macrophylla.*

All the hybrids I have observed, containing well-developed
grains of pollen in their anthers, have been fertile, often to
a high degree, by their ovaries. I have never seen, and I
do not believe it possible to mention a single instance in
which, the ovary being sterile, the stamens have been fertile,
even in the least degree.

The deleterious influence which hybridisation exercises
upon the fertilising apparatus shows itself in different
forms.

The most common, or at least the most remarkable case,
is the direct atrophy of the pollen in the anthers, more
rarely the atrophy of the anthers themselves; but we have
also seen it act on the entire flowers. It is so among all
the hybrids produced by the agency of *Datura Stramonium*,
the flowers in the lowest branches invariably fall without
opening; also among all the individuals of *Luffa acutangulo-
cylindrica* of the first generation,—all the primary male
flowers perish entirely, and also some flowers which begin to
open when the plants are more than full grown, and have
lost part of their vigour. The same phenomenon is observed
in *Mirabilis longifloro-Jalapa*, which loses three-fourths of
its buds, in *Nicotiana rustico-paniculata* and *paniculato-
rustica* of three consecutive generations, &c. In fine, another mode of sterilisation is that effected by the changing
of monœcious male flowers into female, as we have seen in
Luffa hybrids of the third generation.

I have every reason to believe, although I cannot positively affirm so, that the specimen of *Cucumis Figarei*,
so remarkably large, and peculiar by the nearly total
absence of male flowers, which I experimented on in 1856,
and which yielded the results I have mentioned, owed both
its great size and almost female unisexuality to hybridisation.

(2.) *On the Difference of the Fertility of Hybrids.*

Hybrids are self-fertile in all cases in which their anthers contain well-organised pollen ; but if the quantity is very small, it is well not to leave the impregnation to chance, but to aid artificially in fertilising the hybrid with its own pollen.   I have done this in *Luffa acutangulo-cylindrica* of the first generation, which has but few male flowers and a small quantity of good pollen.

In the majority of cases microscopic inspection sufficiently shows the character of the pollen ; the difference in form, size, and transparency distinguishes the good and bad ; and it is easy to judge, at least approximatively, of the relative quantity.   Yet there are some cases, though not very common, where this examination of the pollen is not sufficient to determine whether it is active or inert; for it may happen that it has all the appearance of good pollen without having its qualities.   Such was that of *Mirabilis longifloro-Jalapa*, whose .grains, although unequal, were not deformed, and appeared full of fovilla, notwithstanding their inefficacy upon the stigmas of the two parent plants, as well as upon those of the hybrid.   Perhaps the employment of chemical reagents would better determine their impotency.

There are various degrees of fertility in hybrids by means of pollen.   We have seen *Luffa acutangulo-cylindrica* of the first generation extremely low in this respect, but in the third remarkably productive.   It is the same, and nearly to the same degree, in *Luffa amaro-cylindrica, Nicotiana rustico-paniculata*, and *paniculato-rustica*, and in a great number of the toad-flax hybrids (*Linaria purpureo-violacea*) of the second, third, fourth, and fifth generations.

A greater richness of pollen is seen in *Primula offcinali-grandiflora* of the first, and especially second, generation, and in *Cucumis Meloni-trigonus*, &c.   In fine, there are some hybrids where the pollen is little inferior, if at all, in perfection to the most legitimate species.   This is the case in *Coccinia Schimpero-indica, Datura meteloido-Metel, D. Stramonio-Metel, D. Stramonio-lævis, Nicotiana angustifolio-macrophylla,* N. *texano-rustica,* N. *persico-Langsdorffii, Pe-*

tunia violaceo-nyctaginiflora, &c.; and the same in many of the toad-flax hybrids of the third and fourth generations, already very close to *Linaria vulgaris.*

In a word, as I said at the commencement of this article, hybrids are found of all degrees of fertility, from the extreme case where the ovary only is fertile to that where all the pollen is as perfect as that of the best-established species.

(3.) *Is the Aptitude of Species to cross each other, and the Fertility of the Hybrids which result, proportional to the apparent Affinity of the Species?*

In general this is the case; but there are exceptions, and we have stated some. There are, indeed, some species, closely allied in exterior organisation and physiognomy, which are less disposed to mutual crossing than other species which are far distant in their outward appearance. Thus we have seen three species of eatable gourds, so closely resembling each other that most botanists fail to distinguish them, resist all attempts to cross them; whilst the melon and *Cucumis trigonus*, so very different from one another, easily give origin to very fertile hybrids, though the pollen is a little defective. Such is the case with *Nicotiana glauca*, which, although very distant from *N. angustifolia* and *macrophylla*, yet gives hybrids with them, having very fertile ovaries; whilst *N. glutinosa*, more difficult to cross with them, although belonging to the same section of the genus, only gives one sterile hybrid both by the pollen and ovary. I might also mention the crossing of *D. Stramonium* and *D. ceratocaula*, two species strangers to each other, from which there results a fertile hybrid, although attended by that peculiar kind of partial sterility which consists in the loss of the first flowers.

These exceptions, for which it is probably impossible to assign a cause, do not prevent the affinity of species, as revealed by the exterior organisation, from indicating generally the degree of aptitude to cross, and do not prevent us from forming a conjecture to a certain extent as to the fertility of the hybrids. We have seen the proof in *Datura Meteloido-Metel*, *D. Stramonio-Tatula* and *Tatulo-Stramonium*, D. *Stra-*

monio-lævis, Nicotiana texano-rustica and *rustico-texana, N. angustifolio-macrophylla*,&c., which hybrids, with the marked exception of those of *D. Stramonium*, have perfect fertility. The aptitude of species for mutual impregnation, and the degree of fertility of the hybrids which result, are therefore the true signs of their special affinity as regards generation ; and in the great majority of cases this affinity is indicated by the exterior organisation—in other words, by the physiognomy of the species.

(4.) *On the Physiognomy of Hybrids.*

To give a just idea of the aspect which hybrids present, it is essential to distinguish between the first generation and those which follow.

I have always found in those hybrids which I have obtained myself, and whose origin has been well known to me, a great uniformity of aspect between individuals of the first generation, no matter how numerous, provided they proceed from the same crossing. This we have seen in *Petunia violaceo-nyctaginiflora, Datura Tatulo-Stramonium,* and *D. Stramonio-Tatula, D. Meteloido-Metel, D. Stramonio-lævis, Nicotiana texano-rustica,* and *N. rustico-texana, N. persico-Langsdorffii,* &c.

I do not mean to say that all the individuals of the same crossing are absolutely counterparts of one another ; there are sometimes slight variations between them, but not sufficient to alter the general uniformity in a sensible degree, and it does not appear to me that these differences are any greater than those which are frequently seen between the seeds of legitimate species of the same production. In short, it may be said that hybrids which proceed from the same crossing, resemble each other, in the first generation, as much as, or nearly as much as those which proceed from the same legitimate species.

Must it be admitted, as M. Klotzsch maintains, that mutual hybrids (those which proceed from the two possible crossings between the two species) are markedly different from each other ; for example, the hybrid obtained *from the species* A fertilised by the species B, differs sensi-

bly from that which is obtained from the species B fertilised by the species A ? I cannot deny this in an absolute
manner ; it would be necessary to see the hybrid which
induced M. Klotzsch to make this statement ; but I can
assert, that all the mutual hybrids which I have obtained, as
well between allied species as between distant ones, resembled
one another as much as if they proceeded from the same
crossing.   I have already pointed this out when speaking
of *Datura Stramonio-Tatula* and *Tatulo-Stramonium, Nicotiana paniculato-rustica* and *rustico-paniculata, N. angustifolio-macrophylla* and *macrophyllo-angustifolia, N. texano-
rustica*, and *rustico-texana, N. persico-Langsdorffii*, &c. ;
without doubt it may not be always so, but if the fact is
true, it must be rare, and considered more as the exception
than the rule.

All hybridologists are agreed that hybrids (and it always
applies to hybrids of the first generation), are mixed forms,
intermediate between two parent species. And this is really
what does take place in the great majority of cases ; but it
by no means follows that these intermediate forms are
always at an equal distance between the two species.   On
the contrary, it is often observed that they are frequently
much nearer one than the other. Besides, we may conceive,
that the appreciation of these relations is always a little
vague, and that it is the idea which determines it.   We may
also remark that hybrids resemble sometimes one of the
two species in one character, whilst they resemble the
other in another character.   This is very true, and we have
seen an example in *Mirabilis longifloro-Jalapa*, which is distinctly more like *M. longiflora* in the organs of vegetation,
and *M. Jalapa* in the flowers.   But I think it is wrong to
refer this distribution to the part which the species have
played as father or mother in the crossing whence the
hybrid has arisen. At least I have not seen anything which
confirms this opinion.

M. Regel asserts (*Die Pflanze und ihr Leben*, &c., p. 404,
et suiv.), that when the hybrid proceeds from species of different genera, their flowers bear the essential characters of
those of the father ; but we have seen in the *Datura cerato-
caulo-Stramonium*, proceeding from two nearly generically

different species, the flowers were absolutely like those of
the mother (*D. Stramonium*) ; in *Nicotiana glauco-angustifolia*, and *glauco-macrophylla* obtained from very different
species, they were remarkably more like those of the mother
than those of the father ; whilst in *N. californico-rustica*
and *glutinoso-macrophylla* they were very distinctly intermediate between the parent species.

The rule laid down by M. Regel seems to me therefore
very hazardous, or at least founded upon insufficient data.
For my own part, I believe that these inequalities in resemblance, sometimes very great between the hybrid and its
parents, are maintained chiefly by the marked preponderance which many species exercise in their crossings, whatever may be the part which they act (whether as male or
female).

This we have seen in the hybrids of *Petunia violacea* and
P. nyctaginiflora which have a greater resemblance to the
first than the second ; in *Luffa acutangulo-cylindrica* of
which the forms are far more like *Luffa cylindrica* than the
conjoined species ; and especially in *Datura ceratocaulo-
Stramonium* and *D. Stramonio-lœvis*, of which all the individuals are incomparably nearer *D. Stramonium* than the
other species, although in one case *D. Stramonium* fulfils
the function of the male, and in the other that of the
female.

Commencing with the second generation, the physiognomy of hybrids is modified in a most remarkable manner. Very often the perfect uniformity of the first generation is succeeded by an extreme medley of forms, the one
approaching the specific type of the father, the other that of
the mother—sometimes returning suddenly and entirely into
the one or the other. At other times this recurrence towards the generating types is performed by degrees and
slowly, and sometimes the whole collection of hybrids is seen
to incline to the same side.

I think it is now placed beyond dispute that this dis-
solution of hybrid forms commences, in the great majority
of cases (it may be in all) in the second generation.

(5.) *On the return of Hybrids to the specific forms of the pro-
    ducing species. . What is the cause which determines this
    return ?*

In every hybrid which I have examined, the second
generation presented changes of aspect, and a manifest
tendency to return to the forms of the producing species,
and that under such conditions that it was impossible for
the pollen of those species to have concurred in bringing
them back.   We have seen striking examples in *Primula
officinali-grandiflora,* in all the hybrids of *Datura Stramonium, D. Meteloido-Metel,* the mutual hybrids of *Nicotiana
angustifolia* and *macrophylla, N. persica,* and *Langsdorffii,
Petunia violacea* and *nyctaginiflora,* in *Luffa acutangulo-
cylindrica,* and still more in *Linaria purpureo-vulgaris.*
Among many of these hybrids, from the second generation,
a complete return to one or other, or even both, of the
two parent species has been seen, and approaching them in
different degrees; among many also we have observed
forms continuing intermediate, whilst simultaneously other
specimens of the very same production have effected the
return of which I am about to speak.   Further, we have
stated in some cases (*Linaria purpureo-vulgaris*) that, in
the third and fourth generation, true retrogression towards
the hybrid form takes place ; and sometimes even we have
seen individuals of a plant to all appearance wholly returned
to one of the two species, which seemed to revert almost
entirely into the opposite species.

All these facts are naturally explained by *the disjunction
of the two specific essences in the pollen and ovules* of the
hybrid.   A hybrid is an individual in which two different
essences are found united, each having its particular mode
of vegetation and finality, which are mutually opposed, and
are constantly striving to disengage themselves from one
another.   Are these two essences intimately blended?   Do
they reciprocally penetrate every part, so that each particle
of the hybrid plant, however minute or divided, contains
equal portions of both ?

It may be so in the embryo and first stages of the development of the hybrid ; but it seems to me more probable that

this last, at least in the adult state, is an aggregation of particles, both homogeneous and unspecific when taken separately, but distributed more or less equally between the two species, and mixed in different proportions in the organs of the plant. The hybrid, according to this hypothesis, would be a living mosaic, the discordant elements of which, so long as they remained mixed, would be undistinguishable to the eye ; but if, in consequence of their affinities, the elements of the same species approached each other and agglomerated themselves in small masses, parts and sometimes entire organs, would then be visible, as we have seen in *Cytisus Adami*, and the bizarre group of the orange and citron hybrids, &c. It is this tendency of two specific essences to disengage themselves from their combination, which has induced some hybridologists to say, that hybrids resemble the mother by their leaves and the father by their flowers.

Although the facts may not be sufficiently numerous to conclude with certainty, it seems that the tendency of species to separate, or, so to speak, to localise themselves in various parts of the hybrid, increases with the age of the plant, and is more and more pronounced as the vegetation approaches its term. These disjunctions become more manifest in the highest organisms of hybrids, about the reproductive organs ; in *Cytisus Adami* disjunction shows itself in the flowering branches ; in the orange anomalies and *Datura Stramonio-lœvis* in the fruit itself. In *Mirabilis longifloro-Jalapa* and *Linaria purpurea* the corolla manifests the phenomenon of disjunction, by the separation of the colour peculiar to the producing species. These facts authorise the idea that the pollen and ovules, but especially the former, are precisely the parts of the plant where disjunction goes on with most energy ; and what adds a greater degree of probability to this hypothesis is, that they are at the same time very elaborate and minute organs—a double reason for rendering the localisation of the two essences more perfect. This hypothesis being admitted, and I confess it seems to me extremely probable, all the changes which supervene in hybrids of the second and more *advanced* generations would explain themselves, as it were ;

but if, on the contrary, it be not admitted, they would be perfectly inexplicable.

Let us suppose in the Toadflax hybrid of the first generation, that disjunction takes place both in the anther and contents of the ovary ; that some grains of pollen entirely belong to the paternal species, others to that of the mother ; that in others disjunction has not, or, at least, only just commenced. Again, let us suppose that the ovules are, to the same degree, separated *both in the direction of the male and female parent; what will result when the pollen tubes descend into the ovary to fertilise the ovules ? If the tube of a pollen grain, which has returned to the male parent, meet an ovule separated in the same direction, *a perfectly legitimate* fecundation will be produced, from which will result a plant entirely returned to the paternal species. A similar combination effected between a pollen grain and ovule, both returned to the female parent, the product will return in the same manner to the species of this last ; if, on the contrary, combination is effected between an ovule and pollen grain, separated in opposite directions, they will perform a true crossed fecundation, like that which gave origin to the hybrid itself ; and there will again result a form intermediate between two specific types. The fertilisation of an ovule non-separated, by a pollen grain separated in either direction, would give a quadroon hybrid ; and since disjunctions, as much in the pollen as in the ovules, can take place in all degrees, every sort of possible combination will result as chance may direct. We have seen these multitudinous forms produced in the Toadflax hybrids, and *Petunia* from the second generation.

The retrogression of a hybrid in its course of return to one of the parent species is also easily explained by this hypothesis. I mentioned several examples when speaking of the third generation of *Linaria purpureo-vulgaris*. Thus, for example, among eighty plants sprung from an individual of the second generation, which seemed to be entirely returned to *L. purpurea*, fresh hybrids appeared, which came back to the intermediate form of the first hybrid, and other individuals still more sensibly approached to *L. vulgaris*. The reason is, that the purple-flowered hybrid of the second generation, notwithstanding appearances, still retained some essence of the yellow-flowered Linaria, and this strange particle was sufficient to bring some pollen grains and ovules back either to a mixed state, or altogether to *Linaria vulgaris.*

Similar actions are produced, though less marked, in the descent of hybrids of the second generation, which seem entirely returned to the type of *L. vulgaris,* and even to a certain extent in that of *Dátura Stramonio-lœvis,* where some individuals return to *lœvis,* preserving up to the third generation the accessory characters which belong to that form of hybrids.   All these facts show us that the separation of specific forms allied in hybrids, is not always completed so rapidly as one might be led to suppose, judging from physiognomy and external appearance.

The return of hybrids to the forms of the parent species is not always so sudden as that which we have observed in the Primroses, Petunias, *Linaria purpureo-vulgaris,* D. *Meteloido-Metel,*&c.; it is frequently completed by insensibly minute gradations continued through long series of generations.   We have seen, for example, in *Luffa acutangulo-cylindrica,* even in the third generation, that among forty individuals only one was found which had wholly reassumed the external appearance of *L. cylindrica.*

Hybrids of *Nicotiana persica* and *Langsdorffii,* modify themselves ˙slowly, and ten or even more generations may be insufficient to bring them back entirely to the specific forms.

It is remarkable in the latter case, that the hybrids do not present any appreciable mark of disjunction of the two specific essences, which appear intimately blended together in every part of the plant.   Nevertheless the traits of one of the two species sensibly disappear from generation to generation, as if extinguished by degrees; but it not unfrequently happens that this extinction takes place with such rapidity as to be completed in the second generation.

En résumé, hybrids fertile and self-fertile return sooner or later to the specific types from which they were derived, and this return is effected either by the separation of the *two* mixed essences, or by the gradual extinction of one of the two. In the latter case the hybrid posterity returns entirely and exclusively to one only of the two producing species.

(6.) *Are there any exceptions to the law of return of hybrids to the parent forms? Do certain hybrids become fixed and give rise to new species?*

I have not been long enough engaged in the study of hybrids to have formed any settled opinion on this question. Many botanists of good authority believe that some hybrids, if not all, *can* become fixed and pass to the state of constant varieties, that is to say, true species, intermediate between those of their parents; this is in particular the opinion of M. Regel, who regards it as probable that in the group of willows, roses, and many other genera rich in nearly allied forms, the nomenclature of which is very embarrassing to the botanist, there originally existed but a small number of species (two or three), the fertile crossings of which have given rise to equally fertile hybrids, which, in their turn, crossing between themselves and their parents, have produced, age after age, those multitudes of forms which exist at the present day.

Such may be the case, but it is without proof, and the hypothesis is entirely gratuitous. In my opinion the fact may be explained otherwise in a much more natural and probable manner, viz., by the inherent property of all organisms (at least vegetable) to modify themselves to a certain extent according to the influence of the surrounding medium, in other words, by the innate tendency of what we call *species* to subdivide into secondary species. How can it be admitted, for example, that roses, which are disseminated over the whole extent of the Old World, from Ireland to Kamschatka, from the Atlas and Himalayas to the glacial ocean, which cover all North America, which are often isolated in narrow spaces and different localities, can have met each other to give rise to hybrid forms?

It would be hardly possible to conceive such a fact. Have roses never been subjected to experiment to ascertain how far they can mutually hybridise, and if their hybrids would be fertile or not?   I can affirm this, that I have never obtained a hybrid which manifested the least tendency to form a specific stock.

At present I only know a single instance which might serve as a basis for the hypothesis of fixation of hybrids. Still this fact is doubtful; it is that of Ægilops, closely allied to wheat, which was cultivated at the Museum about ten years, during which the successive generations did not produce any appreciable modification.

It remains to be proved whether the Ægilops cultivated at the Museum (*Æ. speltæformis*, Jord.) is really a hybrid, and that it does not modify itself during a long series of generations: it would be an exception; but this very general rule would not be weakened, at least so long as the fact remained isolated.

(7.) *Is there any precise limit between Hybrids and Crosses?*

Most hybridologists insist on making a distinction between hybrids and crosses, and nothing could be more easy to understand; the hybrid results from the crossing of two distinct species, two true species, as M. Regel says—the crosses from that of two races or varieties.

Theoretically, nothing is more clear; in practice, nothing is more difficult than the application of these two words.

For example, ought the produce obtained by the crossing of the Cantaloup Melon and Netted Melon, that of the Netted Melon and Dudaim, that of Dudaim and *Cucumis Pancherianus*, or even that of *Datura Stramonium* and of *Datura Tatula*, &c., to be called hybrids or crosses? This question gives rise to another, that of the distinction of species, races, and varieties, an everlasting subject of dispute among naturalists, which too often ends in a war of words unworthy of the science; to settle which, it is necessary to turn to the examination of what is understood by the term *Species*, *Race*, and *Variety*.

(8.) *What, therefore, is a Species, Race, and Variety?*

Let us start at the very origin of the notion of species,

and not lose sight of the fact that all our ideas arise from the *contrast of things.*

The man blind from birth has no idea of darkness, because being deprived of the sensation of light he does not perceive the difference between the two ; even one possessed of sight would have no idea of the light which surrounds him if the whole world was luminous and that to the same degree. The notion of species does not escape the common law ; it is more complex, and is formed from more elements, as we shall attempt to elucidate.

If there existed in nature but *one vegetable form,* wheat for instance, always and everywhere alike, without any variation in the innumerable individuals which represent it, we might arrive at the idea of an *individual* and *vegetable,* but not *species ; wheat* and *vegetable* would be confounded in one's mind as one and the same thing.

Let us suppose also that nature had created an indeterminate number of different organisms, and each of them represented on the earth by only a single individual, incapable of multiplying itself, but indestructible and imperishable ; even here we could not arrive at the conception of a species, for each type of organisation would be isolated, and have no resembling individual.

To have a species it is necessary, therefore, 1*st,* To have a *plurality* of similar individuals, that is to say, a group, a collection ; 2*d,* That this group or collection of individuals *contrast in some degree* with other groups of individuals likewise resembling each other, and yet able to approach one another in some common points which render them comparable.

It follows that the idea of species is connected with that of kind or genus (I mean genus taken in a philosophical sense) ; that the one fact always supposes the other ; that, in a word, they are inseparable and unable to exist apart. And as, in the organic world, individuals have a transitory existence, reproducing themselves by generation, it is necessary, 3*dly,* In order that species may have consistence and duration—that the *resemblance of individuals forming a specific collection shall continue in successive series of generations.*

Thus a plurality of similar individuals forming a group, and the contrast of groups among themselves by certain characters common to different groups ; and, *lastly*, The perpetuation of resemblances between the individuals of the same group constitute the elements of species. Species contain nothing more or less.

It is not, therefore, an ideal type, as certain abstract-loving naturalists have suggested ; it is essentially a collection of similar individuals. The abstract ideal type of a common organisation is only, as it were, a tie, which in our mind collects similar individuals in the same bundle, and sums up the contrasts (or differences) which separate their group from every other.

It is necessary, then, to return to the pure and simple definition of Cuvier,—viz., *A species is a collection of individuals descended from one another, or from common parents, and from those which resemble them as much as they resemble themselves.*

Let us remark, in passing, that in thus defining species, Cuvier did not take *races* and *varieties* into consideration.

Everywhere where there is a group of similar individuals, contrasting in some measure with other groups, and pre- serving through a series of generations the physiognomy and organisation common to all the individuals,—*there is a species*.

It is by their contrast that species are distinguished from one another, and it is by comparison that their contrasts appear. Contrasts may be more or less great according to the objects compared. If they are very great and well marked, all the world acknowlege the specific distinction of the compared forms ; if they are very weak, almost inappreciable, opinions are divided ; one party separating the feebly contrasting forms into distinct specific groups, the other collecting them into one, and applying to them in the mean time the qualifications of *races* or *varieties*.

These collections and separations are purely optional, and they can have no other rule than scientific or economic advantage ; in order to determine them it is necessary to be endowed with a certain tact which is ordinarily acquired by experience.

In short, there is no qualitative difference between *species*, races, and varieties ; it is idle to seek one. These three things are formed from one, and the terms which pretend to distinguish them only indicate degrees of contrast between compared forms.

It must be understood that here the question is not concerning simple individual variations, non-transmissible by way of generation, but only forms common to an indefinite number of individuals, and transmitting themselves faithfully and indefinitely by generation.

Contrasts between compared forms *are of all degrees*, from the strongest to the weakest, which simply means that following the comparisons which are established between groups of similar individuals, species are found of all degrees of strength and weakness; and if it was attempted to express these degrees in so many words, the whole vocabulary would be insufficient.

The delineation of species is therefore as I said before entirely optional; it makes them larger or smaller, according to the importance which is given to the resemblances and difference of various groups of individuals taken with respect to each other, and these appreciations vary according to men, times, and phases of science. How many modifications have certain great species of Linnæus and Jussieu undergone during fifty years!

The division of old species, their *pulverisation*, if I may use such a term, seems to have now reached its extreme limits, and many botanists are led by this tendency to complicate the descriptive part of the science in such a way as to threaten to involve the whole life of a man in its minutiæ. Notwithstanding this, if those who have inaugurated these scientific refinements have not committed error by taking individual alterations, non-transmissible and not forming a group, that is to say simple variations, for forms common to an indefinite number of individuals, very constant and very faithfully transmissible in every consecutive generation, there is reason to believe that they have proceeded logically. The whole question is to know if it be advantageous to science to distinguish and enrol in its catalogues, these feebly contrasting species; but it is essentially necessary to be assured that the characters which are assigned them are really specific—that is to say, common to an unlimited num-

ber of individuals, and always faithfully reproduced in every
generation.

But it is more than probable that in a multitude of cases
(in the genus *Rubus*, for example) purely individual variations without persistence, have been taken for common
characters, constant and transmissible.

Does it then follow that the terms *race* and *variety* ought
to be banished from the science? Certainly not, for they
are convenient to designate weak species that ought not
to be enrolled among the official species, but it is proper
to give them their true signification, which is absolutely
the same as that of *species* properly so-called, and to see in
forms designated by these terms some unity of a weak kind,
which might be neglected without inconvenience to the
science.

9. *Can Artificial Hybridisation furnish a mark to determine what it is proper to distinguish as Species?*

I have not the least doubt but that there are some cases
where it would be of a slight assistance, and again a greater
number where it would not be practicable. Here are some
examples of its practical utility.

I have stated before, in speaking of the three species of
eatable gourds, that they but slightly differ in outward appearance, and even by their intimate characters, for most
botanists cannot clearly distinguish them; Linnæus himself
confounded them in one. But these three plants refuse to
give hybrids by mutual crossing; they are then three self-
governed species perfectly distinct.

M. Dunal, in his Monograph of Solanaceæ, combines into
one species *Datura Stramonium* and *D. Tatula*, considering
them as simple varieties of the same species. But the produce of their crossing does not vegetate altogether like
these two forms; it grows much larger and flowers less,
inasmuch as it loses its flower-buds in the seven or eight
first branches. This disturbance caused in the vegetation
of the mixed produce, is an indubitable sign of a difference
in the autonomy of the two parent forms; therefore these
forms ought to be held as distinct species.

Datura Metel and *Meteloides* are' at least as nearly allied to one another as the two preceding ; but, from the second generation, their hybrids cease to resemble them, and a certain number of individuals return to one or other of the two parent forms. Let us therefore conclude that these forms are specific, that they each have their autonomy and deserve, notwithstanding their affinity, to be distinguished from one another.

Nicotiana macrophylla and *N. angustifolia* combined in the "Prodromus" of De Candolle with *N. Tabacum*, give hybrids which, after the second generation, manifest a very appreciable commencement of return towards the producing forms. These last have therefore also their manner of growth proper to each of them. Why do we not admit them as distinct in our botanical catalogues ?

But when the forms are so closely allied to one another that they are with difficulty distinguished, their hybrids must differ still less from one another than they differ between themselves. The data furnished by hybridisation, therefore, here lose their value ; but then it becomes a matter of indifference, whether to separate the two forms as distinct species, or to combine them, by the title of simple varieties, under a common specific denomination.

It follows from all we have said, that the application of the terms hybrid and cross is determined by the rank which may be assigned to the individuals from the crossing of which the mixed forms requiring to be named have been produced—that is to say, it is entirely left to the judgment and tact of the nomenclator.

On Diplostemonous Flowers ; with some Remarks upon the Position of the Carpels in the Malvaceæ. By ALEXANDER DIOKSON, M.D., Edin. (Plate III.)*

It has long been known that in *Geranium* and its allies, the stamens superposed to the petals are external to those superposed to the sepals. That this is the case is very distinctly seen in the adult state, where the dilated bases of the filaments of the outer stamens overlap those of the inner.

* Read before the Botanical Society of Edinburgh, Feb. 11, 1864.

Before these plants were examined organogenically, the outer stamens were, not unnaturally, assumed to be the older; and, as this involved a want of due alternation of parts, it was imagined that a third and outermost whorl of stamens alternating with the petals must have aborted, the idea being held to be countenanced by the frequent occurrence, in these plants, of five glands outside the androecium and alternating with the petals.[*] When, however, the development of the parts was observed, the unsoundness of this theory became evident; for it was found that the outer stamens are the younger; and, moreover, that the glands do not appear until shortly before the time of blossoming.[†]

The fact of the younger whorl of stamens being external to the older one is remarkable, as being exactly the reverse of what one would, *a priori*, have expected. The question as to how stamens should be so arranged, is an interesting one, and derives great importance from the researches of Payer having shown that this arrangement, so far from being uncommon, obtains in the greater number of diplostemonous dicotyledons.

In attempting an explanation of this difficulty, I am fully aware of the delicacy of the questions involved; and I would offer the result of my consideration of the subject, more as a suggestion worthy of being kept in view by those who may examine diplostemonous flowers, organogenically or otherwise, than as a definite solution of the problem. In short, I would submit a *possible* solution, to be substantiated or negatived by more extended and comprehensive observation of the facts.

Of diplostemonous flowers, there are two principal forms which demand our attention :—

1*st*, That in which the younger staminal whorl is the more internal, and the carpels, when of the same number, alternate with the younger stamens. Examples—*Coriaria*,[‡]

---

[*] Maout, Atlas de Botanique, p. 60. Balfour, Class-Book of Botany, p. 783, fig. 1485, with description.

[†] Payer. Organogénie, p. 59; pls. 12 and 13: the development of the glands in *Erodium* is shewn in pl. 12, figs. 17 and 21.

[‡] Ibid., p. 49: pl. 10. *Limnanthes*, in all probability, comes under the same head. Although Payer describes the younger stamens in *L. Douglasii* as *the more* external, his figure (pl. 10, fig. 21) does not seem to bear out

Agrostemma, Cerastium (e.g. *C. triviale*),* *Lasiopetalum* (e.g. *L. corylifolium*),† *Lilium*, &c. (Pl. III., fig. 2.) Comparatively few dicotyledons, but almost all diplostemonous monocotyledons, fall under this head.

2d, That in which the younger stamens are the more external, and the carpels, when of the same number, alternate with the older stamens. Examples—*Geranium, Erica, Malachium*, &c. (fig. 1.) As I have already mentioned, the greater number of diplostemonous dicotyledons fall under this head.

In the first of these two forms of diplostemony (that in which the younger stamens are the more internal), the arrangement requires no explanation, as it corresponds with the ordinary centripetal evolution of successive whorls upon an axis.

The case, however, is widely different where the younger stamens are the more external. Here we seem to have a *centrifugal* succession of parts upon an axis. Have we any analogies to guide us in explaining this apparent anomaly ? What suggests itself most naturally in this regard is perhaps the case of polyadelphous flowers, where the members of each staminal group are usually developed in centrifugal

the statement, for there the circle of the younger stamens appears almost to coincide with that of the older ones, or, if anything, to be somewhat smaller than it. In the more advanced stages, the younger stamens are very distinctly the more internal. At all events, the plant requires re-investigation upon this point.

* I have examined, with great care, the development of the stamens in *Agrostemma Flos-Jovis* and *Cerastium triviale*, and in both of them I can confidently state the younger stamens to be the more internal. I have given figures of young flowers of these two plants (Plate III., figs. 12 and 13). Payer has, on the contrary, stated that the *older* stamens in *Cerastium* are the more internal ; but his figures do not indicate this very satisfactorily. To judge from my own experience of *C. triviale*, I should imagine it to be very difficult to determine which staminal whorl is the more internal, when the flower is viewed so much from above as those represented by Payer (Organogénie, pl. 72, figs. 7 and 8). In all flowers like *Cerastium*, where the receptacle is very convex, it is very necessary to obtain a completely side view, so as to see the difference in *elevation* of the different parts. In flowers with flat receptacles, on the other hand, the view from above is the most advantageous, as the parts are no longer on higher and lower elevations, but on larger and smaller circles.

† Organogénie, pl. 9, fig. 4.—Though the fact is not mentioned in his text, Payer's figure leaves no doubt as to the younger stamens (staminodes) in this plant being on a smaller and more internal circle than the older.

succession upon a cushion-like body which precedes their appearance. Payer has, however, by the most convincing arguments, determined the staminal phalanges of poly-adelphous flowers to be compound stamens, the parts of which, when centrifugally developed, correspond to the basipetally developed leaflets or lobes of ordinary leaves ; so that in these flowers we have no real examples of centrifugal succession of parts *upon an axis*. A curious arrangement, however, is described by Payer as occurring in *Opuntia*, where vast numbers of stamens are developed in centrifugal succession, and apparently distributed uniformly round the receptacle. We may arrive at a comprehension of this re-markable form, if we direct our attention to certain cases which appear to connect it with more easily intelligible forms. In *Brathys (Hypericum) prolifica*, Payer has shown that the staminal cushions (which usually remain distinct in the Hypericaceæ) become fused together, at an early period, into a single nearly uniform annular cushion, upon which the stamens make their appearance centrifugally. Again, in *Cistus*, he has shown that, in the early condition, the centrifugally developed stamens exhibit distinct traces of grouping, although the annular cushion, on which they are developed, is always entire. From *Cistus* we pass at once to *Helianthemum*, where all trace of grouping in the stamens disappears,[*] presenting us with a condition quite analogous to that in *Opuntia*. Such a series of forms leaves us in no doubt that in *Opuntia*, as in *Helianthemum* and *Cistus*, we have merely an extreme case of the fusion of compound stamens, which differs from that in *Brathys* only in being congenital, while in that plant it does not com-mence until a little after the appearance of the staminal cushions.

In connection with the above, I must not omit allusion to Payer's own determination of the signification of the an-drœcium in *Cistus ;* and as the questions suggested by it have no unimportant bearing upon the subject of this paper, I may be excused the apparent digression of commenting upon it. In this plant he· has found that the stamens

---

[*] Organogénie, p. 17, plate iii. fig. 25.

make their appearance in centrifugal succession upon an annular cushion surrounding the centre of the floral axis, and in this wise:—in the first place, a circle of five stamens, superposed to the sepals, makes its appearance on the upper part of the annular cushion; later, alternate with and below these, a second circle of five stamens is developed; still later and lower, ten stamens appear, one on either side of each of the stamens of the second circle; lastly, a great number of stamens continue the centrifugal succession till the annular cushion is completely covered. "From this mode of staminal development," he says, "may we not conclude that the androecium of *Cistus* is composed of only two whorls: the one superposed to the sepals, in which the stamens remain simple and are the more internal; the other superposed to the petals, in which the stamens are grouped in five bundles, the stamens in each bundle appearing from above downwards."*
Here, I cannot but think, Payer has introduced an un-necessary complication into the subject. His interpretation involves at least one serious improbability,—that in the same flower there should be both simple and compound stamens. What induced him to adopt this opinion was no doubt the consideration that, if the five stamens of the first degree, which are superposed to the sepals, were assumed to be the apices of staminal groups, the stamens of the second degree, which are superposed to the petals, must occupy neutral territory between these groups. He seems not to have taken notice of the fact which his figure† plainly indicates, that the same difficulty occurs again lower down, where there are stamens (apparently of the fifth degree) super-posed to the petals, and therefore also on neutral terri-tory. The same thing appears still more strikingly in his figures of *Capparis* (the androecium of which he has recognised as being similar to that of *Cistus*), where it will be seen that in every second or third generation of stamens there are some occupying neutral territory. There seems, therefore, to be no more reason for con-sidering the stamens of the second degree as the apices of staminal groups than for so viewing any of the other sta-

---

* *Op. cit.* pp. 16, 17.   † *Op. cit.* plate 8, fig. 18.

mens of inferior degree which may be superposed to the
petals. It appears to me that in *Cistus* and *Capparis*, all
the stamens superposed to the petals (including of course
the stamens of the second degree) must be looked upon
simply as neutral structures, resulting from the coalescence
of parts at the points of fusion of the contiguous groups,
just as "interpetiolar stipules" are neutral structures, resulting from the coalescence of the stipules of opposite
leaves at the points of fusion of the leaf-bases. This
analogy will at once be admitted as a legitimate one, when it
is remembered that stipules are nothing but lobes of the leaf.

To sum up, I think it is sufficiently evident from the
foregoing considerations, that all the instances of indefinite
stamens exhibiting an apparent deviation from the law of
centripetal succession of leaves upon an axis, may be resolved into cases of compound stamens with development
of lobes from above downwards.

From this conclusion we are naturally led to inquire
whether those diplostemonous forms, where the younger
stamens are the more external, may not in like manner be
found to be merely *apparent* deviations from the ordinarily
recognised laws of leaf-succession ? We are at least bound
to show that the phenomenon is incapable of explanation
by the action of known laws, before we admit it to be an
example of centrifugal succession of leaves upon an axis.*

In the *Geraniaceæ*, which, as I have already mentioned,
exhibit this form of diplostemony, the genus *Monsonia*
presents the remarkable peculiarity of having *ten* stamens
in the younger and outer circle, arranged in five pairs
superposed to the petals.† Payer considered that these five

---

* It may be observed that I have here left out of view the centrifugal evolution of ovules upon central placentas. I have done so, because it is vain to
discuss this subject until we have more definite notions than at present prevail on the morphological value of the ovule itself. If the ovule represents a
modified leaf, the ovular groups will probably fall under the same category
with the staminal groups of polyadelphous plants, the placental elevations corresponding to the staminal bosses or cushions. If, on the other hand, the
ovule is to be viewed as a bud or branch, analogies may be sought for among
contracted centrifugal inflorescences.

† Similarly, in the *Zygophyllaceæ*, the five outer and younger stamens of
Zygophyllum, &c., are replaced in *Peganum* by five pairs of stamens superposed
to the petals.—Payer, Organogénie, p. 69 ; pl. 14, fig. 28.

pairs represented the five single stamens of the outer whorl in *Geranium* congenitally deduplicated.* At first sight, this view seems unexceptionable, since there is no doubt as to the parts being homologous. It appears to me, however, that if we were to invert Payer's statement, and say that in *Geranium* there is a congenital connation, in pairs, of the parts of the outer circle, which are distinct in *Monsonia*, we should thereby be enabled at once to explain the apparent anomaly of a younger whorl being external to an older one.

If, in fact, we adopt a line of argument analogous to that which Payer himself has employed in determining the signification of the epicalyx in the Potentillidæ, the whole difficulty, it seems to me, disappears. " In *Fragaria collina*," he says, " we always observe a calicule, composed sometimes of five leaflets alternate with the sepals, sometimes of ten leaflets grouped in pairs which alternate with the sepals ; and, as this calicule appears always after the calyx, it cannot be doubted that it is formed by the stipules of the sepals." (*Op. cit.* p. 503.) Now, in the Geraniaceæ we have an outer staminal whorl, which consists sometimes (as in *Geranium*) of five stamens alternate with the parts of the inner whorl, and sometimes (as in *Monsonia*) of ten stamens grouped in pairs which alternate with the parts of the inner whorl. Moreover, these outer stamens appear after the inner ones. The parallelism of these two cases, as regards the number, position, and order of succession, of the parts, is quite complete ; for we have in the Geraniaceæ an outer whorl whose relation to the inner one in these respects is exactly the same as that which the apparent outer calycine whorl of *Fragaria* bears to the inner one, or calyx proper.

The presumption, raised by this comparison, that the outer stamens in Geraniaceæ represent the lateral lobes of the inner or primary ones, distinct, or congenitally connate in pairs like interpetiolar stipules, amounts almost to a certainty, when we consider the mode in which the pentadelphous condition of *Monsonia* occurs, where each stamen of the inner whorl is connate with two of those of the outer

* *Op. cit.* p. 60.

—one on either side,—offering the closest analogy to a leaf with lateral lobes, or with adnate stipules. That the presumed interstaminal lobes—if I may so call them—in *Geranium* should so closely resemble, in all essentials, the primary stamens, need not surprise any one who bears in mind instances like *Galium cruciatum,* where the interpetiolar stipules differ in no respects from the leaves between which they are placed.*

It is evident that if the foregoing reasoning holds good as to the Geraniaceæ, we must extend its application to all the numerous cases where a similar diplostemonous arrangement occurs. All such plants, if my view be adopted, must be considered, strictly speaking, as isostemonous, the members of the outer staminal circle consisting merely of the lateral lobes of the primary stamens which form the inner circle.†

We may now proceed to examine what bearing the position of the carpels may have on this question. I have already stated that where the younger whorl of stamens is

---

* It may perhaps be thought that I am begging a question a little, in this allusion to the stipules of the Galiaceæ. Although the stipulary nature of these organs has been admitted by many very eminent botanists, yet I would not thus have assumed their opinion to be correct, had I not satisfied myself on the subject by examination of the course of development in *Galium aparine* where I can positively testify to the appearance, in the first place, of opposite leaves, followed afterwards by the development of intervening lobes, two or three on either side of the axis. I can hardly doubt that the leaf develop ment in the Galiaceæ has been traced by others, but I have not succeeded in finding any references to it.

† It is perhaps worthy of remark that Payer has shown that the "stami-nodes" of *Linum* are not developed until after the fertile stamens are so far advanced as to indicate the distinction between anther and filament, and after the carpels have made their appearance. (See Organogénie, pl. 13, figs. 6 and 7 ; with description, p. 67.) Now, if these staminodes in *Linum* represent, as Payer suggests (*loc. cit.* p. 66), the staminodes in *Erodium,* and these last constitute a true whorl of sterile stamens, it is very difficult to understand their very late appearance. If, on the other hand, we adopt the view stated above, as to the younger and outer stamens being merely the lateral lobes of the primary ones, and analogous to leaf lobes or leaf stipules, this difficulty disappears ; since there is nothing surprising in such structures appearing at a comparatively late period, and it is quite in accordance with what one observes in the case of staminal groups, where frequently the greater number of the stamens (lobes of the compound stamens) are developed *after* the appearance of the carpels.

the more internal (as in *Agrostemma*, &c.), the carpels, when of the same number, alternate with the younger stamens ; but that, where the younger whorl of stamens is the more external (as in *Geranium*, &c.), the carpels alternate with the older stamens.*

The position of the carpels in the first of these two forms requires no explanation, since it is manifestly in accordance with the usual rule of alternation of floral whorls.

In the second form of diplostemonous arrangement (that in *Geranium*, &c.), the case is apparently very different. Here the carpels alternate with the older stamens, and are thus superposed to the stamens developed next to them in order of time. If the outer and younger stamens in this form be regarded as forming a true staminal whorl, and as of equal value with the older whorl, we must admit a very extensive series of exceptions to the rule of alternation of whorls. On the other hand, if we view the younger stamens here as forming a merely adventitious whorl, the symmetry becomes at once intelligible, since the stamens with which the carpels alternate are then the only ones of primary importance. The fact of my interpretation of this staminal arrangement satisfactorily explaining away such a large number of apparent exceptions to the rule of alternation of whorls, is, I think, no small argument in favour of its being well founded.

It is further to be noted that, when, in a group of plants exhibiting this pseudo-diplostemonous arrangement, the outer and younger stamens disappear, the position of the

---

* Some may be inclined to think that the circumstance of the carpels occupying different positions in these two cases is not a point of much importance; that the carpels are only growing out where they have most space for expansion. It is quite true, as matter of fact, that the carpels here do occupy the places where they have most room ; but it appears to me impossible to reflect at all upon the arrangement of the parts of flowers, and admit that this arrangement is primarily dependent upon any such simple law of packing, if I may so express it. That such a law cannot be viewed as the basis of the arrangement of floral parts, must, I think, be apparent, when the not unfrequent instances of superposition of successive whorls are considered ; for in these instances the parts are certainly not developed where there is most room. The fundamental conditions are more likely to be found in the modifications of a contracted spiral, than in the mere influence of surrounding parts upon nascent structures.

carpels is unaffected by such disappearance. This, of course, is only what might have been expected, if the outer stamens are viewed as merely accessory parts. Thus, in the Ericaceæ, we have in most of the species of *Ledum* the apparent diplostemony which is frequent in the family, while in *Ledum latifolium* the younger stamens disappear; and yet in this species the carpels are superposed to the petals just as in the others.* In *Epacris*, so nearly allied to the heaths, we have also an absence of the stamens superposed to the petals; and yet the carpels have the same position as in the Ericaceæ. Contrasted with this, it is striking to observe the consequence of the disappearance of the younger stamens in a group of plants exhibiting what I believe to be a true diplostemonous arrangement—one where the younger stamens are the more internal. I have already mentioned, as an example of this arrangement, a Büttneriaceous plant, *Lasiopetalum corylifolium*, the organogeny of which has been given by Payer. Here, the fertile stamens, which form the outer and older whorl, are superposed to the petals; the inner and younger whorl consists of staminodes alternate with the older stamens; and the carpels alternate with the staminodes, and are thus superposed to the petals.† In *Hermannia*, on the other hand, where only the fertile stamens superposed to the petals exist, the carpels are no longer superposed to the petals, but are now found superposed to the sepals, occupying, apparently, the place of the missing staminodes. In the Dombeyeæ, Baillon has described in *Astrapœa* a single whorl of *staminal groups*, with the carpels similarly superposed to the sepals.‡

* I am indebted to a friend for the facts regarding *Ledum*.

† In the Büttnereæ, moreover, Baillon has shown (*Adansonia*, II. p. 168) that in *Büttneria*, &c., the fertile stamens superposed to the petals are, as in *Lasiopetalum corylifolium*, older than the staminodes which alternate with the petals, and that the carpels are also, as in that plant, superposed to the petals. He has not stated which of these whorls is the more internal; but I can scarcely doubt that further investigation will show that the staminodes are the more internal.

‡ Payer has observed (Organogénie, p. 45) that in *Melhania* the carpels are similarly superposed to the sepals, and I have been able to confirm his observation. I have also found that in *Pentapetes* the carpels occupy the same position. It is probable that this arrangement obtains in the Dombeyeæ generally.

The genus *Melhania* exhibits the simplest form of andrœcium that occurs in

If my explanation of the apparent diplostemony in the Geraniaceæ be admitted, the analogy between such an arrangement and that of polyadelphous stamens will be at once allowed. In this regard, it is not unimportant to inquire whether these two forms may not sometimes pass into one another; and I believe that instances of a passage of this kind do really occur.

the Dombeyeæ. In *Melhania incana* there are five staminodes superposed to the petals, and five stamens which apparently alternate exactly with the staminodes. At first sight, the arrangement of parts in this plant seems very incomprehensible. Here are apparently two staminal whorls, and yet the carpels are superposed to the sepals, as in the isostemonous *Hermannia.* On further examination and reflection, however, I have come to the conclusion that the diplostemony here is only apparent, and that we have merely to do with a much reduced form of the staminal groups which are found in *Astrapæa.* In *Melhania incana,* the stamens and staminodes are connected below into a short tube or ring, which adheres to the petals at points corresponding to the bases of the staminodes. When, however, we detach the corolla from the flower, the staminal ring becomes broken into five parts, a stamen and a staminode coming away with each petal. In those flowers which I have examined, the stamen is always to the left side of the staminode, which, as I have already stated, is superposed to the petal, and adherent to it by its base. From the regularity with which the rupture of the staminal ring takes place, it seems reasonable to infer that the fertile stamens do not exactly occupy the indifferent or neutral position between the staminodes which we should expect, were this a case of two alternating whorls.

In *M. decanthera,* where, instead of one stamen, there are two in each interval between the staminodes, I find that of these two stamens there is always a longer and a shorter one, whose position to right or left as regards each other is constant in the same flower, although differing in different flowers. These facts seem to indicate that the pairs of stamens here have not an indifferent relation to the staminodes between which they are placed.

When we consider how easy the transition is from *Melhania* (through *M. decanthera*) to *Dombeya,* which again is closely allied to *Astrapæa* where Baillon has distinctly traced the origin of the stamens and staminodes to five groups superposed to the petals, we can scarcely doubt that the androecium of *Melhania* is referable to a polyadelphous type, and thus the difficulty as to the position of its carpels disappears.

It will probably be very difficult, in the Dombeyeæ, without organogenic examination, to apportion the fertile stamens to their proper groups, as they appear to vary very widely in their ultimate relation to the staminodes: thus, in *Dombeya viburniflora* (Bot. Mag. *tab.* 4568), the fifteen fertile stamens are collected into five bundles, which apparently alternate with the staminodes: while, in an opposite direction, an example may be found in *Trochetia grandiflora* (Bot. Reg. *tab.* 21), where the stamens and staminodes unite to form five phalanges, each phalanx consisting of a staminode from which four fertile filaments spring, two on either side.

In the Aurantiaceæ, we have, in *Citrus*, staminal groups, of which Payer has fully detailed the development. These groups alternate with the petals. In each group, the successive evolution of the stamens extends, in single line, laterally to right and left from a central oldest stamen superposed to a sepal ; those stamens, therefore, being youngest, which are furthest removed from the central stamen or terminal lobe of the compound stamen.* Payer, moreover, describes the arrangement in *Tiphrasia trifoliata*, in the same family, as diplostemonous with the carpels superposed to the petals. The fact of the carpels here being superposed to the petals is important, as such an arrangement cannot fail to recall that in *Geranium, Erica,* &c., and of course suggests that the diplostemony in *Tiphrasia*, and many other Aurantiaceæ, is of the same spurious character as in these plants. Now, if this is the case, does not the andrœcium of *Citrus* bear to that of *Tiphrasia*, a relation exactly analogous to that which the whorl of apparent leaves of *Galium aparine*, consisting of opposite leaves, with a plurality of intervening lobes, does to that of *G. cruciatum*, where the intervening lobes are reduced to one on either side of the axis ?

Again, in the Philadelphaceæ, there are, in *Philadelphus*, staminal groups, the development of which, as described by Payer, is strikingly like that in *Citrus* ; while we have an apparent diplostemonous form occurring in *Deutzia.* But we seem further to have a form intermediate between *Philadelphus* and *Deutzia*, in *Decumaria*, which is described as having thrice as many stamens as petals, there being single stamens superposed to the sepals, and pairs of stamens superposed to the petals.† Thus, in the Philadelphaceæ, we have—1*st*, in *Philadelphus*, compound stamens with indefinite lobes ; 2*d*, in *Decumaria*, a reduction in the number of the staminal lobes, resulting in a condition apparently analogous to that in *Monsonia* ; and 3d, in *Deutzia*, an apparent diplostemony, probably analogous to that in *Geranium,* &c.‡

* Payer, Organogénie, p. 114, pl. 25.   † Endlicher, Genera, p. 1187.

‡ *Visnea*, in the Ternstrœmiaceæ, is probably another example of a reduced polyadelphous form. In *V. mocanera*, Payer has shown that, of its 15 stamens,

Although we must avoid attaching undue weight to the foregoing facts as to the Aurantiaceæ and Philadelphaceæ—the organogenic evidence being far from complete—yet it may be allowed, I think, that such facts at least heighten the presumption in favour of the justness of my views as to the constitution of the andrœcium in *Geranium*, &c. If my conjectures are well founded, it is possible that the younger stamens in *Tiphrasia* and *Deutzia* may be found to appear on the same circle with the older ones, just as in *Citrus* and *Philadelphus* all the staminal lobes are on one circle, or nearly so.

Having stated my reasons for believing the diplostemony in *Geranium* and the like, to be merely apparent, I would now allude to certain objections which may be urged against that view.

It may be said that such diplostemony may occur in plants whose leaves are neither lobed nor stipulate. In *Malachium*, for example, the leaves are entire and ex-stipulate. Regarding such an objection, I would observe that, in the case of compound stamens, which I believe affords us the closest analogy with this form of diplostemony, there is not only no necessary coincidence between the lobed or compound condition of the stamens and a similar condition of the leaves in the same plant, but there is not even any neces-

---

5 are superposed to the sepals, and 10, in 5 pairs, to the petals. The stamens superposed to the sepals are the first developed, and appear simultaneously. In each of the pairs superposed to the petals, however, there is an older and a younger stamen. From Payer's figure (Organogénie, pl. 154, fig. 25), it would appear that the position of the older and younger stamen, in each pair, to right and left, as regards each other, is uniform; so that each of the primary stamens superposed to the sepals has an older stamen on the one side and a younger on the other. There is evidently here an alternate succession of secondary staminal lobes, analogous, so far as it goes, to what has been described by Payer in *Malvaviscus*, and by Payer and Baillon in *Euphorbia*. The andrœcium of *Visnea* cannot fail to recall that of *Melhania decanthera*, described in a former note, and I have no doubt that the two cases are quite analogous, only, the apices of the reduced staminal groups are represented by staminodes in *Melhania decanthera*, the intervening unequal pairs of lobes being alone fertile.

Aristotelia, in the Tiliaceæ, is evidently also a reduced polyadelphous form, being described as having 5 inner stamens superposed to the sepals, and 10 outer in 5 pairs superposed to the petals (Payer, *Leçons sur les fam. nat.*, p. 278.)

sary correspondence between the mode of succession of the staminal and foliar lobes, when both stamens and leaves, in the same plant, happen to be lobed. Thus, in the Hypericaceæ, Myrtaceæ, &c., we have examples of families characterised by their compound stamens, and yet remarkable for the simplicity of their leaves ; and in *Cajophora* (*Loasa*) *lateritia*, where the stamens are developed in succession from above downwards, or basipetally, upon the staminal cushions,* I find the pinnæ of the leaf to be developed from below upwards, or basifugally. From such considerations it may be inferred that we need not expect any necessary association of lobed or stipulate leaves with pseudo-diplostemonous flowers.† Should any one be inclined to imagine that the facts I have just been stating at all invalidate Payer's determination of staminal groups as compound stamens, I would have it borne in mind that it is no more surprising that there should be entire leaves and compound stamens in the Myrtaceæ, &c., than that in many other plants there should be lobed leaves and simple stamens ; or, again, that in *Cajophora* there should be basipetal development of staminal lobes with basipetal development of leaf-lobes, than that these two modes of development should often occur together in the same leaf, as they do in the so-called mixed leaf-formation.‡

---

* Payer, Organogénie, p. 391, pl. 84.

† I may observe, however, that in some plants with pseudo-diplostemonous flowers, the leaves are not only stipulate, but exhibit a tendency to the formation of interpetiolar stipules. A. P. de Candolle has remarked that " several Geraniaceæ present this peculiarity [fusion of stipules] in a very evident manner." (*Organographie Végétale ;* Paris, 1827 ; tome i. p. 339.)

In some Geraniaceæ I find a very remarkable condition, which, so far as I know, is without parallel in other plants. In *Erodium hymenodes*, for example, where all the leaves are opposite, there are invariably, between each pair of leaves, on the one side of the axis a single, entire or slightly bifid, *interpetiolar stipule*, and on the other side a pair of *free stipules*, the one of which pair overlaps the other, from their bases passing each other. I find a similar arrangement in *E. cicutarium, Pelargonium zonale* and allied forms, *P. scutatum,* &c., whenever the leaves happen to be opposite.

Again, in *Spergula* (which, like *Malachium,* has ten stamens, and five carpels superposed to the petals) there are interpetiolar stipules ; and similar stipules exist in the allied *Lepigonum.*

‡ *Cobæa scandens* affords a very pretty example of the association of basipetal

It has been suggested to me, that if we were to admit the occurrence of accessory stamens, there would be no reason why these might not sometimes be placed on the same circle with, or even internal to, the primary ones, just as stipulary lobes may appear on the same level with the base of a leaf, or, as in the so-called axillary stipule, above it or on its inner face. I have already, when treating of *Tiphrasia* and *Deutzia*, admitted the possibility of accessory stamens being on the same circle with the primary ones. As to accessory stamens being internal to the primary ones, I think it not at all improbable that such an arrangement may also occur ; and, in compound stamens, an analogous phenomenon would be found in the Myrtaceæ, where the staminal lobes appear in centripetal succession as regards the axis.* Now, the possibility of accessory stamens being internal to the primary ones, may be supposed by some to invalidate the morphological distinction of diplostemonous flowers into two forms, which I have endeavoured to estab-

---

with basifugal development of leaf-lobes. Here, I find a succession of lobes, from a point both upwards and downwards. The upper pair of foliaceous pinnæ appears in the first instance ; and, from this, as a point of departure, the cirrhose pinnæ appear in basifugal succession towards the apex of the leaf, while the other foliaceous pinnæ appear successively towards its base. Payer has described an analogous succession from a midway-point upwards and downwards, in the serrations of the leaf-lobes in *Cannabis sativa* (Organogénie, pl. 61, fig. 28 ; with descr. p. 288).

* Organogénie, pp. 460-1, pl. 98. Payer has, somewhat hastily, I think, compared the compound stamens in the Myrtaceæ to leaves with lobes developed from base to apex, or basifugally. (*Op. cit.* p. 718.) His figures however, distinctly indicate that here, as in the ordinary forms of compound stamens, there is a mesial stamen or lobe of the compound stamen, from which, as a point of departure, the evolution of the other stamens extends ; and it appears to me improbable that a basifugal succession of lobes should be initiated by the development of a lobe in the middle line at the base of the compound stamen. The phenomenon seems more naturally explained by supposing that the first developed lobe of the myrtaceous compound stamen corresponds to the first developed or terminal lobe in the ordinary form, in which case the evolution in both forms would be basipetal—the only difference between the two being that, while in the Hypericaceæ, &c., the lobes are developed on the back or outer face of the rachis of the compound stamen (the staminal cushion), in the Myrtaceæ they appear on its front or inner face. In confirmation of this opinion, I may refer to the highly developed staminal groups in *Melaleuca purpurea*, where, in each phalanx, the stamens evidently proceed from the *inner* face of the flattened and elongated rachis.

lish. As to this, I would state that, although the accessory or non-accessory nature of the younger stamens when internal may be very difficult to determine in some cases, where the carpels, from multiplication, or reduction in number, fail to afford any indications, yet, when we consider the relations of the parts in the flowers of *Coriaria, Agrostemma, Cerastium,* &c., where the gynœcium is isomerous with the staminal whorls, and the carpels alternate with the younger stamens, we can have no doubt as to such flowers being truly diplostemonous, and therefore morphologically distinct from those of *Geranium, Erica, Malachium,* &c., where the younger stamens, in being external to the older, occupy a position irreconcilable with the idea of their forming a genuine whorl, and where the carpels alternate with the older stamens.

In the last place, we may consider certain anomalous and somewhat perplexing pseudo-diplostemonous forms, occurring in the Sapindaceæ and Polygalaceæ.

I have constructed, in accordance with Payer's observations, diagrams of the flowers of *Polygala, Kœlreuteria,* and *Cardiospermum.* In these plants, the outer and younger (accessory) whorl of stamens is incomplete.

In *Polygala* (Pl. III. fig. 9), the lower or anterior stamen of the primary, and the upper or posterior stamen of the accessory, whorl are absent.[*] It is worthy of remark, that in this plant, while the disappearance of the anterior primary stamen appears to be the direct cause of a solution of continuity of the staminal tube, the disappearance of the posterior accessory stamen is unaccompanied by any such solution.

In *Kœlreuteria* (Pl. III. fig. 11), the primary staminal whorl is complete, while the accessory whorl is reduced to three stamens alternate with sepals 1 and 4, 4 and 2, 5 and 3.[†] In *Pavia* (*Æsculus*), Payer has described a similar arrangement,—only, the accessory stamen between sepals 1 and 4 (and sometimes also that between sepals 4 and 2) is absent, and that between sepals 5 and 3 is occasionally resolved into two, then resembling those in *Peganum* and *Monsonia.*[‡]

Now, it may seem an objection to my doctrine of accessory stamens, that, in such plants, it requires us to admit that some members of a primary whorl may be provided with accessory lobes, while others are not. Regarding this objection, it is sufficient for me to advert to the remarkable condition of the calyx of the hundred-leaved rose, where only two sepals are provided on both sides with lateral lobes: two of the remaining three being destitute of them, and the other having a lobe on one side only.

In *Cardiospermum* (fig. 10), the number and position of the stamens are exactly the same as in *Kœlreuteria:* the stamens superposed to the petals being reduced to three, and alternating with the same sepals as in that plant. It cannot be doubted that, in the two cases, the andrœcia are essentially the same, although it is to be remarked that, while the three carpels are, in *Kœlreuteria*, superposed to sepals 1, 2, and 3, in *Cardiospermum* they alternate with sepals 1 and 4, 4 and 2, 5 and 3. In *Cardiospermum*, however, the stamens in each whorl, instead of appearing simultaneously, as in *Kœlreuteria*, are developed in a remarkable succession, which I have indicated by the numbers accompanying the stamens in the diagram (fig. 10). In the first place, the two stamens which alternate with sepals 1 and 4, 4 and 2, make their appearance ; next, the two stamens superposed to sepals 1 and 2 ; then, the three stamens superposed to sepals 3, 4, and 5 ; and lastly, the stamen alternate with sepals 3 and 5.[*] Payer has endeavoured to render intelligible the remarkable mode of staminal succession, in this and other analogous cases, by supposing that the irregularity of development, which so frequently manifests itself after the appearance of floral parts, is congenital in such cases ; and it is hardly possible to doubt that his explanation is correct. The anomalous succession is evidently the effect of a disturbing force delaying or arresting, for a time, the appearance of some of the parts, and thus materially affecting the order of staminal evolution. This disturbing force seems to act in quite an arbitrary manner, as it affects different plants in very different manners: thus —to take examples from Payer's work—in *Viola odorata,* the stamens appear successively from before backwards ; in the Resedaceæ, from behind forwards ; while, in *Cardio-*

[*] Organogénie, pp. 150-1.

spermum, the succession may be described, in general terms, as obliquely from side to side. As I have already stated, the two stamens which first appear in *Cardiospermum,* are those which alternate with sepals 1 and 4, 4 and 2. Now, it may be supposed by some to be a formidable, if not fatal, objection to my views, that two of the supposed accessory stamens should appear before the primary ones to which they belong. At first sight, such a mode of appearance seems very improbable ; yet, when we consider the arbitrary manner in which the disturbing force affects the order of staminal succession, we need scarcely be surprised even at such a result. At any rate, it cannot, *à priori,* be said to be more improbable that the appearance of primary staminal lobes should be delayed by a disturbing force until after that of their accessory or lateral lobes, than that the appearance of a normally older staminal whorl should be delayed until after the appearance of some of the parts of a normally younger one, which must be admitted on the ordinary supposition of there being two genuine staminal whorls. I do not think, therefore, that the case of *Cardio-spermum,* although certainly a very strange one, can fairly be urged as invalidating my views.

While engaged in the attempt to determine the morpho-logical constitution of double staminal whorls, I was led, incidentally, to examine the position of the carpels in some of the Malvaceæ.

I have already stated, regarding the Büttneriaceæ, that where, in these plants, there are two staminal whorls, the carpels (as alternating with the younger staminal whorl) are superposed to the petals,—*e.g.* in *Lasiopetalum, Büttneria, Melochia,* &c. ;—but that where the andrœcium is reduced to a single whorl (of simple or compound stamens), the carpels are superposed to the sepals, as in *Hermannia* and the Dombeyeæ.

The researches of Payer leave no doubt that the andrœ-cium of the Malvaceæ consists essentially of a single whorl of five compound stamens, superposed to the petals. With *a staminal* arrangement so closely analogous to what Baillon has described in *Astrapœa*, we may expect to find the carpels, when of, the same number, superposed to the .sepals, as in that plant. Payer, indeed, has stated that the Hermanneæ, Dombeyeæ, and Bombaceæ, in which the carpels are superposed to the sepals, are distinguished thereby from the Malvaceæ, Sterculeæ, and Lasiopetaleæ :* but, as regards the Malvaceæ, I believe it can only have been through an oversight that he has associated them with the Lasiopetaleæ, since he describes the carpels in *Hibiscus* as being superposed to the sepals.† So far as my observations extend, the rule seems without exception, that in 5-carpellary Malvaceæ the carpels are superposed to the sepals, just as in the Dombeyeæ. I have ascertained the occurrence of this arrangement in the following:—(Hibisceæ) *Hibiscus*, *Paritium ;* (Sideæ) *Lagunea.* Moreover, in the Malopeæ, where Payer has shown the gynœcium to consist of five carpellary groups, I find that, in *Malope*, these groups are superposed to the sepals ; so that in this plant we have a similar arrangement to that in *Hibiscus*—only, the five simple carpels of *Hibiscus* are replaced in *Malope* by five carpellary groups (or compound carpels, as they may be termed, being developments evidently of an analogous character with the compound stamens of polyadelphous plants). Payer has described the angles of the pentagon formed by the carpellary groups in *Malope* as superposed to the sepals: but I am quite satisfied that his statement is erroneous. In flowers at or near maturity, there is, sometimes, a slight want of perfect superposition of the carpellary groups to the sepals : but this seems to be never to such an extent as to justify Payer's statement. In the early condition of the ovarian pentagon, the superposition of its *angles* to the petals is quite unmistakeable. The cavity of the staminal tube is five-sided, the sides alternating with the petals ; and the carpellary pentagon, in its origin, is pretty accurately fitted into the bottom of this cavity.‡

The superposition in many Ureneæ of the loculi to the

* Organogénie, pp. 44-5. † Ibid., p. 33.

‡ I have not had an opportunity of examining the position of the carpellary groups in *Kitaibelia ;* and there seems to be considerable confusion in Payer's works, on this point, as these groups are described in the "Organogénie" (p. 34) as alternate with, and in the "Eléments de Botanique" (p. 209) as superposed to, the petals.

petals is only an apparent exception to the rule I have
stated above, as to the position of the carpels in 5-carpellary
Malvaceæ; for the researches of Payer on the organogeny
of *Pavonia* leave no doubt that in this tribe the five loculi
merely represent the fertile members of a circle of ten
carpels, to which the ten styles correspond. In *Pavonia*,
the gynœcium, in its origin, consists of ten carpellary mam-
millæ. Of these, however, only five have loculi developed
in connection with them—every second carpel being, so to
speak, barren. The ten carpels all equally develope styles;
so that in the advanced condition there are five styles prolonged upwards from the loculi, and five continuing the
lines of the dissepiments.* In *Malvaviscus,* Payer describes
the loculi (corresponding to the fertile carpels) as superposed
to the petals.† In *Urena* (*U. americana, U. lobata, U.
scabriuscula, U. sinuata*), I find the same arrangement. In
Pavonia, I have ascertained the remarkable fact that the
loculi are sometimes superposed to the sepals, and sometimes
alternate with them. Thus, in *P. typhalea, P. begoniæfolia*
(Gardner), *P. odorata, P. umbellata,* and *P. zeylanica,* the
loculi are superposed to the sepals; while, in at least one
species, named in the Edinburgh University Herbarium
P. hastata,‡ the loculi are certainly alternate with them, as
in *Malvaviscus* and *Urena.*§

* Organogénie, p. 85, pl. 7.

† Leçons sur les fam. nat. des plantes, p. 281.

‡ I have expressed myself thus guardedly as to the specific name of this
plant, because, by its indefinite stamens, it differs from that to which Baillon
refers as *P. hastata* (Adansonia, II. p. 176), which is described by him as
having only five stamens in the adult state. The Edinburgh plant agrees
with the description of *P. hastata* in Decandolle's "Prodromus" (vol. i. p. 443),
in its lanceolate hastate dentate leaves, axillary unifloral pedicels, and five-
leaved involucre. I cannot say much as to the colour of the petals, except
that a deep red or purplish blotch remains at the base of each. The whole
plant (especially the stem, tho under side of the leaves, the involucre, and the
sepals) is downy, being covered with a short stellate pubescence. The plant
which Payer has examined as *P. hastata* appears to perfect a considerable
number of stamens, as is seen in his representation of the andrœcium " shortly
before blossoming," where there would seem to be 25 stamens, or there-
abouts (Organogénie, pl. 7, fig. 9; with description, p. 38).

§ The position of the fertile carpels seems to offer a much more important
character by which the genus *Pavonia* may possibly be disintegrated, than any
derived from the awned or awnless condition of the fruit, the relative length
of the involucre to the calyx, &c.

In constructing those diagrams which illustrate arrangements in the Malvaceæ (figs. 5, 6, 7, and 8), having found
great difficulty in giving diagrammatic expression to the
staminal groups, I have represented the said groups by
symbols of infinity, which conveniently enough indicate the
indefinite number of the staminal lobes.

Explanation of Plate III.

[Figs. 1–13 are from my own designs. Figs. 14 and 15 are taken from
Payer's "Organogénie," plate xiii. figs. 28 and 32. The diagrams are constructed with the utmost conventional uniformity, being merely intended
to represent the *position* of the parts, not their *form*. In the diagrams, the
posterior aspect of the flower is above, the anterior below, and the stamens
are numbered in the order of their appearance.]

Fig. 1. Arrangement in *Geranium, Malachium,* &c. Younger (accessory) sta
mens external. Carpels alternate with older (primary) stamens.

Fig. 2. Arrangement in *Coriaria, Agrostemma, Cerastium,* &c. Younger stamens internal, probably forming a genuine whorl. Carpels alternate with the younger stamens.

Fig. 3. Arrangement in *Lasiopetalum corylifolium,* and probably in *Büttneria,
Melochia,* &c. Outer and older stamens fertile, and superposed to
the petals ; inner and younger sterile, and alternate with the outer.
Carpels, as in fig. 2, alternate with the younger (sterile) stamens.

Fig. 4. Isostemonous arrangement in *Hermannia.* Fertile stamens, as in the
last, superposed to the petals. The carpels are superposed to the
sepals, apparently replacing the staminodes of the last form.

Fig. 5. Arrangement in *Hibiscus, Paritium,* and *Lagunea.* Same as last form,
except that, instead of five simple stamens, there are five staminal
groups (indicated by symbols of infinity).

Fig. 6. Arrangement in *Malope.* Same as last, except that, instead of five
simple carpels, there are five carpellary groups.

Fig. 7. Arrangement in *Pavonia Typhalea, P. begoniæfolia, P. odorata, P. umbellata,* and *P. zeylanica.* Staminal groups as in figs. 5 and 6, superposed to the petals. Ten carpels ; five fertile, superposed to the
sepals, and five sterile, superposed to the petals. The sterile carpels
are indicated by small circles alternate with the loculi.

Fig. 8. Arrangement in *Malvaviscus, Urena, Pavonia* sp. (*hastata ?*). Same as
last form, except that those carpels which are sterile there, are
fertile here, and *vice versâ.*

Fig. 9. Arrangement in *Polygala,* as described in the text.

Fig. 10. Arrangement in *Cardiospermum,* as described in the text. The sepals
are numbered in the order of their appearance. Sepals 3 and 5
become connate, and the petal (indicated in outline) which alternates
with them aborts. If an oblique line be drawn, as in the diagram,
through sepal 4 and the abortive petal, the parts are arranged
symmetrically on either side of it. This imaginary line, by torsion
of the peduncle, becomes antero-posterior, the abortive petal becoming posterior (superior). See Payer's "Organogénie," p. 168.

Fig. 11. Arrangement in *Kœlreuteria,* as described in the text. As in *Cardio-spermum,* the petal alternating with sepals 3 and 5 aborts.

Fig. 12. Young flower of *Agrostemma Flos-Jovis,* just before the appearance of the carpels. The younger stamens are internal to the older ones. *s,* sepal ; *p,* petal ; *st,* older stamen ; *st',* younger stamen.

Fig. 13. Young flower of *Cerastium triviale,* at same stage as the last. The younger stamens, as in *Agrostemma,* are internal to, or on a higher level than, the older ones. *sa, sp, sl,* anterior, posterior, and lateral sepals ; *p,* petal ; *st,* older stamen ; *st',* younger stamen ; *ax,* convex extremity of the floral axis ; *b, b,* lateral bracts, with secondary floral axes *fl², fl²,* developed in their axils.

Fig. 14. (From Payer). Young flower of *Monsonia ovata.* *s,* sepals ; *p,* petals ; *st°,* older and inner (primary) stamens, superposed to the sepals ; *st°,* younger and outer (accessory) stamens, superposed in pairs to the petals ; *cp,* carpels, superposed, as in *Geranium,* to the petals.

Fig. 15. (From Payer). Andrœcium and pistil from a flower of *Monsonia ovata,* at the time of blossoming. Each of the primary stamens has become connate with the two accessory stamens adjacent to it, one on either side, so that the andrœcium seems now composed of five phalanges superposed to the sepals.

The Classification of Animals based on the Principle of Cephalization. No. I. By JAMES D. DANA. Communicated by the Author.[*]

(*Continued from page* 102).

3. *Classification of Animals.*

1. *Subkingdoms.*—Of the four subkingdoms, first recognised by Cuvier and since by most zoologists, the Vertebrate, Articulate, and Molluscan, are typical, or of the true *animal-type,* and the Radiate is degradational, being *plant-like* in type. Using the terms alphatypic, betatypic, and gammatypic, *simply as a numbering of the grades* of types (see p. 96), their relations are as follows :—

Alphatypic, ...	1. Vertebrates.
Betatypic, ...	2. Articulates.
Gammatypic, ...	3. Mollusks.
Degradational, ...	4. Radiates.

An important dynamical distinction between Mollusks and Articulates has been already suggested by me.

2. *Classes of Vertebrates, Articulates, Mollusks, and Radiates.* —(1.) The classes of *Vertebrates* are four (see page 78),

[*] From the American Journal of Science and Arts, Vol. xxxvi., Nov. 1863.

namely, Mammals, Birds, Reptiles, and Fishes,—three of which are typical, of different grades, parallel with the above.

(2.) The classes of *Articulates* are but three, Insecteans, Crustaceans, and Worms. I have already shown that the three divisions of Insecteans, namely Insects, Spiders, and Myriapods, are distinguished by characteristics analogous to those which separate the divisions of Crustaceans,—Decapods, Tetradecapods, and Entomostracans. The facts on this point are briefly presented on page 97. Insects and Spiders do not, in fact, differ more widely in external form or in structure than Decapods and Tetradecapods.

Insecteans and Birds express in different ways the same type-idea,—that of aerial life, Birds being flying Vertebrates, and Insects flying Articulates; and, in accordance, they are of the same grade of type, both being *betatypic*. This follows, further, from the fact that there are but two grand divisions of Insecteans above the degradational division, that of Worms.

(3.) Among *Mollusks*, there are two well-characterised classes, the *first* including the *ordinary* Mollusks; the *second*, the *Ascidioids*, or the Brachiopods and Ascidians, which are mostly attached species and thus hemiphytoid. Besides these, there are the Bryozoans, which either make a third division under the Ascidioids (Edwards having long since pointed out their relations to the Ascidians); or they constitute a *third* class of Mollusks, characterised by being polyp-like both in external appearance and in being attached, and hence doubly hemiphytoid.

(4.) The *Radiates* are all degradational in their relations to the animal-type. But under the *Radiate-type*, the species of the first two classes are within type-limits, while those of the third are degradational, since almost all are attached and very inferior in type of structure, being the most phytoid of phytoid animals. The grades of structure, as marked in the digestive system, are as follows : (1.) Having approximately normal viscera, as in Echinoderms; (2.) Having, for the digestive system, only a stomach cavity, with vessels, imbedded in the tissues, radiating from it, as in Acalephs; (3.) Having, for the same, no system of viscera or radiating vessels; but only a central stomach surrounded by a cavity more or less divided at its sides by partitions as in Polyps.

The following table presents the relations and the parallelisms of these classes, and of each to the subkingdoms :—

	Subkingdoms.	Vertebrates.	Articulates.	Mollusks.	Radiates.
a.	Vertebrates.	Mammals.			
β.	Articulates.	Birds.	Insecteans.	Ordinary.	Echinoderms.
γ.	Mollusks.	Reptiles.	Crustaceans.	Ascidioids.	Acalephs.
D.	Radiates.	Fishes.	Worms.	Bryozoans?	Polyps.

On the Classification of Animals

Arranging the divisions according to the relations of the groups to the *animal-type*, instead of the special type of each class, the table takes the following form :—

	Subkingdoms.	Vertebrates.	Articulates.	Mollusks.	Radiates.
a.	Vertebrates.	Mammals.	——	——	——
β.	Articulates.	Birds.	Insecteans.	——	——
γ.	Mollusks.	Reptiles.	Crustaceans.	Ordinary.	——
a. D.	——	Fishes.	Worms.	Ascidioids.	——
b. ,,	——	——	——	Bryozoans.	——
c. ,,	Radiates.	——	——	——	Echinoderms.
d. ,,	——	——	——	——	Acalephs.
e. ,,	——	——	——	——	Polyps.

The letters c, *d*, *e*, stand for different grades of phytoid degradational, *b*, hemiphytoid, and *a*, degenerative. The blank interval between Mollusks and Radiates is filled up by the inferior divisions of the higher subkingdoms.

We may now consider the subdivisions under some of the classes; and first, those of Vertebrates.

3. *Higher subdivisions of the class of Mammals.*—The higher subdivisions of the class of Mammals are four in number: Man, Megasthenes, Microsthenes, and Oötocoids, as explained in the preceding volume of the American Journal, Man is shown to stand apart from the Megasthenes on precisely the same characteristic that separates the two highest orders under the classes severally of Insecteans and Crustaceans; for, in passing from Man to the brute Mammals, there is a transfer of the forelimbs from the cephalic to the locomotive series.

Moreover, a study of the Vertebrate skeleton has shown that the forelimbs in the Vertebrate type, as well explained by Professor Owen, are *cephalic appendages*, being normally appendages to the posterior or occipital division of the head. In the Fish, these forelimbs (the pectoral fins) have at any rate an actual *cephalic position* (back of which position they are thrown, by displacement, in other Vertebrates). Now, in Man, they are not only cephalic in normal structural relations, but *cephalic* also in *use*. The transfer of these cephalic organs to the locomotive series, by which the brute structure is made, is a manifest degradation of the type. Man is thus the only Vertebrate in which the Vertebrate-type is expressed in its perfection, and therefore occupies *alone* the sublime summit of the system of life.

Three of the orders of Mammals, namely, Man, Megasthenes, and Microsthenes, are typical of different grades, and one, Oötocoids, is semidegradational.

The Oötocoids may be divided into three groups—a *megas-*

thenic, a *microsthenic*, and a *degradational;* the *first* to include
the genera Phalangista, Dasyurus, Macropus, Diprotodon, &c.;
the *second*, Perameles, Didelphys, Phascolomys, Echidna, &c., or
Marsupial Insectivores, Rodents, and Edentates ; the *third*, Ornithorhynchus.

The following table presents to view the subdivisions of Mammals and its orders. Under Oötocoids, the relations of the two
higher groups are indicated by the above adjectives, without giving them special names :—

	Mammals.	Megasthenes.	Microsthenes.	Oötocoids.
a.	Man.	Quadrumanes.	Chiropters.	————
β.	Megasthenes.	Carnivores.	Insectivores.	Megasthenic.
γ.	Microsthenes.	Herbivores.	Rodents.	Microsthenic.
D.	Oötocoids.	Mutilates.	Edentates.	Ornithorhynchs.

4. *Higher subdivisions of the classes of Birds, Reptiles, and
Fishes.*—(1.) In the class of *Birds*, there are three grand divisions : the first two, as recognised by Bonaparte, are the *Altrices*
(Rapacious birds, Perchers, &c., and other birds that feed their
young until they can fly), and the *Præcoces* (or the Gallinæ, An-
seres, Ostriches, &c., which feed themselves as soon as hatched).
The third includes the Reptilian Birds or Erpetoids (p. 77).
The terms *Pterosthenics* and *Podosthenics* apply equally well
with *Altrices* and *Præcoces* to the two higher divisions of Birds,
as explained on page 83, and have an advantage in their direct
dynamical signification.

The type of ordinary Birds (or Pterosthenics and Podosthenics)
is stated on page 95 to be essentially *limitate*, like that of Insects, while the type of Erpetoids is *multiplicate*, like that of
Myriapods or of ordinary Reptiles ; so that the relation of
Erpetoids to the higher division of Birds is in an important respect analogous to that of Myriapods to the higher division of
Insecteans.

(2.) In the classification of *Reptiles* there are three prominent
types of structure recognised by Erpetologists; (1.) That of the Chelonians ; (2.) That of the Lacertoids (including Saurians, Lizards,
Snakes) ; and (3.) The degradational or hemitypic one of Amphibians. It is now well known that Snakes and Lizards are
alike in type of structure, the two groups graduating almost insensibly into one another, some species ranked as Lizards being
footless like the Snakes. The Snakes constitute the degradational group under the Lacertoids. The Amphibians, constituting the third order, are on the same level with the Erpetoid
Birds and the Oötocoid Mammals, as presented in the following
table.

The three orders of Reptiles—Chelonians, Lacertoids, and Amphibians—make a parallel series with the three lower classes of Vertebrates; the Chelonians representing the Birds, to which they approximate in some points, besides being betatypic like them; the Amphibians representing the Fishes, with a still closer approximation between the two; while the Lacertoids are the typical Reptiles. The Chelonians might be viewed as *hemitypic* Reptiles; not *hypotypic* like the Amphibians, but *hypertypic*, like the Selachians and Ganoids among Fishes.

(3.) *Fishes* are all degradational species in their relations to the animal-type. The two higher groups, or those of Selachians and Ganoids, as already explained (p. 96), are *hypertypic*. The third, including Teliosts, is *typical* if viewed with reference to the Fish-type. Below these, the Dermopters or Myzonts (including Amphioxus, Myxine, &c.), constitute an inferior *hypotypic* or degradational group,—that is degradational in its relations to typical Fishes (p. 94). Thus *typical* Fishes are gammatypic in their relations to other Vertebrates, while the alphatypic and betatypic groups are *hypertypic* orders.

The following table exhibits the relations of the orders in the classes of Birds, Reptiles, and Fishes; and, for comparison, those of Mammals are added :—

	Mammals.	Birds.	Reptiles.	Fishes.
Alphatypic,	Man.	——	· ——	Salachians.
Betatypic,	Megasthenes.	Altrices, or Pterosthenics.	Chelonians.	Ganoids.
Gammatypic,	Microsthenes.	Præcoces, or Podosthenics.	Lacertoids.	Teliosts.
Hemitypic or Degradational,	Oötocoids,	Erpetoids.	Amphibians.	Dermopters.

We pass now to Articulates.

5. *Subdivisions of the classes Insecteans, Crustaceans, and Worms into Orders.*—(1.) The higher subdivisions in each of the classes, *Insecteans* and *Crustaceans*, are three in number, none existing above the betatypic grade, which is that of Articulates among the subkingdoms, and of Insecteans among Articulates.

(2.) *Worms* are of four types of structure. First, *Annelids*, or *typical* Worms, including the Branchiates, Abranchiates, and Nematoids—the last the degradational group, and showing this in the obsolete body-articulations and some internal characters.— Second, *Bdelloids* or *Molluscoid* Worms, including the Hirudines or Leeches, Planarians and Trematodes; characterised by obsolescent or obsolete body-articulations, and by often wanting the nervous ganglia, excepting the anterior; by usually a Gasteropod-like breadth and aspect, an *amplificate* feature; by being in general *urosthenic*, even the highest having a caudal disk for attachment; and in an up-and-down movement of the body in locomotion, *Mollusk-like*, instead of the worm-like lateral movement of the Annelids. The fact of this mode of movement has been recently made known to the writer by Dr Wm. C. Minor, as a distinctive feature of the Bdelloids. Quatrefages remarks that the Planarians and Trematodes may well be regarded degraded forms of the Hirudines, and the three tribes are arranged in one group by Burmeister.—Third, *Gephyreans* (of de Quatrefages), or *Holothurioid* (*Radiate*-like) Worms, including the genera, Echiurus, Sipuncula, &c.*—Fourth, *Cestideans*, or *Protozoic* Worms, including the Cestoids, in which there is no normal digestive system, and the segments are independently self-nutrient.†

The orders of these classes of Articulates are the following:—

	Insecteans.	Crustaceans.	Worms.
Alphatypic,			
Betatypic,	Insects.	Decapods.	Annelids.
Gammatypic,	Spiders.	Tetradecapods.	Bdelloids.
a Degradational,	Myriapods.	Entomostracans.	Gephyreans.
b. ,,			Cestideans.

6. *Subdivisions of the Orders of Insecteans and Crustaceans into Tribes.*—(1.) The orders of *Insecteans* have each three divisions, excepting that of Myriapods in which but two have been recognised. The three of Insects are indicated on pages 83, 98. The fact that Insects are, in type-idea, *flying* Articulates, gives special importance to the wings in classification. The *first* order includes the *Prosthenics*, in which the anterior wings are flying wings, as the Hymenopters, Dipters, Neuropters, Lepidopters, and Homopters. The *second* consists of the *Metasthenics* or *Elytropters*, in which the anterior wings are not used in flying, or but little so, as the Coleopters, Strepsipters, Orthopters, and Hemipters. The Hemipters and Homopters, united in one tribe by most entomologists, are hence profoundly distinct. The *third* tribe, or *Apters*, embraces the Lepismids and Podurellids; the remaining Apterous insects being distributed among the other

* The Holothurioid characteristics are well exhibited by de Quatrefages in Part ii. p. 248 and beyond, of "Recherches Anatomiques et Zoologiques faites pendant un voyage sur les Côtes de la Sicile," &c., in 8 vols. or parts, the second by de Quatrefages. Paris.

† The *Acanthocephali*, according to van Beneden and Blanchard, are Nematoids (with which they agree in form and general structure), although without a digestive system. Blanchard states that there is reason for believing that the digestive system becomes atrophied with the growth of the animal, and mentions that cases of like atrophy occur even in species of *Gordius* and *Nemertes*.

groups, as suggested by different entomologists. The Lepismæ show their degradational character in their larval forms and in other approximations to the Myriapods, and the Podurellids appear to be still inferior in having the abdomen elliptic in some segments.

(2.) The Orders of *Spiders* suggested by the principles of cephalization are in precise parallelism with those of the Decapod and Tetradecapod Crustaceans. They are, first, *Araneoids*, including all the *Pulmonates*, except the Pedipalps; second, *Scorpionoids*, or the Pedipalps from among the Pulmonates, and the Chelifer group from among the Trachearians; third, *Acaroids*.

The Araneoids or *Brachyural* Spiders; the Scorpionoids, *Macrural;* while the Acaroids are *degradational.* The last show their degradational character in having no division between the abdomen and cephalothorax; so that, while Insects have the body in *three* parts, head, thorax, and abdomen, and ordinary Spiders in *two*, cephalothorax and abdomen, the Acaroids have it *undivided* (page 86). Thus, one of the most prominent characteristics marking the descent from Insects to Spiders becomes the characteristic of a further descent among Spiders themselves— illustrating a common principle with regard to such subdivisions. The propriety of making the Acaroids a distinct group appears therefore to be well sustained.

The usual subdivision of Spiders into Pulmonates and Trachearians depends on *internal* characters, which is not the case with any other subdivisions in the table beyond. Moreover, these names, though *seeming* to mean much, are not based on any *functional* difference between the groups. Spiders have many relations to Crustaceans; and it is natural that the subdivisions in both should depend on the same methods of cephalization, the amplificative and analytic (p. 98).

(3.) The two orders of *Myriapods* are examples, one of case *a*, the other of case *b*, under multiplicative decephalization (p. 85).

The close relations between Isopods and the higher Myriapods, suggest that they are of like grade under their respective types, that is, betatypic.

(4.) *a*. Under *Decapod Crustaceans*, the subdivisions are *three*, as already remarked upon by the author.[*]

The Anomurans are only degradational Brachyurans, and do not represent an independent type of structure. The Schizopods, similarly, are degradational Macrurans, with which they should be united. The *third* type is that of the *Gastrurans*, which are peculiar among Decapods, in having the viscera extending into the abdomen, one of the marked degradational features of the type.

---

[*] *See also* Amer. Jour. of Science and Art, vol. xxv. [2], pp. 387, 388.

They are the Stomapods of Latreille; but this author, in his last edition, made the group, in connection with the Schizopods, co-ordinate with that of Decapods. Being co-ordinate with Brachyurans and Macrurans, the change of name is necessary.

b. The *Tetradecapods* include two divisions precisely parallel with the first two of the Decapods, the first literally *brachyural*, the second *macrural.* (See p. 97 of this volume.) The *Anisopods* of the writer, are degradational Isopods, just as the Anomurans are degradational Brachyurans. The Lemodipods (Caprellids, &c.) are only degradational Amphipods, the structure of the two being essentially the same in type. Hence, neither the Lemodipods nor the Anisopods are an independent type corresponding to a *third* division.

The *third* subdivision probably is made up of *Trilobites,* although these are generally regarded as Entomostracans. One of the most prominent marks distinguishing Entomostracans from Tetradecapods is the absence of a series of abdominal appendages. It is highly improbable that the large abdominal (or caudal) plate of an Asaphus, or the many-jointed abdomen of a Paradoxides, Calymene, &c., should have been without foliaceous appendages below; and if these appendages were present, the species were essentially Tetradecapods, although degradational in the excessive number of body-segments.

c. Entomostracans (or Colopods, as they are more appropriately styled) embrace four orders. First, *Carcinoids* (as named by Latreille) consisting of the Cyclops group (Copepods of Edwards), whose species have a strong Macrural or shrimp-like habit; to which should be added the Caligoids, (Cormostomes of the writer, Siphonostomes of others), since they are essentially identical in type of structure with the Cyclopoids, as may be seen on comparing *Sapphirina* of the latter with *Caligus.*—Second, *Ostracoids* (or the Daphnia, Cypris, and Limnadia groups), which have, besides a bivalve carapax more or less complete, a much more elliptic abdomen than the Carcinoids, it being short, incurved, and without a lamellar terminal joint or terminal appendages.—Third, *Limuloids,* which have the abdomen still more elliptic, it being reduced to a mere spine, or nearly obsolete, and which have the mouth-organs all perfect feet and the only locomotive organs. (The joint across the carapax of the Limulus corresponds in position to a suture or imperfect articulation in the carapax of the Caligi, &c.)—Fourth, the *Rotifers,* a low Protozoic grade of degradation, in which all members are wanting, and locomotion is performed by cilia. The Phyllopods are distributed between the first two divisions.

The Rotifers are sometimes arranged under Worms. If they are degradational species of a limitate type, they are Crustaceans;

and if of a multiplicate, they are Worms. The very small number of segments present, when any are distinct, the character of the dentate mandibles (for mandibles are *not* found in the inferior subdivisions of Worms), and the resemblance in the form of some species to Daphniæ and other Entomostracans, sustain the view that they are Crustacean.

The Cirripeds appear to be only attached, amplificate Ostracoids. (See pages 84, 85.)

The subdivisions of the orders of Insecteans and Crustaceans are then the following :—

Insects.	Spiders.	Myriapods.	Decapoda.	Tetradecaps.	Entomostr.
α.					
β. Prosthenics or Ctenopters.	Araneoids.	Chilopods.	Brachyurans.	Isopods.	Carcinoids.
γ. Metasthenics or Elytropters.	Scorpionoids.	Diplopods.	Macrurans.	Amphipods.	Ostracoids.
a. D. Apters.	Acaroids.	?	Gastrurans.	Trilobites. ?	Limuloids.
b. D.					Rotifers.

7. *Subdivisions of the orders of the class of Worms.*—On the true method of grouping the typical (Branchiate and Abranchiate) Annelids, I here make no suggestions. The Cystics are there included with the Cestoids. If any of the *simple* Cystics are really adults, they may possibly make a second subdivision of the Cestideans.

8. *Subdivisions of the classes of Mollusks.*—The ordinary Mollusks include three orders, as usually given: (1.) *Cephalopods,* (2.) *Cephalates,* and (3.) *Acephals ;* of which the first two correspond to different grades of typical Mollusks, and the last is degradational in its relations to the type, the species being imperfect in the senses and means of locomotion.

The Ascidioid Mollusks comprise (1.) *Brachiopods,* and (2.) *Ascidians,* with perhaps the *Bryozoans* as the third order. If the last, however, be made a third *class,* as already suggested (though with hesitation), there is no third order, unless the inferior of the compound Ascidians, having water-apertures to a *group* of individuals instead of to each one, and the mouth-opening of each usually *radiated* (the number of rays *six*), be regarded as the third. This would make the orders, (1) *Brachiopods ;* (2) *Ascidians ;* (3) *Incrustates ;* the first two typical, the last degradational and strikingly hemiphytoid.

4. Conclusions.

The preceding review of zoological classification appears to sustain the following general conclusions.

1. *Number and typical relations of the subdivisions of groups.*

I. The number of subkingdoms, classes, orders, and tribes, in the system of animal life is either *four* or *three*, that is, the division in each case is either *quaternate* or *ternate*.

II. The lowest of the subdivisions in each group is a degradational or semidegradational subdivision, or *hypotypic*.

III. The quaternate division is confined to *six* cases (excepting two or three among inferior types in which there are *two* degradational subdivisions) : 1, the number of subkingdoms ; 2, the number of classes under Vertebrates, the highest of the subkingdoms ; 3, 4, the number of orders under Mammals and Fishes, the highest and lowest classes of Vertebrates ; 5, 6, the numbers of tribes under two of the orders of Mammals.

IV. In *three* only of the six cases of *quaternate* division are the three higher subdivisions all *true typical*, namely : 1, in the division of the animal kingdom into subkingdoms ; 2, of the Vertebrates into classes ; 3, of Mammals into orders. In the last we reach Man. As Man alone is archetypic in the class of Mammals (p. 96), so the Mammal-type is archetypic among Vertebrates, and the Vertebrate-type among the subkingdoms.

b. Below this archetypic level, in the orders of Mammals, the number of *true typical* subdivisions is but *two*—and these are the *betatypic* and *gammatypic;* for the first or alphatypic subdivision in both Megasthenes and Microsthenes, as explained on page 96, is *hypertypic*, and not true typical.

c. Again, of the *four* orders of Fishes only *one* is typical, the *two highest* being *hypertypic* (p. 96).

V. In the rest of the animal kingdom, the number of *true typical* groups, in the classes, orders, and tribes that have been reviewed, is either *two*, the *betatypic* and *gammatypic*, or *one*, the *gammatypic* alone.

2. *Lines of gradation.*—Lines of gradation between groups are lines of convergence or approximation through intermediate species. Before mentioning under this head the deductions from the preceding classification (or VIII. and IX. beyond), two general principles (VI. and VII.), having an important bearing upon them, are here introduced.

VI. The approximations between two groups usually take place, as has been frequently observed, through their *lower limits*, or most inferior species, that is, between the degradational subdivision of the inferior as well as of the superior group.—For example, plants and animals approximate only in their simplest species, the Protozoans and Protophytes ; Birds and Quadrupeds most nearly in the Ornithorhynchus or Duckbill—which, at the same time that it is the lowest of Mammals, is related to a very inferior type of Birds, the Ducks ; Quadrumanes and inferior Mammals through the Lemurs of the former and the Bats and

Insectivores of the Microsthenes, and not through the higher Carnivores or even any of the Megasthenes.

The classes of Reptiles and Fishes may appear to be an exception. But the *Perennibranchs* (or the species with permanent gills) among Amphibians, if referred to the type of Fishes, and especially to the Ganoid type, would rank low, as is obvious from their exsert and loosely-hung gills without gill-covers, the absence of scales, and the general inferiority in all structural arrangements. The Ganocephs, known only as fossils, and generally regarded as Perennibranch Amphibians, have, it is true, a higher . grade of organisation, both as regards gills and scales, being allied in these respects to the highest of Ganoids. And this fact, in view of the above canon, sustains the opinion of Agassiz, that the Ganocephs (or Archegosaurs) are actually Ganoids—having a Reptilian feature in the partial elongation of the limbs, but in little that is fundamental in the structure beyond what belongs essentially to the Ganoid-type.

VII. The lines of gradation between classes, orders, and tribes, are only approximating, not connecting, lines, there being often wide blanks of the most fundamental character. The Ornithorhynchus, although Duck-like in some points, leaves still a very wide unfilled gap between the Mammal and Bird, and the Marsupials a still wider. The species are fundamentally Mammalian, and Bird-like only in points of secondary importance. In a similar manner, there are long blanks between the Oötocoids and higher Mammals; between Myriapods and either Insects or Spiders; between Reptiles and Mammals. The intermediate groups belong decidedly to one or the other of the two approximating groups, and are never strictly intermediate.

VIII. Under any *class, order*, or *tribe*, the lines of gradation run in most cases between the *degradational* subdivision and severally the *gammatypic* and *betatypic* subdivisions, and far less clearly, or not at all, between the gammatypic and betatypic themselves; that is, between D and γ, and D and β, rather than β and γ. For example, in the class of Mammals, the lines run between Oötocoids and either Megasthenes or Microsthenes, and not distinctly between Megasthenes and Microsthenes; in Insecteans, between Myriapods and either Insects or Spiders, and not distinctly between Insects and Spiders; In Crustaceans, between Entomostracans and either Decapods or Tetradecapods, and not distinctly between Decapods and Tetradecapods, &c. There are exceptions to the canon; and still it is a general truth.

IX. Under any *class* or *order*, the line of gradation between the *degradational* and the *betatypic* subdivision (or D and β) is often more distinct than that between the *degradational* and *gammatypic* (or D and γ), although the gammatypic is nearer in grade to the degradational.—Thus, the line between Myriapods and Insects is more distinct than that between Myriapods and Spiders; or that between Entomostracans and Decapods, than that between Entomostracans and Tetradecapods.

There is an exception in the class of Mammals: the Oötocoids seem to graduate towards both Microsthenes and Megasthenes with nearly equal distinctness.

3. *Co-ordinate grades and distinctions in Classification.*

X. The co-ordinate value of subdivisions in the system of classification is brought out to view in the parallel columns of the preceding tables, and evidence is thence afforded as to what groups are rightly designated classes, orders, &c.

a. We thus learn that the subdivisions of the class of Mammals —Man, Megasthenes, Microsthenes—are properly *orders*, if we so call the subdivisions Decapods and Tetradecapods under Crustaceans, or Insects and Spiders under Insecteans.

b. Again, we have a solution of the question whether in each of the classes, Mammals, Birds, and Reptiles, the *hemitypic* division, as so-called on page 76, is a *subclass* co-ordinate with the *typical* division of the same, *or* whether it is an order co-ordinate with the three higher subdivisions of the class. The question appears to be decided (contrary to former views of the writer), that it is correctly made an *order*. These hemitypic divisions actually correspond severally to the degradational division in other columns of the different tables; and, therefore, if in the case of other classes as those of Crustaceans, Insecteans, &c., they are *orders*, so are they in the three classes of Vertebrates mentioned. They have also a relation to the *hemitypic* divisions among Fishes, which are the first and second *orders* of the class.

XI. In an *inferior* or *degradational* group, the distinctions of the subdivisions included are generally much more strongly and obviously exhibited in the structure than among *typical* groups. Thus, the orders of Fishes are based on characters that have nearly a class-value among the higher Vertebrates. In the same manner, Amphibians, or hemitypic Reptiles, differ from true Reptiles more obviously than Oötocoids, or hemitypic Mammals, differ from other Mammals. So, the distinctions among the groups of Crustaceans are very wide compared with those among Insects; and those among degradational Crustaceans far wider than those among the typical subdivisions. The relative force of the life-systems is, in all probability, as great between Oötocoids and typical Mammals as between Amphibians and typical Reptiles, although so unequally expressed in the structure of the high or concentrated groups and the low or lax groups of species. Overlooking this principle has often led authors to allow too great importance to the structural differences among inferior or degradational groups.

XII. Under any class, order, tribe, the *typical* groups are often represented more or less clearly among the subdivisions of the *degradational.* Hence characteristics which separate the typical groups frequently separate only subordinate divisions under an inferior or degradational group. Examples occur in the class of Fishes under Vertebrates, in whose subdivisions the other classes of Vertebrates are partly represented; in the order of Oötocoids under Mammals, which has its megasthenic and microsthenic subdivisions; under Worms, &c.

4. *Distinction between Animals and Plants.*

XIII. This subject well illustrates a fundamental distinction between animals and plants.

a. An animal, as has been stated on page 94, has *fore-and-aft*, or antero-posterior, polarity; that is, it has a fore-extremity and a hind-extremity which have that degree of oppositeness that characterizes polarity.

b. With this fore-and-aft polarity there is also *dorso-ventral* polarity.

c. The dorso-ventral and antero-posterior axes are at *right angles* to one another. In Invertebrates and a large part of Vertebrates the antero-posterior axis is horizontal and the dorso-ventral vertical; and only in Man, the prince of Mammals, is the former vertical and the latter horizontal.

d. An animal, again, has not only oppositeness between the fore-extremity and hind-extremity, but also a *head*, the seat of the senses and mouth, situated at the fore-extremity and constituting this extremity.

e. In addition, the typical animal is *forward moving*.

But in animals of the inferior type of *Radiates*, while there is an anterior and a posterior side, and also, in most species, forward motion, the mouth-aperture—which indicates the *primary centre* in an animal (p. 82)—is not placed at one extremity, but is more or less nearly *central;* and almost precisely central in the symmetrical (and therefore inferior) Radiates. The mouth-extremity and the opposite are at *the poles of the dorso-ventral axis*, and not at those of the antero-posterior; that is, they are at the extremity of the axis which in the inferior animals is normally *vertical.* This is true even in a Holothuria, the mouth of which is not at the *anterior* extremity, but is central, or nearly so, as in an Echinus. A Limulus has been referred to on page 90 as showing an approximation, under the true animal type, to this same central position of the mouth.

We pass now to *Plants*. The plant, in contrast with the fore-and-aft animal, is an *up-and-down* structure, having up-and-down polarity. The axis is *vertical* like the dorso-ventral in the lower animals, to which it is strictly analogous, as is shown from a comparison with Radiates,—Radiates and Plants being alike in type of structure. The primary centre of force is central, in the same sense, in the regular flower and the symmetrical Radiate.

Thus, the structures under the animal-type and plant-type are based on two distinct axial directions, one at right angles to the other: in the *animal-type* the antero-posterior axis being the dominant one, while the two co-exist; and in the *plant-type* the axis at right angles to this being the only one.

In the above way (as well as in its non-percipient nature), the plant exhibits complete decephalization—a condition to which the Radiate only approximates, as it has generally, if not always, an anterior and posterior side, besides other animal characteristics.

Synopsis of Canadian Ferns and Filicoid Plants. By GEORGE LAWSON, Ph.D., LL.D., Professor of Chemistry and Natural History in Dalhousie College, Halifax, Nova Scotia.

(*Continued from the January Number.*)

SCOLOPENDRIUM.

S. vulgare, Smith.—Fronds (in tufts) strap-shaped, with a cordate base, undivided, margin entire, stipe scaly. *Scolopendrium vulgare*, J. E. Smith, Bab., J. Sm., Moore, &c. *S. officinarum*, Swartz, Schkr., Gray Man., p. 593; Torr. Fl. N. Y. ii. p. 490. *S. Phyllitis*, Roth. *S. officinale*, DC. *S. Lingua*, Cavanilles. *Asplenium Scolopendrium*, Linn. Sp. Plantarum, &c. *A. elongatum*, Salisb. *Blechnum linguifolium*, Stokes. *Phyllitis Scolopendrium*, Newman.—Owen Sound, Georgian Bay, Lake Huron, on soft springy ground, amongst large stones, growing in tufts, abundant, 1861, Robert Bell, junior, C.E. This interesting addition to our list of Canadian ferns has been collected in the same place by the Rev. Prof. William Hincks, F.L.S. Mr Bell's specimens agree, in every respect, with the typical European form of the species, which is exceedingly variable. Only one station was previously known for this fern in all North America, viz., limestone rocks along Chittenango Creek, near the Falls, respecting which Professor Torrey observed :— " This fern is undoubtedly indigenous in the locality here given, which is the only place where it has hitherto been found in North America." It was first detected by Pursh, who found it in shady woods, among loose rocks in the western parts of New York, near Onondago, on the plantations of J. Geddis, Esq. This species (he said) I have seen in no other place but that here mentioned, neither have I had any information of its having been found in any other part of North America. (*Pursh.*) Nuttall states that he found it in the western part of the state, without giving the locality ; but according to Dr Pickering, the specimens of Mr Nuttall, in the herbarium of the Academy of Sciences in Philadelphia, are marked, "Near Canandaigua, at Geddis's farm, in a shady wood, with *Taxus canadensis*," Torrey Fl. N. Y. ii. p. 490. This fern occurs throughout Europe, and also in Northern Asia. Mr Moore considers the Mexican *S. Lindeni* as a mere variety of this species. In Europe there are many remarkable varieties, of which Mr Moore has figured and described more than fifty that occur in Britain. The great beauty and remarkable character of many of these render them very suitable for cultivation. None of the abnormal forms have as yet been found in America, probably merely because they have not been looked for.

CAMPTOSORUS.

C. rhizophyllus, Presl.—Frond lanceolate, broad and hastate, or cordate at base, attenuated towards the tip, which strikes root and gives rise to a new plant; hence this fern is called the Walking Leaf; fronds evergreen. *Camptosorus rhizophyllus*, Link, Presl, A. Gray, Eaton, Hooker. *Asplenium rhizophyllum*, Linn. in part (Linnæus's name included *Fadyenia prolifera*, a totally different plant), Michaux, Pursh Fl. Am. Sept. ii. p. 666, Bigelow, Torrey, Beck, Darlington, Lowe's Ferns, vol. v. pl. 14 *a*. *Antigramma rhizophylla*, J. Sm., Torrey Fl. N. Y. ii. p. 494. *Camptosorus rumicifolius*, Link.—On the flat perpendicular face of a rock in the woods, on the Spike's Corners side of the mills at High Falls, township of Portland, C.W., July 1862. In a rocky wood, a mile north-west from the Oxford station of the Ottawa and Prescott Railway, upon a rock slightly covered with mould, B. Billings, jr.; mountain side west from Hamilton, also at Ancaster and at Lake Medad, Judge Logie; Wolfe Island, E. J. Fox; not rare about Owen Sound, Rev. Prof. W. Hincks, F.L.S.; Montreal Mountain, M. L'Abbé Provancher; rather northern in its range in North America, but not common anywhere in Canada. This curious fern has been long in cultivation in the botanic gardens of Europe.

LASTREA.

L. dilatata, Presl.—Fronds spreading, broadly lanceolate, rather pale but vivid green, bipinnate; the pinnules pinnate or pinnatifid with pointed lobes; on the lower pinnæ, the posterior pinnules are longer than the anterior ones; stipe with rather distant pale unicolorous scales; sori small. This description refers only to the commonest form in Canada. It is a very variable species. *Aspidium spinulosum*, Gray. —Abundant in the woods about Kingston, as Collins's Bay, &c., Smith's Falls, Odessa, woods near the Falls of Niagara, Hinchinbrook, Gananoque Lakes, Farmersville, Hardwood Creek, Delta, Upper Rideau Lake, Newboro-on-the-Rideau, Longpoint; Mouth of the Awaganissis Brook, Gulf of St Lawrence, Goulais River, also Grand Island, and at Ke-we-naw Point, Lake Superior, R. Bell, jr.; Ramsay, Rev. J. K. M'Morine, *M.A.*; Prescott, very common, B. Billings, jr.; St John's, St Valentine, and Beloeil, P. W. Maclagan, M.D.; Belleville, very common, J. Macoun; St Joy Woods, W. S. M. D'Urban; Daniel's Harbour, Newfoundland, James Richardson (a peculiar form); Pêche River, Chelsea and Cantley, Hull, D. M'Gillivray, M.D. Of varieties referable to var. *Boottii*, Gray, var. *dumetorum*, Gray, or others, differing from the common (which, however, is perhaps not the typical) form, I have seen specimens from, or obtained information of their having been collected in, the following localities:—Malden, Brighton, Point Rich, Newfoundland, Hamilton's Farm, Murray, Hamilton, &c. These varieties still require careful study, with a view to their identification with European forms, which are now well understood.

β. *tanacetifolia.*—Frond large and very broad, triangular, tripinnate, with the pinnules pinnatifid or deeply incised, lobed. *P. tanacetifolium*, DC. ?—Pointe des Morts, Gaspé, John Bell, B.A. Mr Bell's specimen seems to agree well with Mr Moore's description of var. *tanacetifolia*. The typical *L. dilatata*, with dark-centred scales, so common in Scotland, I have not yet seen growing in the Canadian woods; but a fragment, the upper portion of a frond, from Point Rich, Newfoundland, James Richardson, looks like it.

L. marginalis, J. Smith.—Frond ovate-oblong, a foot, more or less, in length, bipinnate, pale green, somewhat coriaceous, lasting the winter; pinnæ linear-lanceolate, broad at base; pinnules oblong, very obtuse, obsoletely incised; sori marginal; stipe of a pale cinnamon colour when old, with large thin pale scales profuse below. *L. marginalis*, J. Sm., *Aspidium marginale*, Swartz, Pursh, Bigelow, Beck, Darlington, Gray, Eaton, Lowe's Ferns, vol. vi., pl. 6 (a bad figure), Torrey Fl. N. Y. ii. p. 495. *Polypodium marginale*, Linn. *Nephrodium marginale*, Michaux.—This species is as common in the Canadian woods as *Lastrea Filix-mas* is in those of Britain; woods around Kingston, abundant; near Odessa; Newboro-on-the Rideau; along the course of the Gananoque River and lakes, in various places; very fine at Marble Rock; Farmersville; Hardwood Creek; Valley of the Trent, found on the great boulder, &c.; on Judge Malloch's farm and elsewhere about Brockville; on limestone rocks above the Rapids at Shaw's Mill, Lakefield, North Douro, Mrs Traill; Sulphur Spring, Hamilton, Judge Logie; Cedar Island, A. T. Drummond, jr., B.A.; Smith's Falls, and Chippawa, P. W. Maclagan, M.D.; Ramsay, Rev. J. K. M'Morine, M.A.; Prescott, common, B. Billings, jr.; Belleville, in rich low moist woods, common, J. Macoun; above Blacklead Falls, W. S. M. D'Urban; Gatineau Mills, D. M'Gillivray, M.D.; Cap Tourmente, M. L'Abbé Provancher; Harrington, J. Bell, B.A.; London, W. Saunders. This is exclusively an American fern. It varies in size and appearance; in some specimens the pinnæ are wide apart, their divisions small and narrow; in others, the pinnæ overlap each other, and their divisions are broad and leafy, also overlapping, and in such forms they are usually toothed into rounded lobes. Mr Macoun sends a form from Belleville, more deeply serrate than usual.

β. *Traillæ.*—Fronds very large (3½ feet long), bipinnate, all the pinnules pinnatifid.—Lakefield, North Douro, Mrs Traill. This is a very handsome variety, and would form an attractive plant in cultivation.

It has the same relation to the type of *L. marginalis* which *incisa* (*erosa*) has to typical *Filix-mas.*

Lastrea Filix-mas is erroneously referred to in some American works on Materia Medica as a common North American and Canadian fern. It has recently, however, been found on the Rocky Mountains by Dr Parry. Professor Gray says that Dr Parry's specimens are apparently identical with the European plant. Nothing like it occurs in Canada, so far as I can ascertain. Varieties of *L. marginalis* have been sent to me under the name of *Filix-mas.*

L. cristata, Preal.—Fronds erect, rigid, linear-oblong in outline, vivid green, pinnate or slightly bipinnate; pinnæ triangular-lanceolate; pinnules large, oblong, approximate, decurrent; sori large, in a single series on each side of, and near to, the vein; stipe with few pale scales. *Lastrea cristata*, Preal, Moore, &c. *Polypodium cristatum*, Linn. *Aspidium cristatum*, Swartz, Willd., Pursh, E. B., Beck, Torrey Fl. N. Y., ii. p. 496, Gray. *Aspidium cristatum*, β. *lancastriense*, Torrey; *A. lancastriense*, Spreng., Bigelow, Beck, Darlington, Hooker.—Woods around Kingston; near the Pêche River, Gatineau, a tributary of the Ottawa, D. M'Gillivray, M.D.; Three Rivers, St John's, and Chippawa, P. W. Maclagan, M.D.; Sproule's Swamp, east from Belleville (a cedar swamp), not common, J. Macoun; Ramsay, Rev. J. K. M'Morine, M.A.; Prescott, common, B. Billings, jr.; Lake of Three Mountains, W. S. M. D'Urban; Silver Brook, Gaspé, John Bell, B.A.; St Ferreol, M. l'Abbé Provancher; L'Orignal, J. Bell; London, W. Saunders.

L. Goldieana, J. Smith.—Frond very large (3 or 4 feet or more in length), dark green, bipinnate; pinnæ 6 to 8 inches long, narrow, linear-lanceolate, not much attenuated towards the tips; pinnules (12–20 pairs), linear-oblong, approximate, uniformly curved forwards, scythe-shaped, sometimes with an extra lobe at base; sori small, near the midrib; stipe with pale shaggy scales above and larger dark-centred ones below; our largest Canadian fern, usually barren. *Lastrea Goldieana*, J. Smith. *Aspidium Goldieanum*, Hooker, Edin. New Phil. Jour. vi. p. 333, and Fl. Bor. Am., ii. p. 260, Gray. *Nephrodium Goldieanum*, Hook. and Grev. *Aspidium Filix-mas*, Pursh, not of Willd., &c.—Farmersville, in woods near the village, abundant and very fine, forming immense tufts; near Hamilton's Farm and De Salaberry, town line, W. S. M. D'Urban; Beloeil Mountain, Montreal and Malden, P. W. Maclagan, M.D.; Belleville Woods, near Castleton; woods below Heely's Falls, west side, and in Simon Terrill's Woods, Brighton, J. Macoun. Augusta, Robert Jardine, B.A.; about Montreal, Mr Goldie in Hook. Fl. Bor. Am. London, W. Saunders. This fine fern was appropriately named by Sir William Hooker in honour of its discoverer, a successful investigator of Canadian botany, now resident at Paris, C.W. The species belongs exclusively to the American Continent. In Canada we have two sub-varieties :—

α. *serrata*, in which the divisions of the pinnæ are coarsely serrate. Montreal.

β. *integerrima*, in which the divisions of the pinnæ are almost or quite entire. Farmersville.

L. fragrans, Moore.—Frond 8 to 12 inches long, coriaceous, bipinnate, pinnæ triangular, of few (4 or 5 pairs) of pinnules, which are crowded and covered beneath by the large rusty membranous indusia, which conceal the sori. Rachis with profuse, large, palish scales, especially near the base. *Aspidium fragrans*, Swartz, A. Gray.—Rocks, Penokee Iron Ridge, Lake Superior, Mr Lapham, and north-west—Professor Woods, in Class-Book; shaded trap rocks, Falls of the St Croix, Wisconsin, Dr Parry, and high northward, Gray's Manual. I have not yet seen Canadian specimens of this species, which is quite a northern fern, stretching along the northern shores of the Pacific to the Russian Arctic dominions. I have specimens from Repulse Bay, collected by Captain Rae's party while wintering there in 1855. This plant does not appear to be in cultivation in any European garden.

L. Thelypteris, Presl.—Frond erect, lanceolate, mostly broad at base, and narrowed upwards, thin, and herbaceous, or slightly coriaceous, glabrous or downy, pinnate; pinnæ linear, rather distant, deeply pinnatifid; pinnules with revolute margins, veins forked, sori near their middle, becoming confluent. Stipe as long as, or longer than, the frond, and naked. *Lastrea Thelypteris*, Presl, Moore, J. Sm. *Aspidium Thelypteris*, Swartz, E. B. Willd., Pursh, Bigelow, Beck, Darlington, Torrey Fl. N Y. ii. p. 496, A. Gray, Man. *Polypodium Thelypteris*, Linn. *Dryopteris Thelypteris*, A. Gr.—Swamps in the woods, Townships of Hinchinbrook, Portland, Ernestown, &c.; Millgrove Marsh, Hamilton, Judge Logie; Gatineau Mills on the Ottawa, D. M'Gillivray, M D.; Prescott, common, B. Billing, jr.; Temiscouata, Thorold and Malden, P. W. Maclagan, M.D.; Belleville, very common in swamps, J. Macoun; Ramsay, Rev. J. K. M'Morine, M.A.; portage to Bark Lake, and on lumber road through the woods east from Hamilton's Farm, W. S. M. D'Urban; Montreal, Drs Maclagan and Epstein; Hudson's Bay Territories near Red River Settlement, Governor M'Tavish; St Joachim, M. L'Abbé Provancher; L'Orignal, J. Bell, B.A.; London, W. Saunders. In the State of New York this species is common in swamps and wet thickets (Torrey). I have it from West Point, N. Y. In the south, Eaton indicates Florida and northward. Very seldom found with fructification (Pursh). Fertile specimens are not rare with us. The forked veins of the pinnules distinguish this species from the next. In the Canadian plant the outline of the frond is a little different from Scotch and Irish specimens, being less narrowed at base. There are three forms of this species in Canada. The first (α) seems to be the plant of Gray's Manual, the second (β) is more like the *L. Thelypteris* of Europe, and the third (γ) is intermediate between this species and the next.

, *a. pubescens.*—Frond somewhat coriaceous, densely pubescent or downy throughout. Odessa, Hudson's Bay, &c.

β. glabra.—Frond thin, herbaceous, glabrous. Montreal, Chelsea, Hinchinbrook, &c.

γ. intermedia.—Frond narrowed below, glabrous; stipe slightly elongated (veins forked). Gaspé, J. Bell, B.A.

L. Nov-Eboracensis.—Frond lanceolate, narrow at the base, thin and herbaceous, pinnate; pinnæ linear or linear-lanceolate, more or less approximate, deeply pinnatifid; pinnules oblong, usually flat; veins simple (not forked) ; sori never confluent; stipe short, rachis, &c. downy, pinnules more or less distinctly ciliate. *Lastrea Noveboracensis*, Presl ; *Polypodium Noveboracense*, Linn., Schk. *Aspidium thelypteroides*, Swartz. *Aspidium Noveboracense*, Willd., A. Gray, Eaton—Pittsburg near Kingston ; Lakefield, North Douro, Mrs Traill; Mountain side, Hamilton, Judge Logie ; Prescott, common, B. Billings, jr.; Mount Johnson, Montreal, and Beloeil, P. W. Maclagan, M.D.; Ramsay, Rev. J. K. M'Morine, M.A.; near Chelsea, D. M'Gillivray, M.D.; London, but not common, W. Saunders ; L'Orignal, J. Bell. This fern belongs exclusively to the American Continent. It seems to be more abundant and more distinct in the United States than with us. In *Flora Boreali-Americana*, Sir William Hooker observed—" The *Aspidium Noveboracense* is quite identical with *A. Thelypteris*." In the recently published volume of *Species Filicum* (which at present I can only quote at second hand), doubts are still expressed as to its being a species really distinct from *L. Thelypteris*. Mr Eaton and other American pteridologists think it quite distinct. Its most obvious characters are—(1.) The tapering form of the lower part of the frond (although there is also a form of *L. Thelypteris* having this peculiarity ; (2.) sori few, mostly near the base of the pinnules, and not confluent, not overlapped by a recurved margin ; (3.) veins of the pinnules simple, not forked. The outline of the frond must not be depended upon, as the Scotch and Irish *L. Thelypteris* is narrowed at the base like *L. Nov-Eboracensis*. This species is allied to *L. montana*, Moore (*Oreopteris*, Bory).

POLYSTICHUM.

P. angulare, β. *Braunii*.—Frond soft, herbaceous, lanceolate, bipinnate ; pinnules stalked, serrate ; the small teeth tipped by soft bristles ; stipe and rachis scaly throughout. In the Canadian plant the scales of the rachis are larger than in the typical *P. angulare* of England, from which it may be specifically distinct. *Aspidium Braunii*, Spenner. *Aspidium aculeatum* var. *Braunii*, A. Gray, Man. Bot., p. 599, *A. aculeatum*, Provancher ; Harrington, Cap Bon Ami and Dartmouth, N. fork, Gaspé, John Bell, B.A. ; base of Silver Mountain, W. S. M. D'Urban.

P. Lonchitis, Roth.— Frond rigid and shining, linear-lanceolate, simply pinnate ; pinnæ scythe-shaped, auricled, spinose. *Polystichum Lonchitis*, Roth, Moore, J. Sm., &c. *Polypodium Lonchitis*, Linn. *Aspidium Lonchitis*, Swartz, Schk.—Limestone rocks, Owen Sound, C.W., 1859, Rev. Professor William Hincks, F.L.S. Professor Hincks has kindly furnished me with specimens from the above locality. Woods, southern shore of Lake Superior and north-westward, Professor Asa Gray, in Man. Bot., N.S. ; British America, Professor Woods in Class-Book. It will be observed that Professor Hincks's station is the only definite Canadian one with which we are acquainted. Mr T. Drummond found this fern on the Rocky Mountains many years ago.

P. acrostichoides, Schott.—Frond pale green, shining, long and narrow, linear-lanceolate, simply pinnate ; pinnæ long and narrow, linear-lanceolate, shortly stalked, auricled anteriorly at the base, more or less distinctly serrate, with hair-tipped teeth ; fertile (upper) pinnæ slightly contracted, covered beneath by the large confluent sori ; stipe profusely chaffy, with pale scales. *Polystichum acrostichoides*, Schott, J. Sm. *Aspidium acrostichoides*, Swartz, A. Gray, Eaton. *Aspid. auricula-tum*, Schk. *Nephrodium acrostichoides*, Michx.—Abundant in the woods a few miles west from Kingston ; also not rare in the woods of the Midland District of Canada generally ; Upper Rideau Lake ; woods around Toronto, Rev. Dr Barclay ; Stanfold, M. L'Abbé Provancher ; L'Orignal, J. Bell ; London, W. Saunders : Sulphur Spring, Hamilton, Judge Logie ; Prescott, common, B. Billings, jr. ; Nicolet and St Valentine, C.E., and Chippawa, C.W., P. W. Maclagan, M.D. ; Belleville, very common in rocky woods, as in Hop Garden, J. Macoun ; Ramsay, Rev. J. K. M'Morine, M.A. ; bills and woods, portage to Bark Lake, W. S. M. D'Urban ; Gilmour's Farm, Chelsea, D. M'Gillivray, M D. ; Osnabruck and Prescott Junction, Rev. E. M. Epstein. This species is exclusively American.

[β *incisum ;* pinnæ strongly serrate or incised into lobes. *Aspidium Schweinitzii*, Beck. This form, which I have from Schooley's Mountains, &c. (A. O. Brodie), will no doubt be found in Canada.]

CYSTOPTERIS.

C. fragilis, Bernhardi.—Fronds delicate, green, lanceolate in outline, glabrous, bipinnate ; pinnæ and pinnules ovate-lanceolate or oblong ; the latter obtuse, incisely toothed, thin and veiny ; sori large ; stipe dark purple at the base. *Cystopteris fragilis*, Bernhardi, Hook., Bab., Moore, Newm., A. Gray. *Polypodium fragile*, Linn. *Cystopteris orientalis*, Desvaux. *Polypod. viridulum*, Desv. *Athyrium fragile*, Sadler. *Cyathea fragilis*, Sm. *C. cynapifolia* and *C. anthriscifolia*, Roth. *Cystea fragilis*, Sm. *Cyclopteris fragilis*, S. F. Gray.—Rocky woods and cliffs about Kingston, in various places, but not abundant ; Farmersville ; Mountain side, Hamilton, on moist rocks, Judge Logie ; rocks by the bay shore, L'Anse au Cousin, and Dartmouth River, Gaspé, John Bell, B.A. ; Mirwin's woods, Prescott, common, B. Billings, jr. ; Montreal and Jones's Falls, P. W. Maclagan, M.D. ; rocky banks of the Moira, rather rare, J. Macoun ; Ramsay, Rev. J. K. M'Morine, M.A. ; camp at base of Silver Mount, on rocks, also River Rouge, abundant, De Salaberry, west line, and at Black Lead Falls, W. S. M. D'Urban ; St Joachim, M. L'Abbé Provancher ; Grenville, C E., John Bell, B.A. ; London, W. Saunders. In Dr Hooker's valuable Table of Arctic Distribution this plant is indicated as a Canadian species that does not enter the United States, which I presume arises from a misprint, as the species is not uncommon in the Northern States, and extends south to the Mountains of Carolina. The delicate *C. tenuis* is the form known in the south, but in Canada we have the stout typical European form of *C. fragilis*.

β *angustata*.—Pinnules inc.sed, with longish and spreading teeth. *Cyst. frag.* var. *cynapifolia*, J. Lowe.—Gaspé, John Bell, B A. Specimens referable to this form were likewise gathered at Lake of Three Mountains by Mr D'Urban. Mr Bell's specimens agree perfectly with English specimens from Dr John Lowe (*C. f. cynapifolia*). Italian specimens from Professor Caruel of Pisa, labelled " *Cyst. fragilis*," belong to this variety. Mr Bell has a fertile frond from Gaspé with very broad veiny pinnæ, deeply incised, but not pinnate.

C. bulbifera, Bernhardi.—Frond thin, green, lanceolate or linear-lanceolate, bipinnate, bulbiferous towards the apex on the under surface; pinnæ oblong-lanceolate, narrowed at the tips; pinnules oblong-obtuse, incisely toothed; sori small, not very numerous; indusium short. Very variable in the size and form of the frond. *C. bulbifera*, Bernhardi, A. Gray, J. Sm. *Aspidium bulbiferum*, Swartz, Schk., Pursh. *Aspidium atomarium*, Muhl.—Moist swampy woods about Kingston, as Collins's Bay, Kingston Mills, &c.; abundant on Judge Malloch's farm, a mile west from Brockville; Petit Portage. &c., Gaspé, John Bell, B.A.; Wolfe Island, A. T. Drummond, B.A.; Mirwin's woods, Prescott, common, B. Billings, jr. (short form); Beloeil Mountain, P. W. Maclagan, M.D.; rocky banks of the Moira, Belleville, and in cedar swamps and wet woods, very common, J. Macoun; Ramsay, Rev. J. K. M'Morine, M.A.; Mountain side, Hamilton, common, Judge Logie; Black Lead Falls, on limestone rock, W. S. M. D'Urban; Pied du cap Tourmente, M. L'Abbé Provancher; Grenville, C. E., J. Bell; London, W. Saunders. There are two distinct forms or varieties of this species.

α. *horizontalis.*—Frond triangular-lanceolate, broad at base, not more than three or four times longer than broad; pinnæ horizontal. Niagara Falls, within the spray, Collins's Bay, &c.

β. *flagelliformis.*—Frond linear, attenuated upwards, very long and narrow, six or seven times longer than broad; pinnæ less horizontal. Frankville, Montreal, Gaspé, &c.

DENNSTÆDTIA.

D. punctilobula, Moore.—Frond broadly lanceolate, pale green, thin, with a stout rachis, bipinnate; the pinnules pinnatifid; sori minute. usually one on the anterior basal tooth of each lobe of the pinnule, which is reflexed over the sorus; the proper indusium is pale, cup-shaped, opening at top. Rhizome slender, creeping through the soil; whole plant glandular-downy. *Dennstædtia* (Bernhardi, 1800) *punctilobula*, Moore, Index Filicum, p. xcvii. *Dicksonia punctilobula*, Hooker, A. Gray, J. Sm. *D. pilosiuscula*, Willd., Hook. Fl. Bor. Amer. *Nephrodium punctilobulum*, Michx. *Aspidium punctilobulum*, Swartz. *Patania*, Presl. *Dicksonia pubescens*, Schkr *Sitolobium pilosiusculum*, Desv., J. Sm. Gen. Fil.—Pittsburg near Kingston, John Bell, B.A.; River Rouge, W. S. M. D'Urban; Montreal, P. W. Maclagan, M.D.; Prescott, on Dr Jessup's moist pasture land, B. Billings, jr.; New Brunswick, E. N. Kendal, in Hook. Fl. Bor. Amer.; Ramsay, Rev. J. K. M'Morine. Mr Eaton has mentioned to me that the drying fronds have the odour of new hay.

WOODSIA.

W. Ilvensis, R. Br.—Frond lanceolate, usually 4 or 5 inches long, bipinnate, or nearly so, pinnæ approximate, pinnules oblong, obtuse, stipe (red), rachis, and whole lower surface of the frond clothed with chaffy scales, which are rusty at maturity. Sori usually confluent around the margins of the pinnules. First observed in the Isle of Elba (Ilva), hence named, after Dalechamp, *Acrostichum Ilvense,* by Linnæus, whose Phœnix was very wroth thereat ; see English Flora, vol. iv. p. 323. *Woodsia Ilvensis,* R. Br., Hook., Moore, J. Sm., Gray, &c. *Nephrodium lanosum,* Michx.—Abundant on the ridge of Laurentian rocks at Kingston Mills ; Rocks west from Brockville and at Chelsea, B. Billings, jr. ; Mount Johnson and Beloeil Mountain, P. W. Maclagan, M.D. ; mountain gneiss rocks, opposite Rouge River, W. S. M. D'Urban. I have likewise specimens from the Hudson's Bay territories (Governor M'Tavish), but without special locality. On rocks, Canáda, Pursh ; Canada to Hudson's Bay, Hook. Fl. B. A. ; Pied du cap Tourmente, M. L'Abbé Provancher. I think our plant must be much larger and more scaly than the European one. A tuft which I have from Catskill Mountains (A. O. Brodie) has richly fruited fronds a foot long and 2 inches wide. (I find that large American forms of this species have been mistaken for *W. obtusa.* The involucre, which is large and *not* split into hairs in the latter species, serves readily to distinguish it.) Much of the Ilvensis in cultivation in Europe is probably the American form.

β. *gracilis.*—Frond more slender, more hairy and less scaly than the type ; pinnæ rather distant, deeply pinnatifid, or partially pinnate. Dartmouth River, Gaspé, John Bell, B.A. In technical characters, this form agrees better with *W. alpina* (*hyperborea*), but it has quite a different aspect.

W. alpina, S. F. Gray.—Frond small (from 1 to 2 or 3 inches long), broadly linear, pinnate, somewhat hairy without distinct scales ; pinnæ ovate, somewhat triangular, obtuse, pinnatifidly divided into roundish lobes. *Woodsia alpina,* S. F. Gray, Brit. Pl., Moore. *Woodsia hyperborea,* R. Br. in Linn. Trans., vol. xi. ; Pursh. Fl Am. Sept. ii. p. 660.— In the clefts of rocks, Canada, Pursh ; Canada to the Saskatchewan, Hooker. Noticed in Dr Hooker's Table of Arctic Plants as a Canadian species that does not extend into the Ámerican States.

W. glabella, R. Br.—Frond a few (2-4) inches long, linear, bright-green and glabrous on both sides, simply pinnate ; the pinnæ short, rounded or rhombic, cut into rounded or wedged lobes. Stipe with a few scales at the base only. *Woodsia glabella,* R. Br., Hook. Fl. Boreali Americana, tab. 237 ; Gray. Canada, Professor Woods in Cl. Bk. Sir W. Hooker, in the Fl. B. Amer., gave Great Bear Lake as the only station then known for *W. glabella.* Mr D. C. Eaton has kindly furnished me with specimens from Willoughby Lake, Vermont (Goodale leg.), and Professor Gray notices its occurrence on rocks at Little Falls, New York (Vasey), and " high northward."

β. *Belli.*—Frond larger (6-7 inches long) ; pinnæ more elongated, pinnatifidly incised into rounded lobes (bright green, glabrous). Gaspé, on the Dartmouth River, twenty miles from its mouth, John Bell, B.A.

W. obtusa, Torrey.—Frond nearly a foot long, linear-lanceolate, glandulose, bipinnate ; pinnules slightly decurrent, oblong, obtuse, crenate, or somewhat pinnatifid ; indusium large, enveloping the sorus, torn into a few marginal lobes ; stipe with few scattered, pale, chaffy scales. *Woodsia obtusa,* Torrey, A. Gray, J. Sm. *Aspidium obtusum,* Willd. *Physematium obtusum,* Hook. Fl. Bor. Am. *Woodsia Perriniana,*

Hook. and Grev. Ic. Fil. *Polypodium obtusum*, Swartz.—An impression prevails that this plant, which is said to be common in the Northern States, especially towards the west, grows also in Canada. Mr D. C. Eaton, in the kindest manner, cut out of his own herbarium a specimen for me, from near High Bridge, New York city, in an excellent state for examination, which has enabled me to understand the species and to ascertain that we have as yet no satisfactory evidence of its occurrence in Canada. Large forms of *W. Ilvensis* have in some cases passed for it. (I introduce this notice of the plant with a view to promote farther inquiry.)

Osmunda.

O. regalis β. spectabilis.— Fronds erect, pale-green, glabrous, bipinnate ; pinnules oblong-lanceolate, oblique, shortly stalked, very slightly dilated at the base, nearly entire ; fertile pinnules forming a racemose panicle at the summit of the frond. *Osmunda spectabilis,* Willd., J. Smith. Farmersville ; Hardwood Creek. Hinchinbrook, and other places in rear of Kingston, usually in thickety swamps, by corduroy roads, &c. ; Millgrove Marsh, Hamilton, Judge Logie ; Ramsay, Rev. J. K. M'Morine, M.A ; woods near the Hop Garden, Belleville, not common, J. Macoun ; Prescott, common, B. Billings, jr. ; around Metis Lake, &c.; opposite Gros Cap ; also Sou-sou-wa-ga-mi Creek and Schibwah River, R Bell, jr. ; near Montreal, Rev. E. M. Epstein and W. S. M. D'Urban ; mountain, Bonne Bay, Newfoundland, on rocks 1000 feet above the sea, James Richardson (a small form) ; Welland, J. A. Kemp, M.D. ; Osnabruck and Prescott Junction, Rev. E. M. Epstein ; Nicolet, Wolfe Island and Navy Island, P. W. Maclagan, M.D. ; Lake St Charles, M. L'Abbé Provancher ; Caledonia Springs and L'Orignal, J. Bell ; Portland, Thos. R. Dupuis, M.D. ; Bedford ; London, W. Saunders. The fronds of our plant are a little more drawn out than those of the European one ; the pinnules are often distinctly stalked, and the overlapping auricles either altogether absent or only slightly developed. This is *O. spectabilis*, Willd. ; *O. regalis, β.* Linn. Sp. Pl. Some botanists distinguish two American forms, one agreeing with the typical *regalis* of Europe ; but it is difficult to do so. The typical *O. regalis* is a larger, more robust, and more leafy plant, with more widely spreading or divergent pinnæ, and more leafy auricled sessile pinnules, more or less pinnatifid at the base ; in our Canadian plant they are quite entire. The divisions of the fertile portion of the pinnæ are also more widely divergent in *α regalis*. The frond, moreover, is of a darker colour.

O. cinnamomea, Linn.—Sterile and fertile fronds distinct, the former ample, broadly lanceolate, pinnate ; the pinnæ rather deeply pinnatifid ; lobes regular, entire ; fertile frond contracted, erect, in the centre of the tuft of sterile fronds, and not at all foliaceous. Sporangia ferruginous. Fertile frond decaying early in the summer. *Osmunda cinnamomea*, Linn , Gray, J Sm. *O. Claytoniana*, Conrad, not of Linn.—Fairfield farm and elsewhere about Kingston, not uncommon ; Millgrove Marsh, Hamilton, Judge Logie ; Sandwich and Montreal, P. W. Maclagan, M.D.; opposite Gros Cap ; also Two Heart River, Lake Superior, R. Bell, jr., C E ; Belleville, swamps and low grounds, common, J. Macoun ; Ramsay,

Rev. J. K. M'Morine, M.A.; St Joy Woods, on the river shore, near Gatineau Mills, D. M'Gillivray, M.D.; Newfoundland, Miss Brenton, in Hook. Fl. Bor. Am.; Prescott, common, B. Billings, jr.; Nicolet, M. L'Abbé Provancher; L'Orignal, J. Bell; near London, W. Saunders.

O. Claytoniana, Linn.—Frond narrowly lanceolate, pinnate; pinnæ lanceolate, about three pairs of pinnæ near or below the middle of the frond contracted and fertile; sporangia brown, with green spores. This species, when fresh, has a strong odour, resembling that of rhubarb (Pie-plant) stalks. *O. Claytoniana*, Linn., Gray, J. Sm. *O. interrupta*, Michaux.—Between Kingston and Kingston Mills, in wet swampy places by the roadside; Little Cataraqui Creek; Waterloo; banks of the Humber, near Toronto; Princes Island, Hamilton, Judge Logie; Ramsay, Rev. J. K. M'Morine, M.A.; Ke-we-naw Point, in wet soil, R. Bell, jr.; Belleville, low rich grounds, not rare, J. Macoun; Prescott. common, B. Billings, jr.; Round Lake, W. S. M. D'Urban; Lake Settlement, and on the river shore near Gatineau Mills, D. M'Gillivray, M.D.; Newfoundland, Miss Brenton, in Hook. Fl. Bor. Am.; Osnabruck and Prescott Junction, Rev. Dr Epstein; on Judge Malloch's farm and elsewhere about Brockville; Dartmouth River, Gaspé, John Bell, B.A.; St Ferreol, M L'Abbé Provancher. Abundant on uncleared land along the Bedford Road, where the dried fronds are used by the farmers as winter fodder for sheep. Augmentation of Grenville, C. E., J. Bell, B.A.; near Komoka, C.W., W. Saunders. This fern is common also in the Northern States. I have a lax form, with long stipes and remarkably short somewhat triangular pinnæ, from Schooley's Mountains.

Schizæa.

[*S. pusilla*, Pursh.—Newfoundland, De la Pylaie. I have no further information respecting its occurrence in British America. Professor A. Gray indicates its distribution in the United States thus :— " Low grounds, pine barrens of New Jersey, rare," which is not at all favourable to its being found in Newfoundland or Canada. Mr Eaton has sent me beautiful specimens from sandy swamps in Ocean County, New Jersey.]

Nat. Ord. OPHIOGLOSSACEÆ.

Botrychium.

B. virginicum, Swartz.—Barren branch sessile, attached above the middle of the main stem, thin, delicate, veiny, tripinnate, lobes of the pinnules deeply incised; fertile branch bi- or slightly tri-pinnate. Very variable in size, usually a foot or more in height, but sometimes only a few inches. *Botrychium virginicum*, Swartz, A. Gray, J. Sm. *B. virginianum*, Schk. *Osmunda virginica*, Linn. Sp. Pl. *Botrypus virginicus*, Michx.—Not uncommon in the woods about Kingston and the surrounding country, as near Odessa, in Hinchinbrook, &c.; Delta; Toronto; Sulphur Spring, Hamilton, Judge Logie; Prescott, in woods, common, B. Billings, jr.; Nicolet, Montreal, Wolfe Island and Chippawa, P. W. Maclagan, M.D.; Belleville, rich woods, very common, J. Macoun; Ramsay, Rev. J. K. M'Morine, M.A.; River Marccuin, St Lawrence

Gulf, also opposite Grand Island, Lake Superior, R. Bell, jr., C.E ; Marsoni, Riviere Rouge, and De Salaberry, west line, W. S. M. D'Urban ; Montreal, Osnabruck, and Prescott Junction, Rev. E. M. Epstein ; Hill Portage above Oxford House, Governor M'Tavish ; Newfoundland, Miss Brenton, in Fl. Bor. Am. ; Lake Huron to Saskatchewan, Hook. Fl. Bor. Am. ; Gaspé, John Bell, B.A. ; Stanfold, M. L'Abbé Provancher ; Grenville, C. E., J. Bell ; London, W. Saunders.

β. *gracile.*—Very small (5 or 6 inches high), fertile branch less divided. *B. gracile*, Pursh. Hill Portage, above Oxford House, Governor M'Tavish.

γ. *simplex.*—Barren branch oblong, pinnatifid, the lobes ovate, incised, veiny. *B. simplex*, Hitchcock. Grenville, C.E., John Bell, B.A.

B. lunarioides, Swartz.—Barren branch long-stalked, arising from near the base of the main stem, thick and leathery, bipinnate, the pinnules slightly crenate ; fertile branch bipinnate. Root of long thick tuber-like fibres *Botrychium lunarioides*, Swartz, Gray. *B. fumarioides*, Willd., Provancher. *Botrypus lunarioides*, Michx.—Gananoque Lake, May 1861 ; Plains near Castleton, and woods near the Hop Garden, Belleville, rare, J. Macoun ; Three Rivers, C.E., P. W. Maclagan, M.D. ; Waste places west from Prescott Junction, rare, B. Billings, jr.: St Joachim, Provancher ; L'Orignal, J. Bell ; English's Woods, W. Saunders ; in the Northern States this species grows in dry rich woods, " mostly southward," according to Professor Gray's Manual.

B. obliquum (Muhl.), appears to be chiefly distinguished by its larger size, more compound fertile frond, and the narrower oblique divisions of the barren one. *B. obliquum* (Muhl.), Pursh. Fl. Amer. Sept., vol. ii. p. 656. Newfoundland, Dr Morrison in Hook. Fl. Bor. Am. ; " Wesleyan Cemetery, London," W. Saunders.

B. Lunaria, Swartz.—Barren branch sessile, arising from the middle of the stem, thick and leathery, oblong, pinnate ; pinnæ lunate or fanshaped, slightly incised on the rounded margin. *Botrychium Lunaria*, Swartz, Schk., Hook., Moore, J. Sm. *Osmunda Lunaria*, Linn.— Nipigon, 1853, Governor M'Tavish ; N.E. America, Dr Hooker's tab. ; Newfoundland, Saskatchewan, and Rocky Mountains to Behring's Bay in N. W. Am., T. Moore, Hbk. Brit. Ferns.

OPHIOGLOSSUM.

[*O. vulgatum*, L., which is widely distributed throughout Europe and Northern Asia, and grows also in the Northern United States, although there " not common," is to be looked for in Canada. In one of its forms (*O. reticulatum*, Linn.), it extends to the West Indies.]

Nat Ord. LYCOPODIACEÆ.*

PLANANTHUS.

P. Selago, Pallisot-Beauvois.—Stem dichotomously branched, erect,

* In this order the arrangement of A. M. F. J. Pallisot-Beauvois is adopted, as it seems to afford the best basis for a readjustment of the genera of *Lycopodiaceæ*, which is much required. For P.-B.'s genus *Lepidotis*, I have thought it better to substitute the name *Lycopodium*, an old name that should not be discarded.

fastigiate; leaves in about 8 rows, more or less convergent or spreading, lanceolate, acuminate, entire; sporangia in the axils of the common leaves (not in spikes). *Lycopodium Selago*, Linn., E. B., Bigelow, Beck, Hook. and Grev., Torrey Fl. N. Y. ii. p. 508, Gray.—Labrador, Hudson's Bay to Rocky Mountains, Hook. Fl. B. A.; shore of Lake Superior and northward, Professor A. Gray, Man. Bot., N. S., p. 603. I have not seen Canadian specimens of this plant. The stations known show that it encircles Canada, and some of them are probably within our limits. Principal Dawson obtained the alpine variety on the White Mountains, Herb. Bot. Soc. Canada. It is a rare plant in the United States. There are two forms of this species (both of which are figured by Dillenius). a. *sylvaticus*, leaves convergent, almost appressed. β. *alpinus*, leaves widely-spreading, stems shorter.

P. lucidulus. Stem dichotomously divided into long erect branches; leaves bright green, in about 8 rows, reflexed, linear-lanceolate, acute, denticulate; sporangia in the axils of the common leaves (not in spikes). *Lycopodium lucidulum*, Michaux, Pursh, Bigelow, Torr. Fl. N. Y. ii. p. 508, Gray, Beck, Darlington, Hook. and Grev. Bot. Mis. *L. reflexum*, Schk. *Lycopodium suberectum* of Lowe, a Madeira plant. *Selago americana, foliis denticulatis reflexis*, Dill. Hist. Mus. t. lvi.— Gananoque Lakes, Collins's Bay, Newboro-on-the-Rideau, woods in rear of Kingston, &c.; Prescott, common, B. Billings, jr.; Nicolet, C.E., St Catherine's and Grantham, P. W. Maclagan, M.D.; Belleville, in swamps and cold woods, rather common, J. Macoun; River Ristigouche, St Lawrence Gulf, R. Bell, jr., C.E.; L'Orignal, J. Bell, B.A.; London, W. Saunders; Ramsay, Rev. J. K. M'Morine, M.A. This species is stated by Professor Torrey to be rather common in New York State. " Frequently bears bulbs instead of capsules," Pursh.

[*P. alopecuroides*, P. Beauv.—The habitat " Canada" is given for *Lycopodium alopecuroides*, Linn., in the " Species Plantarum," ed. 3, vol. ii. p. 1565; but it is probably not a Canadian plant.]

P. inundatus, P. Beauv.—Stems prostrate, adherent to the soil, the fertile ones erect; leaves secund, yellowish green, lance-awl-shaped, acute; sporangia in distinct, terminal, leafy, sessile, solitary spikes. *Lycopodium inundatum*, Linn., E.B., Michaux, Pursh, Beck, Tuckerman, Torr. Fl. N. Y. ii. p. 508, Gray. *Plananthus inundatus*, Beauv. *L. alopecuroides*, Linn., in part?—In cedar swamps and overflowed woods, Canada, Pursh. Professor Torrey notices its occurrence in the north-western part of the State of New York. Professor Gray observes, that the leaves are narrower in the American than in the European plant, and suggests that it may be a distinct species. I have not yet seen Canadian specimens.

LYCOPODIUM.

L. clavatum, Linn.—Stems robust, and very long, prostrate, rooting, forked, with short ascending branches; leaves pale, incurved, linear-awl-shaped, tipped with a white hair point; sporangia in scaly catkins, which are usually in pairs on common peduncles. *Lycopodium clavatum*, Linn., E. B., Michaux, Pursh, Bigelow, Beck, Darlington, Spring, Hook.,

Torrey, Gray. *L. tristachyum*, Pursh ? *L. integrifolium*, Hook.
L. aristatum, Humboldt.—Occasionally found in the woods in rear of
Kingston, but not common; Newfoundland, Hook. Fl. Bor. Am.;
between Thessalon and Missisaugi Rivers, Lake Huron, R. Bell, jr.;
Prescott, common, B. Billings, jr.; Three Rivers, Temisconata, and
Wolfe Island, P. W. Maclagan, M.D.; Seymour, in pine woods, rare,
J. Macoun; Ramsay, Rev. J. K. M'Morine, M.A.; River Ristigouche,
St Lawrence Gulf, R. Bell, jr.; London, W. Saunders, C.E.; L'Orignal
and L'Anse au Cousin, Gaspé, J. Bell; Belmont. The spores, chiefly of
this species, constitute *pulvis lycopodii*, which is used by apothecaries,
and was at one time employed for making artificial lightning in the
theatres.

L. annotinum, Michaux.— Stems very long, prostrate, creeping, forked,
with ascending branches; leaves bright green, spreading or slightly
deflexed, in about five rows. linear-lanceolate, mucronate, serrulate;
sporangia in scaly catkins, which are sessile, solitary, oblong-cylindrical,
thick. *Lycopodium annotinum*, Michaux, E. B., Pursh, Beck, Tuckerman, Torrey, Fl. New York State, ii. p. 509.—Pine forests in Hinchinbrook; rocky woods in Pittsburg, on the north bank of the St Lawrence,
near Kingston; Gananoque Lakes; L'Anse au Cousin, Gaspé, John Bell,
B.A.; Prescott, common, B. Billings, jr.; Rivière du Loup, Nicolet,
Montreal, and Kingston. P. W. Maclagan, M.D.; Belleville, in cool
woods, common, J. Macoun; Ramsay, Rev. J. K. M'Morine, M.A.;
Priceville, C. I. Cameron, B.A.; Newfoundland, Hook. Fl. Bor. Am.;
St Augustin and Cap Tourmente, M. L'Abbé Provancher. Frequent in
New York State, according to Professor Torrey. Of this species there
are two forms, only one of which, the normal one, or type, I have as yet
observed in Canada. The var. β *alpestre*, Hartm. Scan. Fl., having
broader, shorter, paler, less spreading leaves, I have from the Dovrefieldt
(T. Anderson, M.D.), Lochnagar (A. Croall), and entrance to Glen Fee,
Clova, where I found it growing with the typical form.

L. dendroideum, Michx.—Stems upright, bare below, bushy above
(giving the plant a tree-like aspect), arising from a long creeping rhizome,
leaves more or less appressed; sporangia, in scaly catkins, which are
sessile, cylindrical. *Lycopodium dendroideum*, Michx., Pursh, Bigelow, Hook., Beck, Darlington. *L. obscurum*, Linn., Bigelow, Oakes.—
White cedar woods near Bath, abundant, and throughout the woods
generally in rear of Kingston; Gananoque River; Priceville, C. I.
Cameron, B.A.; Prescott, common, B. Billings. jr.; Nicolet, Mount
Johnson, and Montreal, P. W. Maclagan, M D.; Seymour and Cra-
mahe, in cold moist woods, J. Macoun; River Ristigouche, Gulf of
St Lawrence, R. Bell, jr.; Ramsay, Rev. J. K. M'Morine, M.A.; New
Brunswick, Hook. F. B. A ; Osnabruck and Prescott Junction, Rev. E.
M. Epstein; London, W. Saunders; Harrington, L'Orignal, and Gaspé,
John Bell, B.A.; St Joachim, M. L'Abbé Provancher.

L. complanatum, Linn.—Stems rhizome-like with ascending branches,
which are dichotomously divided, flattened; leaves short, in four rows,
those of two rows imbricated, appressed, of the other two somewhat
spreading; sporangia in scaly cylindrical catkins, in twos, threes, or
fours, on a common peduncle. *Lycopodium complanatum*, Linn., Gray,

Blytt. *L. chamæcyparissias*, Braun. *L. sabinæfolium*, Willd.—Not uncommon ·in the woods about Kingston, and in rear ; Newboro-on-the Rideau ; Gananoque River ; River Ristigouche, St Lawrence Gulf, and St Joseph's Island opposite Campment D'Ours, Lake Huron, R. Bell, jr. ; Ramsay, Rev. J. K. M'Morine, M.A. ; pine grove near Blue Church Cemetery and woodlands west from Brockville, not common, B. Billings, jr. ; Three Rivers and Temiscouata, C.E., P. W. Maclagan, ·M.D. ; sandy woods around Castleton, sterile hills Brighton and Murray, J. Macoun ; L'Orignal and L'Anse au Cousin, Gaspé, J. Bell, B.A. ; Trois Pistoles, M. L'Abbé Provancher ; London, W. Saunders. To this species is referred *L. sabinæfolium*, Willd, *L. chamæcyparissias*, A. Braun, with branches more erect and fascicled. Professor Asa Gray remarks :—The typical form of *L. complanatum*, with spreading, fan-like branches, is abundant southern (in N. States), while northward it passes gradually into var. *sabinæfolium*." I have only one rather imperfect specimen of the European *L. chamæcyparissias*, collected at Bonn, on the Rhine, by my friend Professor G. S. Blackie, which does not differ in the branching from ordinary Canadian forms of *L. complanatum*. It appears to be quite a common species in the States, for I have it from a great many places.

SELAGINELLA.

S. spinulosa, A. Braun.—Small, prostrate, leaves lanceolate, acute, spreading, spinosely toothed ; fertile branch stouter, ascending spike sessile. *Selaginella spinulosa*, A. Braun, Blytt, Norges Fl. *Lycopodium selaginoides*, Linn., Pursh Fl. Am. Sept. ed. ii. p. 654. *Selaginella spinosa*, Beauv. *Selaginella selaginoides*, A. Gray, Man. Bot. N. States, p. 605.—Gaspé, John Bell, B.A. ; Canada, Michaux ; Lake Superior and northward, pretty rare, Professor Asa Gray in Man. Bot. N. States ; Canada, Pursh, who observes : " The American plant is smaller than the European."

STACHYGYNANDRUM.

S. rupestre, P. Beauv.—Much branched, leaves slightly spreading when moist,. appressed when dry, carinate, hair-tipped ; compact and moss-like, growing on bare rocks. *Selaginella rupestris*, Spring, A. Gray, Eaton. *Lycopodium rupestre*, Linn., Pursh Fl. Am. Sept. ed. ii. p. 654.—On the perpendicular faces of Laurentian rocks, along the north bank of the St Lawrence, in Pittsburg, and on the Thousand Islands at Brockville, &c. ; Longpoint on the Gananoque River ; near Farmersville, C. W., T. F. Chamberlain, M.D. ; rocks in pine groves two miles west from Prescott, near the river, and on rocks west from Brockville, not common, B. Billings, jr. ; Ramsay, Rev. J. K. M'Morine, M.A. ; Beloeil and Mount Johnson, C.E., P. W. Maclagan, M.D.

DIPLOSTACHYUM.

D. apodum, P. Beauv.—Stems creeping, branched ; leaves pale vivid green, of two kinds,—the larger spreading horizontally, ovate oblique, the smaller appressed, acuminate, stipule·like. Forms compact tufts. *Lycopodium apodum*, Linn., Pursh. Fl. Am. Sept. ed. 2, ii. p. 654.

Selaginella apus, Gray, Eaton.—Abundant on low wet ground east of Front Street, Belleville, below the hill, where it was pointed out to me by Mr J. Macoun, July 1863. In September 1863, I found it sparingly but fertile, on grassy flats by the river side at Odessa. Near London, W. Saunders; Detroit River, C. W., P. W. Maclagan, M.D. Apparently not common in the United States. I have it from Schooley's Mountains. This is a very small, compactly-growing moss-like species, well adapted for cultivation under a glass shade. It was a great favourite with the late Dr Patrick Neill, in whose stove, at Canonmills, Edinburgh, I first saw it many years ago.

Nat. Ord. MARSILEACEÆ.

AZOLLA.

A. Caroliniana, Willd.—Pinnately branched with cellular, imbricated leaves; plant reddish, circular in outline, $\frac{1}{2}$–1 inch in diameter; leaves ovate obtuse, rounded and roughened on the back (Eaton). Resembles a floating moss or Jungermannia (Torrey). Gray, Man. Bot., t. 14. Floating on the waters of Lake Ontario, Pursh Fl. Am. Sept., ed. 2, ii. p. 672. In the adjoining states, Professor Asa Gray notices it as occurring in pools and lakes, New York to Illinois and southward, and observes that it is probably the same as *A. magellanica* of all South America.

SALVINIA.

[*Salvinia natans* = *Marsilea natans,* Linn. Sp. Pl. "Floating like lemna on the surface of stagnant waters, in several of the small lakes in the western parts of New York and Canada."—Pursh Fl. Am. Sept. ed. 2, ii. p. 672. Professor Asa Gray states, that it has not been found by any one except Pursh, and he therefore omits it from his Manual of Botany of the Northern States.]

ISOETES.

I. lacustris, L.—Beloeil, C. E., P. W. Maclagan, M.D.; Saskatchewan, Hook. Fl. Bor. Am. This plant is spoken of by Pursh as growing in the Oswego River, near the Falls; and Professor Gray and others allude to it as not rare in the New England States. It should be carefully looked for in the numerous lakes and creeks of Upper Canada. It grows in muddy bottoms, forming green meadows under water. Much interest is attached to the genus *Isoetes,* since Professor Babington has shown that instead of one there are many species, or at least distinct races or forms, in Britain. In the United States four are known:— *I. lacustris,* Linn.; *I. riparia,* Engelm.; *I. Engelmanni,* Braun; and *I. flaccida,* Shuttlew., the last a southern form. Professor Babington is certain of the existence of at least eight European species:—*I. lacustris,* L.; *I. echinospora,* Dur.; *I. tenuissima,* Bor.; *I. adspersa,* A. Br.; *I. setacea,* Del.; *I. velata,* Bory.; *I. Hystrix,* Dur.; and *I. Duriæi,* Bory. As yet we know of only one Canadian species, which is here rendered, rather uncertainly, *I. lacustris.* The American species are described in Gray's Manual, the British ones in the new Journal of *Botany,* London.

Nat. Ord. EQUISETACEÆ.

EQUISETUM.

The *Equiseta* having been described in a previous paper, it will be sufficient to give here a mere list of the species, with some additional notes obtained since the former paper was written.

E. sylvaticum, Linn. Newfoundland and New Brunswick, Hook. Fl. Bor. Am.

E. sylvaticum, β. *capillare*. Much branched; branches very long. straight, and exceedingly slender (capillary). Farmersville.

E. umbrosum, Willd. Belmont.

E. arvense, Linn. West from London, W. Saunders. The rhizome bears large spherical pill-like nodules, which are, however, more conspicuous in var. β. *granulatum*.

E. arvense, β. *granulatum*.

E. Telmateja, Ehrhart. Shores of Lake Ontario, Beck.

E. limosum, *Fries* —The great value of this species and of *E. arvense* as fodder plants, is confirmed. On the western prairies horses are said to get " rolling fat" on Equisetum in ten days ; and experienced travellers tell me, that their horses always go faster next day after resting at night on Equisetum pasture. The horses do not take to it at first ; but after having a bit of Equisetum put occasionally into their mouths, they soon acquire a liking for it, and prefer it to all other herbage. Near Komoka, W. Saunders.

E. hyemale, Linn. Lake Huron, Hook. Fl. Bor. Am. ; St Joachim, M. l'Abbé Provancher ; London, W. S.

E. robustum, Braun. Stems much thicker than in *E. hyemale*, the ridges with one line of tubercles ; sheaths shorter than broad, with a black band at base, and a less distinct one at the margin ; teeth about forty, three-keeled. *E. robustum*, Braun, A. Gray. Grenadier Pond, on the Humber River near Toronto, 3d June 1862. It is difficult to decide whether this and other forms are really distinct from *E. hyemale;* certainly that species varies in size, in roughness, and other characters. In *robustum* the teeth are twice as many as in *hyemale*, but even this is perhaps not a constant character.

E. variegatum, Weber and Mohr. ; St Joachim, M. L'Abbé Provancher.

E. scirpoides, Michaux.

E. scirpoides, β. *minor*.

E. palustre, Linn.—" Canada, from Lake Huron, Dr Todd, Mr Cleghorn, Mrs Perceval, to the shores of the Arctic Sea, Dr Richardson, Drummond, Sir John Franklin, Captain Back."—Hook. Fl. Bor. Amer. Professor A. Gray speaks of " the European *E. palustre*," " attributed to this country (the N. American States) by Pursh, probably incorrectly." Dr Hooker indicates its existence, without doubt, in Arctic West America and Arctic East America. The name of the plant has occasionally appeared in Canadian lists, but I have as yet seen no Canadian specimen. It remains for Canadian or Hudson's Bay botanists to trace its southern limit on the American Continent. In Europe and Asia it has no tendency to Arctic limitation.

The Hypothesis of Molecular Vortices.

TO THE EDITOR OF THE EDINBURGH PHILOSOPHICAL JOURNAL.

SIR,—As the article on Thermodynamics in the "North British Review" is perhaps the most complete history of that science which has yet appeared, and is written with a scientific precision which is unusual in journals not specially devoted to science, I wish to correct an oversight that the Reviewer has committed in describing the "Hypothesis of Molecular Vortices," or "Centrifugal Theory of Elasticity," as proposed by me in 1849.* He speaks of atmospheres of ether surrounding nuclei of ordinary matter; whereas in the hypothesis, as I put it forward, the nuclei perform the functions of ether, and the atmospheres those of ordinary matter. Radiance is supposed to consist in oscillations of the nuclei, transmitted in waves by means of the forces which they exert on each other at a distance; and thermometric heat is supposed to consist in an agitation of the atmospheres, producing outward pressure according to the known laws of centrifugal force. Emission of radiance takes place when the atmospheres whirl faster than the nuclei oscillate, so that the nuclei are undergoing acceleration, and the atmospheres retardation; absorption of radiance takes place when the nuclei oscillate faster than the atmospheres whirl, so that the nuclei are undergoing retardation and the atmospheres acceleration. In perfect gases, the nuclei oscillate with little impediment from the atmospheres, and the transmission of radiance is rapid; in substances in a more dense condition, each nucleus is, as it were, loaded with a part of its atmosphere (like a pendulum in a resisting medium), and the transmission of radiance is slower. It is this peculiar view of the respective functions of the nuclei and the atmospheres, that constitutes the main distinction between the hypothesis put forth by me and other hypotheses involving atomic nuclei and atmospheres (as that of Mossotti), or accounting for the phenomena of heat by molecular motions (as that of Mr Herapath).

Of course a mechanical hypothesis does not form an *indispensable* part of Thermodynamics, more than of any other physical science; but if a hypothetical theory of Thermodynamics is to be used, it appears to me that its fundamental principles must be such as I have described.—I am, Sir, your most obedient servant,

W. J. MACQUORN RANKINE.

GLASGOW, 1st *March* 1864.

* Transactions of the Royal Society of Edinburgh, 1850–51.

REVIEWS AND NOTICES OF BOOKS.

Journal of the Scottish Meteorological Society, New Series, No. I. William Blackwood and Sons, Edinburgh and London. January 1864.

The Council of the Scottish Meteorological Society having recently resolved to publish their proceedings in the form of a quarterly journal, the first number of the new series, got up in a handsome form by the Messrs Blackwood and Sons, has just made its appearance. The Society will now be better able to carry out the important objects it has in view, and which cannot be better stated than in the words of the prospectus, namely,—" to investigate Scottish meteorology, and particularly to ascertain the leading features of the climate of different districts of the country; to point out the bearings of meteorology on public health, and on the prevalence of diseases affecting crops and live stock; to investigate the origin, progress, and recurrence of storms in Scotland; to point out the difference between the climate of Scotland and that of other countries; to ascertain the peculiar causes of the climate of Scotland in summer as well as in winter; to investigate the general laws regulating atmospheric changes, the discovery of which may lead to a knowledge of the coming weather; and to disseminate meteorological information by the circulation and publication of interesting papers."

For the first few years of the Society's existence, its efforts were almost exclusively devoted to the collection and reduction of meteorological data, and to the establishment of observatories over the country, of which there are now seventy-two. As it is now eight years since the Society began its operations, a sufficient time has elapsed to warrant conclusions being drawn regarding the climate of Scotland in its relations to pressure, temperature, humidity, rain, and wind. Accordingly, various papers have of late appeared in their proceedings, some of which we noticed in this Journal at the time of their appearance, giving a *resumé* of these elements of the weather, and pointing out the influence they severally exercise on the health of the people and on vegetation. In these papers the chief principles have been educed that give its peculiar character to the Scottish climate, and in this way an amount of reliable information respecting its advantages and disadvantages in the different seasons of the year has been disseminated among the people. The good services thus rendered to science and the public service are now, we are happy to see, begun to be recognised and appreciated, as evinced by the comparatively large accession to its membership the Society has lately received. But considering the importance of the objects aimed at, the support yet given is not nearly such as might have been expected; and there is no doubt that many more will be induced to join the Society when they are made aware of its claims, and the easy terms of admission to membership. The claim to greater support is strengthened by the consideration that in the case of meteorological investigations, numbers and extensive combination are indispensable to insure anything like success. We hope the time is not far distant when Government will recognise the claims this and other scientific bodies have on its support.

The first paper in the present number of the journal is a very interesting one by Mr Buchan on the " Weekly Extreme Temperatures and Rainfall in Scotland for March, April, May, and June, on an average of seven years ending with 1863." From the valuable tables which accompany this paper, the following highly interesting and important conclusions are drawn regarding the occurrence of frost over Scotland :—

" 1*st*, That up to the end of May frosts may be expected to occur every week somewhere in Scotland. This remark is applicable in its fullest extent to the inland parts of the country taken as a whole.

" 2*d*, That in every part of the country, frost will occur some time or other in the months of March and April; in inland and eastern districts nearly every year some time in May; but in the islands and in the west, the probability of its occurring or not occurring in May is about the same.

" 3*d*, That in June frosts do not occur except in high situations situated in hollows and surrounded by hills.

" 4*th*, That at each particular station frost may be expected to occur every week in inland and eastern districts up to about the end of the first week in May, but at western and island stations only till the middle of April.

" 5*th*, That from the second week in May till the end of the month, frost is as likely to occur as not in inland districts on the east coast till the middle of May, and in the west and islands from the middle of April to the end of the first week in May.

" 6*th*, That at inland places frost rarely—that is, practically— ceases to occur from the beginning of June; and at all other places in the west and east, and islands, this takes place at least a fortnight earlier, or in the middle of May.

" 7*th*, That at inland places in elevated situations and in hollows, such as Braemar, Thirlestane, and Stobo, frost cannot be considered as ceasing to occur till about the middle of June, or a full month later than in the islands and along the sea-coast.

" 8*th*, That at elevated stations among hills, such as Castle

Newe, frost may be regarded as of certain occurrence every week up to the end of the first week of May; in the west and islands to the end of March; and elsewhere to the middle of April.

"9*th*, That at Glasgow and Baillieston, frost may not be expected to occur after the first week of May, being a fortnight earlier than the other inland stations. Query—Is this due to the proximity of the Atlantic? and is the effect of that ocean with respect to frosts really felt in so marked a manner in localities at so great distances from the west coast?

"These conclusions will be at once recognised to be of the utmost value in their bearings on the character of the climate of the different parts of the country, and in determining how far each locality may be suitable in the treatment of different classes of diseases, and in supplying valuable information in deciding whether particular crops may be cultivated with a fair prospect of success. To farmers and gardeners they are invaluable, as indicating the proper and safe time for planting the potato and other tender exotics grown in the open air, inasmuch as the time when frosts most commonly cease to occur is thereby made known."

If the temperature during the day repeatedly rise to $65°·0$, the heat thus received by the crops is sufficient for the growth and ripening of oats and the coarser sorts of grain; it is also sufficient for the growth of wheat and the finer cereals up to the period of flowering. The observations made respecting the occurrence of this important degree of temperature over the country are thus summarised:—

"1*st*, That as regards the months, in the central and eastern districts, a temperature of $65°·0$ will occur some time in May and June; in April the chances of its occurring or not occurring are about equal; and in March it is not likely to occur at all;—in the western districts these occur nearly a month later; and in islands nearly two months later.

"2*d*, That as regards the weeks, $65°·0$ occurs so seldom, that it need not be expected on the mainland till the end of April, and in a few places there not till a week or a fortnight later; and in the islands till the second week of June.

"3*d*, That in these different localities, for four weeks after the above times, it occurs just so often as to make it a matter of equal probability, whether in any of these weeks it happen or not.

"4*th*, That, from and after the last week of May in the central and eastern districts, from the first week of June in the western districts, it will probably occur every week; and in the islands during the whole of June, the chance of its not occurring is the greater.

"5*th*, That the districts where $65°·0$ is of most frequent occurrence, are Strathmore, Strathearn, the Lothians, and the Merse in Berwickshire; and where it occurs least frequently are the northern islands."

The temperature of 70°·0, the occasional occurrence of which is necessary for the proper ripening of wheat, is shown to have occurred only once in Orkney, and twice in Shetland and the outer Hebrides during the period. It is of most frequent occurrence in Berwickshire, the Lothians, Fife, Strathmore, and along the Moray Firth; in other words, in the best wheat-producing counties.

The second paper, also by Mr Buchan, on the "Isothermals of the British Isles in January and July," is of great value, and is accompanied by a beautiful map in colours, engraved by the Messrs Johnston, Edinburgh. The results are given in the following extract:—

" With regard to the lines of equal summer heat, there are one or two points of some interest which may be pointed out. The most noticeable of these points is the general slope of the lines from north-east to south-west, thus indicating a higher summer temperature in the east as compared with the west of the British Isles. This is occasioned by the general direction of the wind in July being from the Atlantic Ocean, which at this time of the year is colder than the land, and also by the greater amount and less height of the clouds in the west as compared with the east. The curving northward of the lines in the centre of England is deserving of attention. This is owing to the circumstance, that it is only this part of the island which is broad enough to allow of the central parts acquiring so much of a Continental summer climate as to give a decidedly northern flexure to the lines. The southern part within the isotheral of 64°·0, of which London may be said to form the centre, is interesting as being undoubtedly the hottest district in the British Isles during summer, and consequently that part best fitted for the culture of exotics which require a high summer temperature. The colder summer climate of the south coast, as compared with the district immediately to the north of it, just referred to, is caused by the moderating influence of the sea which flows round it. The whole of the British Isles south of the Dornoch Firth and Skye may be considered as suited to the successful cultivation of wheat and the finer cereals, except those districts where the rainfall is excessive or the elevation too great.

" The most noticeable point with respect to the lines of equal winter temperature is the crowding of the high temperatures in the western, and particularly the south-western, parts of the British Isles. Ireland and the peninsula of Cornwall, being, as it were, more completely bathed in the warm waters of the Gulf-Stream during the winter months, exhibit for this time of the year a markedly high temperature for their position on the globe, being no less than 26°·0 above their normal winter temperature.* The winter temperature, 39°·0, of the west coast of Great Britain is the same from the most northern of the Shetland Isles as far south as North Wales; and the winter temperature, about 37°·0, of the east coast, from Kinnaird's Head in Aberdeenshire to the mouth of the Thames. Hence in winter the west coast is generally advantaged at least 2ᶜ·0, and the south-west of Ireland as much as 6°·0 of mean winter temperature, as compared with the eastern parts of England in the same latitude; whereas the eastern coast of Great Britain receives no benefit whatever, as regards its mean winter temperature, from the North Sea, since the cold of winter is quite as low in these as in the central parts of the island. Does not the higher winter temperature of the west coast of Ireland, as compared with that of Great Britain, supply additional proof of the Gulf-Stream reaching our shores, as it would be difficult to account for the decrease in the temperature of the Atlantic as *we proceed from west to east, unless on the supposition of a slow gradual translation of its whole waters eastward, under a cold wintry atmosphere?* The higher winter temperature of the west coast in winter is owing to these causes :—(1.) The warm south-west wind, which loses some of its heat before reaching the east coast. (2.) The larger amount of vapour deposited in rain on the west coast—being nearly double that in the east—which thus liberates a very large quantity of latent heat. (3.) The larger amount of vapour in the west, which obstructs radiation, not only when in a visible state in the form of clouds, but also when dissolved through the atmosphere in an invisible state, in which state, as shown by the recent researches of Professor Tyndale, it serves as a covering to the earth both from its own radiation by night, and from the sun by day, which would otherwise burn up everything in the fierceness of his heat."

The next paper is an "Address on the Importance of Medical Climatology," by Dr Scoresby-Jackson, the convener of this department, lately added to the Society's operations. The address is judicious and able, and contains much useful information, and many suggestive practical remarks. The subject is one of great importance and complexity, and will require for its elucidation the active and willing co-operation of medical men and meteorologists for a series of years. The little that is yet known of the

---

* The excess of the actual temperature of January over the normal temperature is even more striking in the case of Shetland. Thus Lerwick should only possess a January temperature of 3°·0, if it received no more heat than is due to its position on the globe in respect of latitude, whereas, owing to the causes mentioned in the text, its temperature for that month is 39°·0, or 36°·0 higher than it would otherwise be.

subject, is given by Dr Jackson with a modesty and candour truly
admirable. The committee on this subject, indeed, find it almost
necessary to commence investigation at the very beginning, and to
limit themselves very much to the accumulation of accurate observations of well-attested facts. It is to be hoped that the
medical faculty will at once come forward and give their hearty
support to the Society in carrying out this object.

Flora of Ulster and Botanist's Guide to the North of Ireland.
By G. DICKIE, A.M., M.D., F.L.S., Professor of Botany,
Aberdeen. Aitchison, Belfast, 1864. 24mo, pp. 176.

We have here a most useful and accurate botanical guide to the
flora of the north of Ireland, written by one who is already known
as the author of an excellent flora of Aberdeenshire. The district
included in the work lies to the north of the 54th parallel of latitude, extends due west from Dundalk, and therefore includes
nearly the whole of Ulster, and the northern portions of Leitrim,
Sligo, and Mayo, belonging to Connaught. As to geological structure, it may be said that " Silurian formations occur in the south-
east; granite in the Mourne Mountains; in the east there is an
extensive mass of basalt,—chalk, greensand and oolite being here
and there exposed; metamorphic and granite rocks appear on the
north and north-west, carboniferous limestones in the south-west,
and Devonian rocks in parts of the interior. The higher mountains
are chiefly grouped in the east, north-west, and west. The extreme highest points are,—in County Down, Slieve Donard, 2796
feet; in Donegal, Muckish and Erigal, respectively 2190 and
2460 feet; in Mayo, Nephin, 2646 feet."
The climate, in the northern parts of Ireland, which are in-
dented with arms of the sea, is comparatively mild and moist, and
the extreme ranges of temperature are moderate. The temperature of the Atlantic is about 3° above that of the North Sea, for
the six months beginning with October and ending with March;
and there are many seaweeds, which indicate the milder temperature of the sea on the Irish coasts.
Dr Dickie follows in his work Hooker and Arnott's British
Flora, and he adopts Watson's types as regards the distribution of
species. The general characters of the flora are thus stated: plants
of Germanic and Highland types, especially the former, constitute
a very insignificant part of the Ulster flora; in the more northern
counties a few of the Scottish type are plentiful, as *Empetrum
nigrum* and *Ligusticum scoticum;* a third of the entire number belong to the Atlantic type; plants of the English type form a
decided character in the flora. The author mentions 570 dicotyle-

dons, 192 *monocotyledons*, and 43 ferns and their allies, making in all 805 species. The *Pteris aquilina* ceases at a lower level in the north of Ireland than it does in Scotland, and so does *Erica Tetralix*. We want the true alpine flora in Ulster, as none of the mountains attain a sufficient elevation. We have only such sub-alpine plants as *Saxifraga stellaris, Salix herbacea, Carex rigida, Lycopodium Selago, Saxifraga umbrosa, Arbutus Uva-Ursi, Rubus Chamæmorus, Alchemilla alpina*.

With the Latin name of the species there is given the English name, the time of flowering, and the range, with the various localities in Ulster in which it has been found. In a supplement, lists are given of species doubtfully native, and of others which are not strictly indigenous.

We have great pleasure in recommending the book as being well executed, and as being a most valuable pocket companion for students who are examining the botany of the north of Ireland. To botanical pupils in the College of Belfast, the work supplies a long-felt desideratum; and to all who are examining the distribution of plants in Great Britain and Ireland, the facts in this volume must prove highly useful.

A Hand-Book of Descriptive and Practical Astronomy. By GEORGE F. CHAMBERS, F.R.G.S. Murray, London, 1861. 12mo, pp. 514.

While we have abundance of manuals in various departments of science in this country, there is undoubtedly a deficiency as regards astronomy. Mr Chambers has produced a book which fully supplies this want. The work is designed to occupy a middle place between purely elementary · works and advanced treatises. He has rendered it attractive to the general reader, useful to the amateur, and valuable to the professional astronomer as an occasional book of reference. The author first gives a sketch of the Solar System—the sun and the various planets, each of which are discussed in separate chapters. He next considers Eclipses and their associated phenomena,—eclipse of the sun and moon, transits of the inferior planets, and occultations. The next subject taken up is the Tides. Then follow miscellaneous astronomical phenomena, as the Obliquity of the Ecliptic, Precession, Nutation, Aberration, Parallax, Refraction, Twilight, Zodiacal light, &c. The subject of Comets occupies the whole of the fifth book; and, after making some interesting general remarks, the author proceeds to treat of Periodic comets, remarkable comets as Donati's, the comets of 1811, 1843, and 1860, cometary statistics and historical notices. Book sixth is occupied with Chronological Astronomy. time in its various phases—hours, days, years, measurement of time, solar and zodiacal time, &c. Book VII. is entitled The Starry Heavens, embracing the pole star, double stars, variable stars, clusters and nebulæ, the milky way, and the constellations. The various astronomical instruments are then noticed and described—telescopes, the equatorial, the transit instrument, &c. A sketch of the history of astronomy is given in Book IX., and the author concludes in Book X. with Meteoric Astronomy, including the consideration of aerolites, shooting stars, and other meteors. The whole is illustrated by an extensive series of excellent woodcuts, and references are given to the principal authorities. Amidst such a multiplicity of subjects it is not possible to select passages for quotation. Suffice it to say, that the work is one of standard merit ; that it will completely repay a careful perusal ; that it is carefully executed, and that it gives the account of most interesting phenomena in a pleasing and popular style. We do not know a better astronomical text-book for a student.

PROCEEDINGS OF SOCIETIES.

Royal Society of Edinburgh.

Monday, 7th December 1863.

Professor Innes, one of the Vice-Presidents, delivered the following Opening Address :—

GENTLEMEN,—The opening of our Session requires that I should lay before you the state and prospects of our Society, which I hope may to some extent be considered the criteria of the state and prospects of the sciences which it cultivates.

The Society has lost since the commencement of last Session by death, six Fellows, viz.,—Robert Allan, Esq., Beriah Botfield, Esq., Dr James Keith, Dr David Boswell Reid, Professor Connell of St Andrews, Professor Mitscherlich of Berlin; and by resignation, two, the Rev. G. V. Faithfull and D. R. Hay, Esq.

In room of whom the Society has elected twenty-five new Fellows, viz.,—Professor Blackie, William Brand, Esq., W.S., Robert Campbell, Esq., advocate, Dr Hugh F. C. Cleghorn, India, Charles Cowan, Esq., W. Dittmar, Esq., Dr J. Matthews Duncan, the Right Hon. Lord Dunfermline, Professor Everett, Nova Scotia, James Hannay, Esq., William Jameson, Esq., India, Hon. Lord Jerviswoode, Charles Lawson, Esq., Hon. G. Waldegrave Leslie, *G. R. Maitland,* Esq., W.S., Edward Meldrum, Esq., Rev. Dr

Nesbit, Hon. Lord Ormidale, David Page, Esq., Dr A. Peddie. James Sanderson, Esq., Deputy-Inspector of Hospitals, Dr John A. Smith, Dr Murray Thomson, Dr J. G. Wilson, Dr John Young.

Our roll, therefore, stands thus :—The number of Fellows in 1862 was 258, of which we have lost by death 6, by resignation 2 = 8, leaving 250. To which add the new Fellows, 25, making the whole number of the Fellows of our Society 275, a larger number than has appeared on the list for many years.

I am enabled, chiefly through the active kindness of our Secretaries, to offer a few notices of the members we have lost, during the past Session.

ROBERT ALLAN, son of Mr Thomas Allan, a banker in Edinburgh, a Fellow of the Society, and for many years Curator of its Museum and Library, and well known as an early and successful collector of a fine cabinet of minerals, was born in 1806, and educated at the High School and University of Edinburgh. He inherited his father's taste for minerals, and while still a youth followed out the study in extended travels in company with Professor Haidinger, who introduced him to the acquaintance and to the cabinets of all the chief foreign mineralogists—among others, Berzelius and Mitscherlich.

Mr Allan passed advocate in 1829, but never practised, and was admitted a Fellow of this Society in 1832. He was also a member of the Geological Society of London.

Mr Allan published in 1834 a Manual of Mineralogy, the classification founded on the external character or natural historical arrangement.

In 1837 he edited a fourth edition of "Phillips' Mineralogy," in which he added notices of 150 new minerals.

On his return from an excursion to the volcanic district of Italy and Sicily, Mr Allan presented to this Society a set of specimens of volcanic rocks of the Lipari Isles, with a descriptive notice, an abstract of which is in our Transactions, of date 16th January 1831.

He communicated an account of a visit to the Geysers and Hecla to the British Association at Glasgow, in 1855.

Mr Allan died in consequence of a fall in his garden.

BERIAH BOTFIELD was of a Shropshire family, in which county his grandfather, Thomas Botfield, made his large fortune as a manager and lessee of the Dawlay Collieries. Thomas's third son inherited Norton Hall, near Daventry, in Northamptonshire, and lived the life of an English sporting squire. He married Charlotte, daughter of William Withering, M.D., F.R.S., the author of "The Botanical Arrangement of British Plants." The only child of that marriage was Beriah, the subject of the present notice, who, in addition to his father's property, inherited the estates of both his uncles, and had become before his death a man of very large fortune.

Beriah was born 5th March 1807, and succeeded his father in 1813. He was educated at Harrow and Christ Church, where he took his Bachelor's degree in 1828.

After leaving Oxford he made a tour in the Highlands of Scotland, a journal of which he printed for private circulation,—printed at Norton Hall, 1830, 12mo.

He was High Sheriff of Northamptonshire in 1831.

In 1840 he was elected Member for Ludlow, and again in 1841.

In 1847 he was beaten by the Whig candidate.

In 1857 he was solicited to stand again, and he sat in Parliament for Ludlow for the rest of his life.

Mr Botfield was a member of the Royal Society of London, the Royal Geographical Society, Royal Institution, Society of Arts, of the Antiquaries of London, Scotland, and Copenhagen, of the Royal Irish Academy, l'Institut d'Afrique, and of all the principal Societies in the Kingdom, and of a great number of literary Clubs,—as the Roxburghe, Bannatyne, Maitland, Spalding, Surtees, Abbotsford, Camden, Percy, Ælfric, Hakluyt, Cheetham; to most of which he gave valuable contributions, his part being generally to defray the expense.

In addition to these, and some smaller tracts printed for private circulation, Mr Botfield published "Notes on the Cathedral Libraries of England," from a personal examination, 1849; "Prefaces to the First Editions of the Greek and Roman Classics, and of the Sacred Scriptures," 1861. Large 4to.

Another work, for which he was making collections when he died, and which would have been of great interest and value, was intended to illustrate the history of the old monastic libraries of England. A collection of the extant catalogues and inventories of these was already in type, to which he meant to add the catalogues of other Middle Age libraries. His collections, made for these objects will, it is feared, be lost to the world by his death. He had previously edited (in 1838), for the Surtees Society, catalogues of the Library of Durham Cathedral, at various periods.

In 1858, Mr Botfield printed, for private circulation, *Stemmata Botevilliana*, a large volume illustrating the descent and antiquities of all the Bottevilles, Thynnes, and Botfields.

He was a liberal collector of pictures, and was also known as a zealous book-hunter.

Mr Botfield married Isabella, daughter of Sir Baldwin Leighton, Bart., but left no family; and has entailed a considerable part of his property on the second son of the Marquis of Bath, in respect of a very old but perhaps real connexion between his family and the Thynnes.

JAMES KEITH, second son of William Keith of Corstorphine Hill, accountant in Edinburgh, was born 29th November 1783, and was educated at the High School and University of Edinburgh. He was apprentice to Messrs· Bell, Wardrope, and Russell; went to London in 1804, and attended the London Hospital and Guy's. Was surgeon of the Berwickshire Militia for two or three years, which he resigned on entering into partnership with Dr Andrew Wardrope, which connection terminated by Dr Wardrope's death in 1822.

Mr Keith took the degree of M.D. in the University of Edinburgh in 1804, and he became a Fellow of the College of Surgeons in 1810. He was physician to the Deaf and Dumb Institution for many years. From the extreme shyness of his disposition, his worth and ability were known only to a limited circle of intimate friends. He died 12th May 1863. His widow and two sons survive—William Alexander, M.A. Oxon., and Charles Maitland.

DAVID BOSWELL REID was the second son of Dr Peter Reid, physician in Edinburgh. His mother, Christian Arnot, was the eldest daughter of Hugo Arnot of Balcormo, advocate and antiquary, well known to the last generation by his book on the history of Edinburgh and his collection of Scotch criminal trials—and perhaps still better by the extraordinary attenuated, almost skeleton, figure of the old gentleman preserved to us in Kay's Portraits. Dr Peter Reid (whose mother was a Boswell of the Balmuto family) was the editor of Dr Cullen's "First Lines of the Practice of Physic," 1802. A new edition was published, with supplementary notes, in 1810. He was also the author of a little duodecimo volume, entitled "Letters on the Study of Medicine and on the Medical Character, addressed to a Student," Edin., 1809. Besides the subject of my present notice, Dr Peter Reid had two sons,—Dr William, a lecturer in Edinburgh on the practice of Medicine, and Dr Hugo, well known as the author of several popular works, the last of which is a modest and temperate memoir of his distinguished brother, to which I beg to acknowledge my obligation.

David Boswell Reid was educated at the High School and University of Edinburgh. At the former, Mr Pillans, the rector, has mentioned him as "among the head boys of the Rector's class." While a medical student he became a member of the Royal Medical Society, of which he was chosen senior president in 1826-27, his junior being James Kay, now Sir James Kay Shuttleworth.

In 1827, Mr David Reid commenced a course of practical chemistry, which was very useful and very popular. He aimed at enabling each student to familiarise himself, by experiments made under the directions of a teacher, with the properties of the chief chemical substances, and the phenomena attending their action on each other.

After much approval in his extra-mural lecture-room, he joined Dr Hope in the College, and was again quite successful in the object of his course. But the Professor and Assistant had some misunderstandings, which led Mr Reid to leave the College, and renew his independent lectures, which were highly appreciated—attended by all classes,—the young ambitious student,—the veteran philosopher and man of science,—the man of intelligence feeling the want of science. On his benches met Dr Chalmers and Sir John Leslie, Professors George Joseph Bell and Pillans, Dean Ramsay and Mr Combe.

After the burning of the Houses of Parliament, and in contemplation of a new building, when a committee of the Commons was inquiring on the subject of its ventilation and acoustics, Dr Reid was examined as a witness, from having devoted much attention to those subjects, and having shown excellent examples of his skill, first in his own lecture-room, and, later, in the great temporary edifice, erected 15th September 1834, in the High School ground, for the Edinburgh dinner to Lord Grey, at which 2768 persons were present, and 240 ladies in the gallery, and each individual speaker was distinctly heard.*

The result of his examination was, that Dr Reid was employed to direct the ventilation and acoustics of the temporary House of Commons in 1836. It is not pretended that his plans gave universal satisfaction to the 700 members, each of whom had a different notion, and of course a peculiar constitution of body to be suited. But, after ten years' experience, in 1846, a fair committee of the House reported as to " the great improvement effected," and " concurred in the general opinion in its favour."

In 1840 arrangements were made for Dr Reid settling in London, and, while taking charge of ventilating the temporary House of Commons, superintending also the ventilation of the new building then in progress. This brought Dr Reid necessarily into close contact with the architect of the new palace, Mr Barry, and unfortunately they did not agree. The difference got worse and worse, till in 1845 they were no longer on speaking terms, and every detail of such extensive operations had to be settled by correspondence,—a state of things which could not be allowed to last. The quarrel broke out in some strong expressions of Dr Reid,—a prosecution for libel by Mr Barry,—a pretty general attack on Dr Reid by the public press, and a Reply by him to " The Times" newspaper [1845-47].

In 1852 a negotiation was entered into, by which the Government proposed to secure Dr Reid's services permanently, and to throw the ventilation of the whole buildings of the Houses under his charge,—one part of which, the House of Lords, had hitherto been managed by Mr Barry on a different system,—but " these negotiations were abruptly broken off." In fact, Dr Reid was turned off, after sixteen years successful service, and, as his brother tells us, " a small sum was given to him as some compensation for the loss which he had sustained. His friends who knew his whole career, and the proceedings connected with his removal to London, to take the charge of ventilating the Houses of Parliament, were of opinion that the sum awarded was totally inadequate to compensate for the sacrifices he had made."

Dr Reid went to New York in 1855. He delivered lectures in the Smithsonian Institution there, and at Boston. In the beginning of this year (1863), he received the appointment of Inspector of Military Hospitals, but soon after, while engaged in an official journey, he died suddenly at Washington, on 5th April 1863.

Dr Reid's system of ventilating great buildings, where crowds habitually assembled, consisted in forcing in a current of air by

* The Pavilion was 113 feet in length by 101 feet in breadth.

means of a powerful engine—the air being previously washed to free it from dust and to give it the requisite moisture. Some of his experiences are curious.

" The house is heated to 62° before it is opened, and maintained in general at a temperature between 63° and 70°, according to the velocity with which the air is permitted to pass through the house. This velocity is necessarily regulated by the numbers present, the temperature to which the air can be reduced in warm weather, and the amount of moisture which it may contain when the quantity is excessive. Some members are much more affected by an excess or deficiency of moisture than by alterations of temperature. In extremely warm weather, by increasing the velocity, air even at 75° may be rendered cool and pleasant to the feelings."

He goes on to say—" The temperature may always be advantageously increased and the velocity diminished before the usual dinner hour. After dinner, other circumstances being the same, the temperature should be diminished, the velocity increased, and the amount of moisture in the air reduced. During late debates, as they advance to two, three, four, or five in the morning, the temperature should be gradually increased as the constitution becomes more exhausted, except in cases where the excitement is extreme."

Next to the Houses of Parliament, Dr Reid's greatest and most successful undertaking of ventilation was the St George's Hall at Liverpool, in which immense building, on some occasions, there have been as many as 4500 persons for about ten hours; the air during all that time having been supplied to all that multitude in a pure state, and in a comfortable and agreeable condition as to temperature and moisture.

Dr Reid superintended while in this country the arrangements for ventilating the royal yacht, " The Victoria and Albert," and the steamships used in the expedition to the Niger, in both instances to the entire satisfaction of his employers; and since going to America, he was employed in the ventilation of a Russian frigate, " The Grand Admiral," built at New York

ARTHUR CONNELL, eldest son of Sir John Connell, Judge of the Admiralty Court, and author of a well known work on the Law of Scotland respecting Tithes, entered the High School of Edinburgh in 1804, and the University of Edinburgh in 1808, where he studied under Playfair, Leslie, Dugald Stewart, and Hope. From Edinburgh Mr Connell went to Glasgow College, where he studied under Jardine and Young, and, having obtained a Snell exhibition, went to Balliol College, Oxford, in 1812.

In 1817 Mr Connell passed advocate at the Scotch Bar, but he had from boyhood a remarkable turn for science, especially botany and chemistry, and he ultimately devoted himself exclusively to the latter science.

In 1840 he was presented to the Chair of Chemistry in the University of St Andrews.

In 1843 Mr Connell was candidate for the Chemistry Chair at

Edinburgh, vacant by the death of Dr Hope, and though not successful, produced a collection of testimonials of the highest character. Most of these were the more worthy of attention as not made for the occasion and so in some degree influenced by private friendship. They are for the most part notices in the published works of eminent chemists and in scientific journals, of Mr Connell's chemical labours, and the papers in which these were announced and described.

Having failed in this object of his ambition, Mr Connell continued to study and teach his favourite science at St Andrews till 1856, when the fracture of a limb, and its effects upon a constitution already long enfeebled, completely incapacitated him from active duty.

Mr Connell became a member of this Society in 1829, from which time till 1843 he contributed to the Transactions, or published in the pages of the "Edinburgh Philosophical Journal," memoirs to the number of 29.

His chief merit lay in his skill and unrivalled accuracy as a mineral analyst. To him we are indebted for several new mineral species—for the discovery in the minerals Brewsterite and Harmotome of the earth barytes in combination with silicic acid—that earth previously having been found combined only with the sulphuric and carbonic acids; while his ascertaining the constitution of the mineral Greenockite, *on one grain* of the substance, displayed a dexterity seldom if ever surpassed.

Mr Connell also engaged in somewhat more ambitious researches on the voltaic decompositions of alcohol, ether, and other liquids, and has presented us with an instrument for ascertaining the dew point, superior in several respects to that generally used.

Mr Connell was of a very retiring nature, modest, gentlemanly, and gentle in disposition. He expired peacefully on 31st of October last.

Eilard Mitscherlich, born 7th January 1794, at Neurede, in the Grand Duchy of Oldenburgh, where his father was a minister of the Lutheran Church, was educated at Heidelberg and Paris, and studied afterwards at Göttingen. His first objects of study were language and ethnology. Later in life he devoted himself more to natural science, and especially chemistry. He assisted Berzelius at Stockholm for some years.

In 1821 he was appointed Professor of Chemistry in the University of Berlin, and attached to the Friedrich Wilhelm Institut. His lectures were held in high estimation, and attended by numerous classes of students.

In 1828 he was elected an Honorary Member of this Society, and in 1829 was awarded a Medal by the Royal Society of London for his discourses " regarding the laws of crystallization and the properties of crystals."

In 1852 Mitscherlich was elected an Associate Member of the Institute of France. His great European reputation is founded on his studies on crystallization and some ingenious adaptations of instruments for practical chemistry. His text-book—*Lehrbuch der Chemie*—has gone through a great many editions.

Mr Mitscherlich died in the present year.

His experiments and disquisitions tended to establish the rule that bodies crystallizing in the same shape (isomorphous) have an analogous chemical composition—throwing great light on chemical classification, and giving us one of the greatest generalizations (after the Atomic theory) which chemistry has gained by the researches of philosophers.

When I have laid before you these slender memorials of our deceased brethren, I may claim to have discharged the real duty of my office to-night. If indeed I were worthy to fill the chair in which your favour has placed me,—if I had, like some of our distinguished Fellows, a knowledge of all science, or even a special acquaintance with any *one*,—it would be my duty to submit to you a survey, or at least some outline, of the progress of science among us and among our neighbours. But for such a task you know me to be ill qualified. I should not venture to speak in the language of science anywhere, and least of all in the presence of the men whom I now see around me.

There are subjects, however, in which scientific men and men of no science feel an equal interest—which must engage the attention of every person of common intelligence.

Among these is the great step recently made in African geography—the discovery of the head of the Nile. No other geographical discovery can ever compare with this. It is not the solution of a puzzle in the Geographical Society. It is removing the "*Impossible*"—the very type of impossibilities—from our books. It is opening to the whole world the mystery which was a mystery even to the initiated. Poets have lost a topic! What philosophers and historians guessed and speculated about, is now written down plain on the map. That is now clear which has been wondered at since men began to ask the meaning of anything. We have lost the oldest subject of curiosity in the world!

A grave, prosaic mind loses its equanimity, and gives way to the charm of romance at the thought of the veil being raised that has for so many thousands of years covered the head of the great mysterious river which was worshipped of old—not more for its beneficent overflowings, regular as the seasons, yet unaccountable, than because of its unknown, unapproachable source.

I do not mean that the facts which our travellers have brought to light run counter to the conclusions of former geographers. On the contrary, I think the body of history on the one hand, the speculations of science on the other, had prepared the world for such a discovery. Glancing at the ancient, I mean the classical authorities, without arraying them before you, I may say that among innumerable fables and much unphilosophical reasoning, they almost concur in giving the Nile its source in a mighty lake—some say two immense lakes—fed by periodical rains,—fed also, say some, by subterraneous streams flowing from the west (these subterraneous rivers were favourites with the wonder-loving naturalists of old).

This great lake was further believed to lie at the foot of lofty, snow-covered mountains, named the Mountains of the Moon. Herodotus indeed demurs to the snow. The Reservoir Lakes become immeasurable marshes in some of the accounts. Indeed I should despair of producing a *catena* of witnesses for any single point of the statement; but such as I have described was nearly the mind of ancient Greece and Rome, speaking on the information obtained in Egypt.

It is more remarkable to find a similar shadow of the truth from a different quarter, and perhaps of an earlier date. The ancient inhabitants of India seem to have felt the same interest, and to have had an equal glimmering of the course of the Nile. In a well-known paper by Mr Wilford, in the Asiatic Researches, we have a sort of abstract of the ancient Indian belief concerning the Nile, drawn from the Puránas and other Hindu or Sanscrit books.

The name of the river in those most ancient books is *Kali*, black. (Though Homer names the river *Aegyptus*, it was known to ancient Greeks as Μέλας.) According to the same authorities, that famous and holy river takes its rise from the lake of the gods, thence named Amara or Deva, Saróvera in the region of Sharma or Sharmasthan, between the mountains of Ajagara and Sitanta, part of Soma-giri, or the Mountains of the Moon, the country round the lake being called Chandristhan or Moon-land. The Hindus believed in a range of snow-covered hills in Africa.

From thence the Kali flows into the marshes of the Padma-van, and through the Nishada Mountains into the land of Barbara; whence it passes through the mountains of Hemacáta; then entering the forests of Tapas (or Thebais) it runs into Kantaka-desa, or Mitha-sthan, and through the woods emphatically named Aranya and Atavi into Sanchabdhi (or our Mediterranean).

From the country of Pushpaversha, it received the Nanda or Nile of Abyssinia, the Asthimati or smaller Krishna, which is the Takazzi or little Abay, and the Sanchanaga or Mareb.

The Ajagara Mountains, which run parallel to the eastern shores of Africa, have at present the name of Lupata, or the back-bone of the world. Those of Sitanta are the range which lies west of the lake Zambre or Zaire, words not improbably corrupted from Amara or Sura. This Lake of the Gods is believed to be a vast reservoir which, through visible or hidden channels, supplies all the rivers of the country.

The Hindus, for mythological purposes (says Mr Wilford), are fond of supposing subterranean communications between lakes and rivers, and the Greeks, we know, had the same leaning.

We really had made little progress beyond these ancient guesses, till in the year 1858 Captains Speke and Burton saw and sailed upon the great lake Tanganyika, 600 miles from the coast at Zanzibar. The lake is narrow, but 300 miles long, and 1800 feet above the level of the sea. Very soon after, Captain Speke alone had the glory to see and bear witness to the great inland sea which he has named Victoria. Having only seen this mighty lake, and *being* obliged to leave it unexplored, Captain Speke made haste to return to it, and this time in company with his old comrade and brother-in-arms Captain Grant, and through toils and dangers which men like these love almost for their own sake, they, together, reached in 1861 the Victoria Lake, which Speke had discovered three years earlier.

It happened (and such coincidences are frequent in science) that at the very time when Speke and Grant were fixing the bearings and heights of the great lake and its mountains, Baron von Decken and Mr Thornton measured and estimated the altitude of *Kilima Nearo*, one of a mountain range to the eastward of our travellers' route, at 20,000 feet, while the snow line descended below 16,000.

At present our information is necessarily meagre, but on the testimony of these two veteran travellers, furnished as they were with instruments for observation, we have some actual certainty, and room for infinite speculation.

The Victoria Sea of fresh water is about 150 miles square. The equator line runs through it, though nearer its north shore. Its waters are 3563 feet above the sea level. It is skirted, if not quite surrounded, by ranges of mountains of 10,000 feet high. Without farther evidence, independent even of the high authority of Captain Speke's opinion, we receive as certain that in the Victoria great lake is the source, or rather the great reservoir of the Nile, for of course the lake is fed by numerous streams, in fact by a stream from every valley among the surrounding mountains, and then it follows that the White Nile, not the Blue Nile as Bruce believed, is the chief of the two streams that join at Kartom, lat. 15° 30'.

Thus was the mystery cleared up that had defeated the ingenuity and enterprise of philosophers and travellers, of kings and Cæsars, since the days of Herodotus.

Captain Speke thinks very highly of the country he has explored in a commercial and agricultural view. He found the people not all savage, but capable of intelligent interest and quite awake to kindness and friendship. But the country is everywhere thinly peopled, and productive much beyond the wants of the population. Along the equator, at heights varying from 6000 to 12,000 feet, the travellers found a delicious climate, with abundance of water, and no excessive heat, full of cattle and corn. In the kingdom of Karagwé (Lat. 1° 40', elevation 5100 feet), the temperature for five months ranged from 60° to 70° at 9 morning. From what they could learn of the country to the westward of the lake, it preserves the same character for several hundred miles, and I know that Captain Speke believes there is a continuance of that which he calls the *Fertile Zone* almost to the coast of the Atlantic. He tells his friends he has "discovered a great fertile zone there, caused principally by the Mountains of the Moon, situated close to the equator, in the midst of the continent of Africa. These are great rain condensers. Round them are the sources of several rivers, the Nile on one side, the Tanganyika and the Congo on the other. The rains falling all round make that a fertile zone—the most fertile in the world. There is nothing in India or China to equal it."

It is in that direction the indefatigable traveller proposes to make his next expedition, and let us hope that in two years more we shall welcome Captain Speke returning from the mouths of the Congo.

I know not whether to congratulate or condole with the Society upon another advance in science, or whether that is to be called an advance which some consider a double trespass, a breaking down of the boundaries between geology and archæology, and overleaping the ancient landmarks which divided natural science from sacred history.

Certain well-known discoveries of hand-shaped weapons and implements, found along with the remains of some extinct animals, in undisturbed beds of a very ancient alluvial deposit both in France and in England, led the antiquary, whose department is limited to the human period, to seek to extend that period into what had hitherto been the exclusive province of the geologist; and the geologist again, driven to admit that these flint spear-points have been shaped by man's hand, and used upon (or among) the *Elephas primigenius*, the *Rhinoceros*, and other extinct animals whose teeth and bones now bear them company, has to seek for an extension of the period hitherto allotted for the operations and deposits which the race of man has witnessed.

This only brought out more palpably what geologists had for some time taught—had taught indeed almost as early as geology took the dimensions of a science—that the globe itself was immeasurably older than the age assigned for man.

That period—the creation of man—the age of man on the globe—had been early, and nearly unanimously fixed, by calculations based upon the data afforded by the Mosaic books.

Such calculations were necessarily more or less conjectural, founded on interpretations of archaic forms of language, and of words which might have different meanings. Numbers and figures were to be read in varying manuscripts, often from faulty copies; and although great men like Newton had satisfied themselves that the received age of the world and its inhabitants was the true one, new facts, of a science unknown to Newton, had shaken that opinion, and it seemed probable that the Biblical scholar, the student of sacred history, in the view of geological facts, would, in the first place, abandon the position that the age of the creation, the antiquity of the earth, was to be determined by the interpretation of the Mosaical books; and, *secondly*, that he would not shut his eyes to new evidence offered upon the questions, whether the Mosaical books intended to affirm the age of man upon the globe, and whether the interpreters of those books had accurately and precisely and definitely ascertained their meaning and intention in that matter.

I should perhaps do better in using the terms of the latest authority on this subject, which comes with "Oxford" on its title page to vouch its orthodoxy, and with the sound sense of our friend Dr Hannah to commend it to our acceptance:—*

* Dr Hannah's Bampton Lectures. Oxon., 1863.

· " It is surely mere misapprehension to suppose that the revelation with which Moses was really entrusted could traverse the path of the modern geologist, or contain any thing that would either confirm or contradict his readings of those buried rocks. From whichever side the error comes, we are bound to shake ourselves free from it, not by saying with some that God cared not though His instruments should make mistakes on scientific subjects, but by pointing out that there can be no error where there is no assertion, and that a purely theological revelation contains no assertion which falls within the proper sphere of science."

I say then the two parties, the scientific inquirer and the Mosaical scholar, both earnest for truth, would have come to some understanding, not surely to conceal or shut out the truth, but to give each full license to inquire and experiment, and to draw all legitimate inferences from facts discovered; for after all, the disputes between theologians and geologists relate rather to inferences from facts than to the reality of the facts themselves. The theologian infers certain truths from the words of the first chapter of Genesis; the geologist infers certain notions from what he sees in an open quarry. The inferences are mutually contradictory; but as the theologian and the geologist are both capable of drawing false inferences, such inferences may be contradictory, and neither may be true. A new light on the meaning of the word "Day," in the Mosaic language, might end the controversy; so might some evidence that the best instances of hand-formed flint implements found in ancient drift were fictitious and fraudulent.

We must suppose that a candid student of the Divine books will take what help is in his power for explaining their difficulties, and, be sure, he will not neglect the testimony of the rocks—the history of creation written in other letters but by the same Author. So a candid geologist, who reflects that the purpose of Moses was clearly not to teach natural philosophy, but to inculcate and enforce the worship of the true God, will acknowledge that the order of creation given in Genesis does agree marvellously with the inverse order of the fossils actually found—plants, marine or aquatic animals, birds, mammals, man.

I say these disputants might have come to terms—explaining the Scripture history of the creation by the help of a careful and reverent study of the created universe. But a third party has lately rushed among the combatants, and now fight with two-edged weapons. These are theologians too—at least they are churchmen, and Hebraists, and mighty arithmeticians; but, with a singular view of their duty to their Church, they cavil at the foundations of its history and doctrine, and think it necessary to tell the world so. These critics insist, that no interpretation, construing of a phrase, word, or numeral of the Mosaical books shall be admitted—that all shall stand or fall together; and then, having picked out some words, especially some numbers, which they judge erroneous—though not affecting a single point of doctrine or morals, or the

essentials of history—they say the books ascribed to Moses are
devoid of authority, and must be abandoned!

That is not the way in which we are accustomed to read any
ancient history; and, though different canons are used for criti-
cising the inspired writers from those applied to other historians,
yet, as to the mere text, the books of Moses are entitled beyond
others to a fair and liberal construction, as the most ancient books
in the world, and as having passed through an infinite number of
transcriptions and translations.

But I must declare my entire concurrence with Dr Hannah, that
"it is a dangerous and mistaken policy to raise these disputes to
adventitious importance, by treating them as though they necessarily involved the issue of our highest interests."

For the persons of tender conscience, who feel themselves constrained " to build up those scattered fragments of difficulty into a
coherent edifice of doubt," they would themselves surely feel easier,
as it would be a relief to the world, who are judging in the quarrel,
if they could cease to be members of a Church which founds so
confidently on the Mosaical history. They would *assail* with more
satisfaction if they had not promised to *defend*.

For the geologist, if my voice were wanting to encourage him,
I would bid him go forward, cautiously, reverently, yet without
fear. Let him test the evidence with all care before publishing a
discovery. He must consider he has everything to prove, and he
should assert nothing without evidence, and take nothing for
granted. We want proof of the antiquity of the Drift-deposit, and
of the fossils contained in it belonging to the extinct animals
named. We want proof that the flints are hand-wrought, and not
chipped accidentally in the rolling drift. Much more, we desire
proof that they *were* found there, and not *placed* to be found by
some cunning quarryman. It is not only the flint instrument but
its manufacture, its chipping into shape, that must be tested. Is
the fracture of the flint such that it might have been made many
thousand years ago?

Farther, the geologist should publish to the world the evidence
of his facts; for the inquiry is one that concerns the public, and in
which the public take an interest. But why should I intrude my
advice upon men who have shown they know well what is required
at their hands in a momentous inquiry? Nine of the most eminent
geologists of France and England met in friendly conference at
Paris, and, later, at Abbeville, to compare specimens, to test the
evidence, to do everything for ascertaining the truth; and they
published the *procès-verbal* of their proceedings in the " Natural
History Review" of last August, with the sanction of Dr Falconer's
name, and others equally well known. It seems hardly to be doubted,
that numerous frauds have been perpetrated upon the naturalists.
When specimens are well paid for, they become plentiful, both in
England and in France, but there may be means of detecting the im-
positions, and these means our geologists are using with all care.
The iron horse-shoe, lately put forth among the primeval relics, has

been, as I understand, withdrawn; the bones of elephant and other animals, bearing marks of human hands, are not yet accepted by these naturalists. As to the Abbeville jaw-bone of a man, whose jaw must have ceased chewing long before the flood, there is but one opinion in England, which I am informed by Mr Evans is also gaining ground in France—that the whole thing was an impudent imposition. Mr Prestwich, who was once a believer, published his recantation in the last Quarterly Journal of the Geological Society.

It may be permitted me, perhaps, as one of the public, to offer one more advice to the naturalist. He must take care not only that his reasoning is logical, his inferences cautious and careful, but he will do well to avoid even the appearance of disputing for victory. Science has no enemies if its votaries do not raise them up by indiscretion and intemperance. ·

Monday, 21*st December* 1863.—Dr CHRISTISON, V.P., in the Chair.

The following Communications were read :—

1. On the Morphological Relationships of the Molluscoida and Cœlenterata and of their leading members, *inter se.* By John Denis Macdonald, R.N., F.R.S., Surgeon of H.M.S. "Icarus." Communicated by Professor Maclagan.

2. On the External Anatomy of a New Mediterranean Pteropod. By John Denis Macdonald, R.N. Communicated by Professor Maclagan.

3. On the Limits of our Knowledge respecting the Theory of Parallels. By Professor Kelland.

The Author has in this paper traced to its consequences the assumption, as if it were an axiom, of the proposition " That the angles of a triangle are together less than two right angles." The results as regards the theory of parallels are such as to imply that such lines would have most of the properties of equal circles exterior to one another.

Professor Tait reminded the Society that, at the close of last session, he and Balfour Stewart, F.R.S., of the Kew Observatory, had deposited with the Secretary a sealed packet containing the coincident results of certain investigations which they had separately carried on from totally distinct points of view, and which appeared to lead to a new principle in Natural Philosophy.

Experimental attempts at verifications of this principle have since been made by them in various ways, and others are in progress. Meanwhile, the authors desire to put on record that it appears probable, from their experiments, that the viscosity, &c., of

The following Communications were read:—

1. On the great Drift-Beds with Shells in the South-west of Arran. By the Rev. Robert Boog Watson, B.A., F.R.S.E., Hon. Mem. Naturw. Ver. Lüneburg.

These beds, as examined by the author, lie in the Torlin or Kilmorie Burn basin, in the Scoradale or Slidry Water basin (more strictly in the first north or north-west tributary of each, reckoning upwards from the sea), and in the Clachan Glen,—all in the south-west of Arran. They are of great extent and depth; at certain points they contain boreal shells in considerable numbers. They are divisible into two classes, (1.) underlying fine dark sands and clays; and, (2.) overlying coarse red clay with striated stones, probably boulder clay.

They are interesting, because,

1. They present, in a striking form, proof of the immense destruction of the surface of the land.

2. They afford unusually good sections, from the rock on which they rest, upwards.

3. They throw some light on the formation of the boulder clay.

4. They present sea shells, at one point land plants, and also at one point a later lake basin.

The special information they give is,—

1. That *all* the latest geological changes have not materially affected the relations of hill and valley.

2. That the valleys were largely excavated by ice.

3. That the ice covered the land till it was submerged.

4. That the depression of the land below the sea was continuous, and ultimately attained 1000 feet at least.

5. That the depression was, at one point at least, sudden.

6. That this sudden fall did not begin later, at least, than the time at which the present 90 feet line above the sea-level reached the level of the sea.

7. That this sudden subsidence could not have amounted to less than 200 feet.

8. That it could not have much exceeded 300 feet.

9. That under obvious limitations, the beds which lie *nearest the sea-level and deepest below the surface*, are the *oldest*, and that those are *contemporary* which occupy the *same relative position to the sea-level and the underlying rock*.

· 2. On the Agrarian Law of Lycurgus, and one of Mr Grote's Canons of Historical Criticism. By Professor Blackie.

3. On the Occurrence of Amœbiform Protoplasm and the Emission of Pseudopodia in the Hydroida. By Professor Allman.

The author described the contents of the small tubular appendages, named Nematophores by Busk, which are developed upon certain definite points of the hydrosome in the *Plumularidœ*. These contents were shown to consist of a granular protoplasm, with occasionally a cluster of large thread-cells embedded in it.

The protoplasm has the property of emitting pseudopodia, which are very extensile and mutable in shape, and exactly resemble the pseudopodial prolongations, whose occurrence among the *Rhizopoda* is so eminently characteristic of this group of *Protozoa*. The contents of the nematophores, indeed, except alone in the presence of thread-cells, are indistinguishable in structure, and in the phenomena presented by them from the sarcode or protoplasm, which forms the substance of an amœba, a difflugia, or an arcella.

Monday, 18*th January* 1864.—His Grace the DUKE of ARGYLL, President, in the Chair.

The following Communications were read :—

1. Description of the Lithoscope, an instrument for distinguishing Precious Stones and other bodies. By Sir David Brewster, K.H.

   The Instrument was exhibited.

2. On the Temperature of certain Hot Springs in the Pyrenees. By R. E. SCORESBY-JACKSON, M.D., F.R.C.P.

In the year 1835, Principal Forbes very carefully observed the temperatures of certain springs in the Pyrenees, with the view of ascertaining to what extent changes of temperature occur in them. Observations previously made were, for several reasons, of uncertain worth, and Principal Forbes was desirous of then fixing "data for future observers with a degree of accuracy hitherto unattempted."

The author having determined to spend his autumn holidays in the Pyrenees, believed that a careful repetition of such observations, after the lapse of twenty-eight years, would afford results of some interest. He furnished himself with accurate thermometers made for the purpose, and, during the month of August, carefully observed the temperatures of several of the springs visited by Principal Forbes in 1835. In the tables which are distributed through the paper, the temperatures of the springs are given as recorded by different observers at various periods between the years 1835 and 1863. From these records it would appear that whilst there is perhaps in no instance a general or permanent change of temperature, neither is there in any an undeviating temperature. It is probable that the temperatures of the springs in the interior of the globe have undergone no change, and that the changes observable upon the surface of the earth are due to superficial causes, such as external temperature, the infiltration of cold surface water, &c. To a certain extent, an allowance must be made for inaccuracies; for it is scarcely to be supposed that all the observers dipped their thermometers exactly at the same points, nor does the author know that in all cases the instruments employed were without errors.

3. On Human Crania allied in Anatomical Characters to the Engis and Neanderthal Skulls. By Wm. Turner, M.B., Senior Demonstrator of Anatomy in the University.

The Author compared the above crania with various human skulls which had come under his observation. He exhibited a skull brought by Mr Henry Duckworth, F.G.S., from St Acheul near Amiens, which in its general contour presented a striking resemblance to the Engis skull. The St Acheul skull was somewhat smaller, being probably that of a female. It might almost have been regarded as a reduced copy of the Engis skull. There was no evidence that the skull from St Acheul was of an earlier date than the Gallo-Roman period of French history. The Neanderthal skull was compared with several modern crania, mostly British, especially with reference to the projection of the supra-orbital ridges, the retreating forehead, and the slight convexity of the occipital region. He exhibited several skulls which were closely allied to it in one or other of these features. It was shown also that the Neanderthal skull, although below the European mean in its internal capacity, yet exceeded the dimensions of some normal modern European crania which had been carefully measured—its large transverse parietal diameter compensating for the brain space lost by the retreating forehead and flattened occiput.

As the history and geological age of the Neanderthal skull were both unknown, and as many of its most striking anatomical characters were closely paralleled in some modern European crania, the Author considered that great caution ought to be exercised in coming to any conclusion, either as to the pithecoid affinities or psychical endowments of the man to whom it originally appertained.

4. Notice of a Simple Method of Approximating to the Roots of any Algebraic Equation. By Edward Sang, Esq.

5. Notice of the State of the Open-Air Vegetation in the Edinburgh Botanic Garden, during December 1863. By J. H. Balfour, A.M., M.D., F.R.S., F.L.S.

The state of the vegetation in the open ground of the Botanic Garden during the month of December 1863 was so very remarkable that I have been induced to submit a notice of it to the Royal Society. The number of phanerogamous species and varieties in flower during the month amounted to 245; of these 35 were spring-flowering plants which had anticipated their period of florescence, while the rest were summer and autumn flowers which had protracted their flowering beyond their usual limits.

The following are the details as given by Mr M'Nab :—

Plants in flower in the Royal Botanic Garden, Edinburgh, from 1st to 31st December 1863.

Annual plants, chiefly summer and autumn flowering species and varieties, ...	36
Perennial plants, chiefly summer and autumn flowering species and varieties, ...	133
Trees and shrubs, chiefly autumn flowering, . .	41
Spring-flowering trees, shrubs, and perennial herbaceous plants, ...	35
Total flowering plants, . .	245

The state of vegetation in December was much influenced by the nature of the weather during the preceding months of October and November. No marked check was given to it during these months, and the temperature was such as to stimulate the action of the cells and vessels of the plants. In December the comparatively high temperature continued. I have asked my friend, Mr Alexander Buchan, Secretary of the Scottish Meteorological Society, to draw up for me a tabular view of the temperature of the three months during the last seven years. I have supplied the lowest daily temperatures, as observed in the Botanic Garden; and for want of sufficient data in the garden, he has selected for the other data a station in Fife (Balfour) where the temperatures in general resemble much those noticed at the Botanic Garden. (I hope to be able to make arrangements in future for a full series of thermometrical, barometrical, and hygrometrical observations being made at the Botanic Garden.)

1. *Mean Temperatures at Balfour, in Fife, in the months of October, November, and December, during seven years ending with 1863. A comparison of 1863 with the Means of the previous six years is also given.*

YEAR.	OCTOBER.			NOVEMBER.			DECEMBER.		
	Mean of Day.	Mean of Night.	Mean Temp.	Mean of Day.	Mean of Night.	Mean Temp.	Mean of Day.	Mean of Night.	Mean Temp
1857	54·0	41·8	47·9	48·1	36·0	42·0	48·7	39·0	43·8
1858	50·8	38·3	44·5	45·0	33·8	39·4	43·3	33·9	38·6
1859	51·7	39·9	45·8	42·8	32·7	37·8	38·1	30·8	34·4
1860	52·7	42·1	47·4	42·5	37·6	40·0	36·3	29·0	32·6
1861	55·4	43·7	49·6	43·7	34·0	38·8	41·2	32·8	37·0
1862	54·3	41·1	47·7	42·9	30·6	36·8	46·2	38·6	42·4
Mean of 1857–62	53·2	41·1	47·2	44·2	34·1	39·1	42·3	34·0	38·1
1863	53·6	42·5	48·0	49·4	37·3	43·4	46·6	36·5	41·6
Excess of 1863 over the Mean of the previous 6 years . .	0·4	1·4	0·8	5·2	3·2	4·2	4·3	2·5	3·5

II. *The Highest Temperature of the Day observed at Balfour, in Fife, in each week of the three months October, November, and December, during seven years ending with 1863.*

YEAR.	OCTOBER.					NOVEMBER.				DECEMBER.				Oct.	Nov.	Dec.
	1	8	15	22	29	5	12	19	26	3	10	17	24			
1857	63·0	56·0	59·5	56·0	57·0	51·0	51·5	52·5	54·0	56·0	50·5	53 5	51·5	63 0	57·0	56 0
1858	61·0	59 0	51·5	54·0	52 0	47·0	46·0	46·0	56·0	52·0	45·0	50·5	46 5	61 0	56·0	54·0
1859	65·6	61·1	57·6	39·6	45·9	51·6	47·9	46·4	45 6	46·5	45 4	40 1	43 6	65·6	51·6	46·6
1860	60·1	56·6	55·6	59·1	48 6	47·6	46·6	45·1	45·6	45·1	44·6	37·6	36·1	60·1	48·1	45·6
1861	59·6	63·6	56·1	59 6	50·6	49·6	44·6	51·6	54·6	44·6	54·1	47·6	41·6	63 6	54 6	54·6
1862	65·6	65 1	59 6	55·6	54·6	47·6	46·1	43·1	44·1	52·6	49·6	51·6	49·6	65·6	54·6	52 6
1863	61 0	57·0	58·0	58 0	55·0	47·5	56·0	57·0	58·0	52·0	53·0	55·0	50·5	61·0	58 0	55 0

III. *The Lowest Temperature of the Night observed at Balfour, in Fife, in each week of the three months October, November, and December, during seven years ending with 1863.*

YEAR	OCTOBER.					NOVEMBER.				DECEMBER.				Oct.	Nov.	Dec.
	1	8	15	22	29	5	12	19	26	3	10	17	24			
1857	31·0	44·0	42·0	31·0	31·0	32·5	33·0	26·0	26·0	36·0	36·0	33·0	28 0	31·0	26·0	28·0
1858	35·0	34·5	33·0	30·0	30·0	31·5	28 0	22·0	36·0	26·5	25 0	31·5	31·0	30·0	22·0	25·0
1859	41·3	38·5	29·5	24·5	24·3	24·5	27·5	29·5	28·5	26·5	23 5	15 0	26·5	24·3	24·5	15·0
1860	41·5	30·5	40·0	39·0	37·5	38 5	27·5	26·5	30·0	39·5	30·5	18·0	1·5	30·5	26·5	1·5
1861	36·0	37 5	34·0	35·5	31·5	25·5	23·5	22·5	25·5	29·0	31·5	26·5	22 5	32·5	22·5	22·5
1862	34·5	37·0	36·5	35·0	29·0	27·0	20·5	20·5	22·0	35 5	33·0	34·5	31·5	29·0	20·5	31·5
1863	29·5	45·5	40·0	34·0	28·5	26·5	32·5	35·0	30·0	31·0	38·5	30·0	23·5	29·5	26·5	23·5

IV. *Number of Nights each Week on which the Thermometer, exposed in the Botanic Garden, four feet above the ground, fell to freezing (32·0). An asterisk (*) is put to indicate the nights on which it fell to at least 6·0 below freezing.*

Year.	October.					November.				December.			
	1	8	15	22	29	5	12	19	26	3	10	17	24
1857	1		1				1	3*	2				2
1858				4*		5	5*	6*		2	5*	1	
1859			2*	5*	3	2*	4	3*	5*	3	5*	6*	6
1860		2			3*	4	4*	2*	2*		3	7*	7*
1861			2		3	5*	3*	3*	2*	1	1	4*	7*
1862					2	5	6*	7*	5*			1	
1863	1			2	4*	4*	1	1	5*	3		3	5*
Lowest each week in 1863, . . }	28·0	40·0	35·0	31·0	26·0	23·5	7·0	32·0	26·0	30·0	35·0	25·0	19·0

Report on the Weather of October, November, and December 1863, as compared with the previous Six Years. By Mr ALEXANDER BUCHAN.

The first three tables present a detailed statement of the temperature in October, November, and December, during the seven years ending with 1863, as observed at Balfour, near Markinch, in Fife, one of the stations of the Scottish Meteorological Society. It is the nearest station to Edinburgh at which full and well-authenticated observations on temperature have been made for so long a period, and, besides, its position is such as to represent fairly both sides of the Forth.

Table I. gives the mean monthly temperature of the day and of the night, and the mean temperature of these months for the past seven years, and a comparison of 1863 with the means of the previous six years. From this Table, we learn that the peculiar features of the weather of October, November, and December last, as respects temperature, were as follow :—

In *October*, the mean temperature was nearly a degree (0°·8) above the average of the month ; but whilst the mean temperature of the day was less than half a degree (0°·4), that of the night was about a degree and a-half (1°·4) above the average, thus indicating a cloudy sky and comparative absence of frost. At the Botanic Garden, the thermometer fell only three times to freezing, the lowest being 28°·0 on the night of the 6th. This frost continued but for a short time, and very little damage was done except to Heliotropes; dahlias were only slightly affected.

In *November*, the mean temperature was 4°·2 above the average, which increase was very unequally distributed between day and night,—the mean temperature of the day being 5°·2, and of the night only 3°·2 above the mean of the month. This temperature is not only greatly above the average, but it is also about a degree and a-half higher than any previously recorded November, and 6°·6 higher than the November of 1862.

In *December*, the mean temperature was 3°·5 above the average, and the manner of its distribution between day and night similar to November,—the mean of the day being 4°·2, and of the night 2°·5 above the average.

Hence the characteristic feature of the weather of this period is the unprecedentedly high temperature during the day in November and December,—a point to which special attention is directed.

Table II. gives the highest temperature of the day, and Table III. the lowest temperature of the night, in each week of the period under consideration.

Table IV. gives the number of nights each week on which the temperature at the Botanic Garden fell to freezing or lower, and an asterisk is put to mark those cases when it fell to at least 6°·0 below freezing.

These tables furnish the data from which an explanation may be had of the remarkable vegetation of December last, in so far as that depended on the character of the then current weather. The explanation is twofold—*first*, the high temperature during the day in November and December; and, *secondly*, the comparative absence of frost during the night.

This remarkably high temperature was preceded by a period of cold weather, extending from the 29th of October to the 12th of November, during which frosts were of frequent occurrence. At Balfour the thermometer, four feet above the ground, and protected, fell to 26°·5; and at the Botanic Garden, four feet above the ground, but exposed, it fell to 23°·5, and indicated freezing on ten nights. Though dahlias and other plants were destroyed, yet many survived, owing, it is supposed, to the remarkably dry state of the weather, and to the very brief periods during which the severity of the cold in each instance lasted. This cold period also contributed to the remarkable growth which followed, since by playing the part of winter, though in a modified degree, it arrested the vital functions, and gave plants the benefit of a fresh start with the warmth which succeeded.

The unprecedentedly high temperature began on the 13th of November, and continued with scarcely any interruption till Christmas : see Table II. Of this period the warmest part extended from the 13th of November to the end of the month, during which the mean temperature of the day was 52°·5, or 9°·0 above the average. This day temperature usually prevails about the end of April or beginning of May, when the temperature of the night is much lower than obtained in November last.

In the end of November and beginning of December (see Tables

III. and IV.), the temperature of the night declined occasionally to freezing. At Balfour it fell to 30°·0, and at the Botanic Garden (exposed) to 26°·0,—a degree of frost insufficient to damage those autumn flowers which had stood the more severe frost in the beginning of November, or check the growth of the spring flowers rapidly coming into bloom.

This anomalous weather sufficiently accounts for the strange spectacle of *sweet peas* and *Hepaticas* blooming together.

Monday, 15th February 1864.—Dr CHRISTISON, V.P., in the Chair.

The following Communications were read :—

1. On the Influence of the Refracting Force of Calcareous Spar on the Polarization, the Intensity, and the Colour of the Light which it Reflects. By Sir David Brewster, K.H., F.R.S.

In the "Philosophical Transactions" for 1819, the author had shown that the doubly refracting force of calcareous spar extended beyond the sphere of the reflecting force, producing a change in the polarising angle varying with the inclination of the incident ray to the axis of the crystal, and producing a deviation of the plane of polarisation from the plane of incidence and reflexion, when the reflecting force of the crystal was reduced by contact with oil of cassia and other oils. These experiments were made on the face of the primitive rhomb.

In the present paper, the author gives an account of the results which he obtained upon other natural and artificial faces of calcareous spar, inclined 0°, 5¼°, 12°, 22½°, 67¼°, and 90°, to the axis of the crystal. On all these surfaces, when the reflecting force is reduced by contact with oil of cassia or other oils and fluids, the intensity and colour of the reflected pencil, and the deviation of the plane of polarisation from the plane of reflection, experiences remarkable changes, depending on the inclination of the incident ray to the axis of double refraction.

2. On the Most Volatile Constituents of American Petroleum. By Edmund Ronalds, Ph.D.

It was shown by this paper that the gases dissolved in American petroleum, and which gave to it such a high degree of inflammability, were composed of the lower members of the marsh gas series, having the general formula,

$$C_n H_{2n+1}, H,$$

and to which the liquid products have already been referred.

The gases evolved from the Pennsylvanian oil were collected at a temperature of − 1° Cent., as they floated, mixed with air, over the surface of the liquid in the casks in which it is imported into this country and the hydrocarbons were shown by eudiometrical analysis to have the composition of a mixture in nearly equal proportions of the hydrides of ethyl and propyl.

The first portions of incondensible gas evolved on warming the most volatile product of the distillation of petroleum on a manufacturing scale were also found to contain a mixture of these hydrides, while portions of gas collected at a later period of the operation approached more closely to the composition of pure hydride of propyl, or were mixtures of the hydrides of propyl and butyl; the last gas collected being nearly pure hydride of butyl.

The liquid condensed by a mixture of ice and salt during the collection of these gases gave, upon redistillation, a considerable portion boiling between 0° and 4° Cent.; this, as well as that which passed over as high as 6° Cent., was shown by analysis to be nearly pure hydride of butyl having the composition C_4H_9, H.

This liquid has a specific gravity of 0·600 at 0° Cent.; it is consequently the lightest liquid known. Its vapour density was by experiment found to be 2·11. It is colourless, possesses a sweet, agreeable smell, is soluble in alcohol and ether, but not in water. Alcohol of 98 per cent. dissolves between 11 and 12 times its volume of the gas at 21° Cent. The liquid and the gas are not preceptibly affected by sulphuric or nitric acid, nor by bromine; mixed with twice its volume of chlorine in diffuse daylight, the gas is converted into liquid chloride of butyl, while the original three volumes become condensed into two volumes of hydrochloric acid.

3.—On the Action of Terchloride of Phosphorus on Aniline. By Magnus M. Tait, F.C.S.

More than a year since my attention was directed to a statement of Hofmann's, that the action of terchloride of phosphorus on aniline yielded a white substance of crystalline character, as the investigation of this compound so produced seemed likely to be of interest, I began its examination, but circumstances prevented me from completing it at that time. The publication of Schiff's papers, however, on the metal-anilides again drew my notice to the subject, and I considered it a duty to myself to publish the results of my experiments, more especially as the reaction which forms the subject of this paper appears to have escaped the attention of that chemist.

Terchloride of phosphorus was added, drop by drop, to the aniline, which required to be kept cool by ice, as the reaction tends to be of a rather violent character, great heat being produced. In a short time the whole solidified into a soft granular mass, which dissolved readily in water, alcohol, and ether. The mass was dissolved in hot water, and, on cooling, the excess of aniline rose to the surface as an oily layer, and was separated by passing it through a moistened filter. The watery solution was evaporated at ordinary temperatures over sulphuric acid, and when it had reached a syrupy consistence it slowly solidified into a mass of fine needle-shaped crystals, which were the hydrochlorate of a new base, to which I give the name of Phosphaniline.

The crystals were well pressed between folds of filter paper, and then being placed on a filter, were washed with a very small quantity of alcohol and ether ; again dissolved, and evaporated as before, the crystals were pure. The substance so obtained dissolves easily in water, alcohol and ether, and is neutral to test-papers. Gently heated, it sublimes, and gives fine prismatic crystals. Treated with solution of potash it is decomposed. Strong sulphuric acid expels hydrochloric acid, and gives a colourless solution. Nitric acid oxidises it, and gives a coloured solution.

On analysis, it yielded the following results :—

		I.	II.	Theory.
Carbon,	. .	45·12	...	44·68
Hydrogen,	. .	4·94		4·84
Phosphorus,	.	...	...	
Nitrogen,	. .	...	..	
Chlorine,	. .	24·82	25·89	25·40

This analys's shows that the substance is produced by the direct union of the chloride of phosphorus and aniline. It is, however, a hydrochlorate, and is formed from three equivalents of aniline, in which three equivalents of hydrogen are replaced by phosphorus: thus—

$$\left. \begin{array}{c} C_6 H_6 \\ C_6 H_6 \\ C_6 H_6 \end{array} \right\} P''' \left\{ \begin{array}{c} N \\ N \\ N \end{array} \right\} + 3HCl.$$

$$C_{18} H_{18} PN_3 + 3HCl.$$

Platinochloride of Phosphaniline.—A portion of the original salt was dissolved in water, the solution acidified with hydrochloric acid, and bichloride of platinum added, in a short time crystals began to appear ; these, after a sufficient quantity had formed, were placed on a filter and washed with a small quantity of alcohol and ether, and dried over strong sulphuric acid, until their weight was constant. They were in the shape of small granular crystals of a light yellow colour, soluble in alcohol and water, but not in ether. On analysis, the following numbers were obtained :—

	I.	II.	III.	Theory.
Carbon,	23·52	...	...	23·38
Hydrogen,	2·40			2·35
Phosphorus,	...			
Chlorine,		...	...	
Nitrogen,		...	...	
Platinum,	...	32·23	32·73	31·98

The platinum in the third column was estimated by direct ignition, the high result obtained was probably owing to the formation of a little phosphide of platinum.

The analysis corresponds with the following formula :—

$$C_{18} H_{18} P''' N_3 \, 3HCl + 3PtCl_2$$

Zincochloride of Phosphaniline.—Hydrochloric acid was added to a portion of the solution of the hydrochlorate in water, and then some fragments of pure zinc were thrown in. After the zinc had dissolved the fluid was evaporated at a gentle heat (about 200° Fabr.), and filtered from a few greenish flakes which had formed, as even that low temperature appears to decompose the salt. The clear fluid was then put over sulphuric acid and left for some days, when crystals of a zinc salt were obtained, but under the same conditions as the hydrochlorate, as the solution was very concentrated before the salt appeared, and then it solidified completely. The crystals were pressed between folds of filter paper, and washed with a mixture of alcohol and ether, and dried in vacuo over strong sulphuric acid. They were white needles, slightly deliquescent, and soluble in alcohol. In ether it does not dissolve, but it becomes liquid, having the same appearance as a drop of oil in water.

The chlorine only was estimated, the analysis gave—

	I.	Theory.
Chlorine, ...	84·57	84·28

which agrees with the following formula :—

$$2 \ (C_{12} \ H_{18} \ P''' \ N_2 \ 3HCl) \ 3Zn'' \ Cl_2$$

Bromine water immediately precipitates a brownish-coloured substance from an aqueous solution of the hydrochlorate. This precipitate was washed with water, and a portion of it boiled for some time, and found to be insoluble in water; but a substance appeared to have volatilized along with the vapour of the water, as the neck of the flask and a glass rod which was held over the mouth of it were covered with white feathery crystals. The fluid was filtered, and, after drying the brown residue, it was put into a beaker covered with filter-paper, and left over a water-bath, when it nearly all sublimed in crystals corresponding to those obtained when attempting to dissolve it. A few of these crystals were dissolved in alcohol and bichloride of platinum added, but no precipitate was formed, and on evaporation the original substance crystallized out. These circumstances indicated the body to be Tribromaniline. The filtrate from the brown substance obtained originally was treated with bichloride of platinum also, but no precipitate forming, it was presumed no bromaniline or bibromaniline had been formed.

Cadmium Salt.—Chloride of cadmium gives, with strong solutions of the hydrochlorate, scaly crystals of a double salt, moderately soluble in water.

Copper Salt.—On adding chloride of copper to a solution of the hydrochlorate, and evaporating over sulphuric acid, small granular crystals of a beautiful green colour are obtained.

Mercury Salt.—If a strong solution of chloride of mercury is added to a concentrated solution of the hydrochlorate, beautiful white scaly crystals precipitate out immediately ; but if the solutions are dilute no precipitation takes place, if the solution is now warmed, a crystalline substance is thrown down. This appears the more curious, when it is known that the other salts cannot be heated without decomposition.

When the hydrochlorate is heated with potash the phosphaniline, at the moment of separation, appears to undergo decomposition, for the smell of aniline is apparent even in the cold, but no precipitation takes place, so that phosphaniline must be itself soluble in water. An attempt was made to obtain it in the separate state by acting on the hydrochlorate with oxide of silver, a precipitate of chloride of silver was formed immediately. The filtered fluid was alkaline to test paper; it clearly contained phosphaniline, but on evaporating the fluid it became coloured, owing to the decomposition of the base, which is very changeable, and cannot be obtained in the pure state.

4. On Fermat's Theorem. By Professor Tait.

The author stated that in consequence of Legendre's work, the proof of Fermat's Theorem is reducible to showing the impossibility of

$$x^m = y^m + z^m,$$

when m is an *odd prime*, x, y, z being integers.

Talbot has shown that in this case x, y, z are necessarily *composite* numbers.

The author shows, among other results of very elementary processes, that if numbers *can* be found to satisfy the above equation, x and y leave the remainder 1 when divided by m; and that z has m as a factor. Many farther limitations are given on possible values of x, y, z—the process being based on the consideration of their prime factors, and on Fermat's Elementary Theorem $N^m - N = Nm$.

5. Professor Archer called attention to a curious binocular telescope, bearing the following inscription :—

PETRVS PETRONVS
SAC : CÆS⁴⸴ ET CAT⁴
MAIES¹⸴ OPTICUS
MEDLANI 1726

The instrument belongs to the Royal Institution of Liverpool, and is supposed to have been part of a collection of rarities, made by Wm. Roscoe, in Italy. As a telescope, it is of great power; the focus is adjusted by one portion of the case acting as a draw-tube within the other part.

Royal Physical Society.

Thursday, 25th November 1863.—JAMES M'BAIN, M.D., R.N., President, in the Chair.

The PRESIDENT delivered an Opening Address " On the Theory of a Central Heat."

The following Communication was read :—

On the Strata discovered in making the East of Fife Extension Railway ; with Special Remarks on the Brick-Clay Beds at Elie, Fife. By the Rev. WALTER WOOD, Elie.

———

Wednesday, 23d December 1863.—JAMES M'BAIN, M.D., R.N., President, in the Chair.

The following Communication was read :—

On the Central Heat of the Earth. By Dr STEVENSON MACADAM, Ph.D., &c., Lecturer on Chemistry.

Dr J. A. SMITH Exhibited a male and female of the *Anser Egyptiacus,* the Egyptian Goose, recently shot near Dunbar.

———

Wednesday, 27th January 1864.—WILLIAM TURNER, Esq., M.B. , President, in the Chair.

The following Communications were read :—

I.—*Deductions from the Hypothesis of the Internal Fluidity of the Earth.* By WILLIAM STEVENSON, Esq., Dunse.

II.—*Remarks on Dr Macadam's " Spheroidal Theory" of the Interior of the Earth.* By T. STRETHILL WRIGHT, M.D.

———

Wednesday, 24th February 1864.—THOMAS STRETHILL WRIGHT, M.D., President, in the Chair.

The following Communications were read :—

I.—*A Wernerian Examination of the Six Points of Pluto-Huttonism.* By Professor W. MACDONALD, St. Andrews.

II.—*Some Objections to the Nebulo-Geological Hypothesis, as stated in Dr James M'Bain's Opening Address to the Royal Physical Society.* By PATRICK MACFARLANE, Esq., Comrie.

III.—*On the Irregularities of the Earth's Surface, and the probable Mean Line of the Terraqueous Circumference.* By WILLIAM RHIND, Esq.

The periphery of the earth's surface consists of land and water at different levels, and the question may be suggested—whether is the surface of the ocean or the mean level of the land the true line of the earth's circumference, or in other words, what is really the medium line of the irregularities of the earth's surface?

Though our knowledge of the sea-bottom is yet very limited, yet the investigations of recent years have added very considerably to this knowledge. The soundings of Sir James Ross in various parts of the ocean have disclosed interesting facts concerning its depth and temperature ; these were followed by other British and American navigators, and the whole has been collected and published by the labours of Lieutenant Maury. From these and other researches we find that the earth's superficies both under the ocean and above it presents an exceedingly irregular form, consisting of a series of elevations and depressions. On this irregular surface are diffused the waters of the ocean, spreading over and concealing from view more than two-thirds of the superfices, while the higher portions only, amounting to somewhat less than one-third, appear as dry land. The greater amount of soundings have been made in the Atlantic Ocean, and thus we have become better acquainted with its bottom than with that of the other oceans of the globe. A section from the Cape de Verd Islands, on the coast of Africa, to the mainland of South America, gives soundings of 17,000 and 22,800 feet. A section south of Newfoundland gives the greatest depth yet authentically ascertained as 27,180 feet. Further north in the line of the late electric telegraph from Ireland to Newfoundland, the depths are 10,000, 11,000 and 12,000 feet. On taking a mean of twenty-seven soundings in various parts of the Atlantic, and rejecting a few doubtful ones, the mean depth of this ocean is indicated as 13,100 feet, or two and a half miles. Soundings in the Pacific, also, indicate depths equal to the above in some positions, but the probability is that the central portions of that vast ocean, occupied by extensive coral reefs and innumerable islands, are of less depth than the mean of the Atlantic. The Mediterranean exhibits depths of 5000, 10,000, and south west of Malta, 15,000 feet. The North Sea is shallow. On the whole an approximate mean of the ocean depth may be estimated at two miles. Now, if we turn to the elevations of dry land we find that a few mountain peaks attain heights equal, if not surpassing, the extreme depressions of the ocean, in the Andes, 20,000 to 23,000 ; in the Himalaya, Kin-Kinchunga, 28,000 feet; and Mount Everest, 29,000 feet. But the mean elevation of land is far inferior to the mean depth of ocean. According to Humboldt's calculations, were the whole surfaces of the continents of Asia, America, and Europe reduced to a uniform level, that mean level would stand at only 1000 feet above the sea-level. From recent explorations of Africa, by Beke, Livingstone, and Speke, we find that extensive table lands of 2000 to 3000 feet occupy the central portions. Some mountain peaks attain a height of 20,000 feet, and these, contrasted with the vast level- deserts and low lying river valleys and shores, would seem to indicate not a higher general level for Africa than that of the other continents. Similar recent explorations in Australia also indicate that that region may be also comprehended in a general mean elevation of 1000 feet of the whole dry land of the globe. We thus find, that while the extremes of elevation of land about equal the extreme depressions of ocean, the mean depth of the ocean is 10,560 feet, or two miles, while the mean elevation of land is only 1000 feet. If we add these two means together, we have 11,560 feet as the mean of the irregularities of the earth's surface. Now, it will be perceived that the ocean surface does not stand at the mean or half section of these irregularities, but on the contrary, stands at the base of the dry land, or 1000 feet from its mean upper surface, and nearly 5000 feet above the line which we have indicated as the mean of the earth's irregularities ; for if the ocean were entirely awanting, the line would, in reality, be the mean line of the earth's irregularities, and consequently the true circumference and central line of gravity regulating the invariable diurnal revolution of the spheroid. Or to extend the expression of the formula used by Humboldt, if the whole irregularities of the earth's surface were levelled down to one uniformity, the earth's circumference line would occupy a position nearly 5000 feet below the present ocean level. But as water is only about half the specific gravity of the materials composing the superficial strata of the earth, twice the volume of water is thus necessary to fill up the depressions and to bring about that equilibrium which is required. In looking, therefore, at a section of the earth's surface, constructed according to the above measurements, it will at once be seen how small a proportion the mean elevation of land bears to the mean depth of the ocean—that while the extremes of both are nearly equal, the extreme elevation of mountains equalling the extreme depressions of ocean—the ocean surface occupies a level half way between these extremes, and thus becomes the actual line of circumference. But then its surface is thus raised from the true central line of the earth's inequalities, in consequence of the less relative specific gravity which water bears to the superficial strata of the earth's crust. There would appear also to be this general arrangement in the seeming irregularities of the earth's surface, that the depressions of the sea bottoms accompany and compensate the elevations of continents, and thus preserve the due equilibrium of the rotating spheroid. Thus the Atlantic is a great hollow basin between the elevated continents of Asia and Africa on the east and America on the west; while the deepest portions of the Pacific are on the west side of the American continent, and its central portions are comparatively shallow. Even in inland seas, the depths of the Mediterranean, ranging from 5000 to 15,000 feet, correspond with the elevations of Mont Blanc and the Alpine range, while the shallow Baltic and North Sea are surrounded by lands of no great elevation. We thus find in the great operations of nature that adaptation of means to ends which pervades the whole works of creation, and which are as perfect in the arrangements of the mechanism of worlds as in the minutest objects which exist on their surfaces.

Botanical Society of Edinburgh.

Thursday, 14th January 1864.—Professor BALFOUR, President, in the Chair.

The following Communications were read :—

I. *New Researches on Hybridity in Plants.* By M. CH. NAUDIN. Part I. Translated from the French, and communicated by Mr GEORGE M. LOWE.

(This paper appears in the present number of the Journal.)

II. *Letter from* ROBERT BROWN, Esq., *Botanist to the British Columbia Association.* Communicated by Professor BALFOUR.

Valley of the Ses-haat Indians, Barclay Sound,
Vancouver Island,
Lat. 48° 47′ 28″, Nov. 4, 1863.

Though I will be in Victoria about ten days after this date, when I will write you a full account of my transactions for the last three months, yet the politeness of the master of a trading schooner enables me to save a mail, and inform you that I am still in life and at work. Since I last wrote I have been to Washington Territory, U. S., British Columbia, and I have (first of white men) reached the head of the "great" central lake of Vancouver Island—that *ignis fatuus* of the local geographers of this far western portion of her Majesty's dominions. I have obtained seeds of between one and two hundred species of plants (in almost every case in quantities sufficient to allow of a complete division), including some very pretty and previously unintroduced species of herbaceous plants, shrubs, and forest trees. Among the last are some good coniferæ—including *Juniperus*, three species—one, a large tree, *Taxus*, n. sp. : fine lot of the rare *Cupressus nutkanus*, Lamb ; *Thuja Craigana* ; a large quantity of *Abies Bridgei*, not yet introduced into England ; about 100 ounces of the finest seed of *Abies Douglasii* ; *Pinus*, species undescribed ; *Abies grandis*, Dougl. ; 20 or 30 ounces *Abies Menziesii* (is this *Pinus Sitchensis*, Bognard Veg. Sitch; in the St Petersburg Academy Transactions ?) &c., and among non-coniferæ, *Quercus garryana ?* *Arbutus Menziesii*, Pursh ; *Spiræa*, sp. n. ; *Oreodaphne*, sp. n., a fine shrub ; and, what I think justly entitles the expedition to the credit which it originally laid claim to—viz., national importance,—about three pounds weight of a fine pasture grass from the Upper Fraser, which survives all winter, and is accounted by the Cariboo muleteers superior in fattening qualities to hay, and certainly much more valuable to them, as the cattle and the mules have nothing else to subsist on during the long bleak winter, when the ground is covered with snow, and hard as iron. I have gathered, and am still gathering, many particulars of its properties from the muleteers and teamsters, which in due time I will submit to you. I believe that it is superior to the much-lauded "Tussac grass" of the Falkland Islands, and might be introduced with great advantage into some of the bleak islands of the Hebrides, or of Orkney and Shetland, where I have known cattle to die off in the winter from want of proper fodder. My attention was originally drawn to this grass by Colonel Moodie, R.E., to whom, therefore, all credit is due ; and I believe that I am at liberty to say that he coincides fully in the above statement. You may remember that it was he who introduced the Tussac grass, and obtained a gold medal for his discovery. The great region for coniferæ I have found, since my present summer's travels have terminated, to lie south of this latitude, and to it therefore, with your permission, I shall in future devote more attention. My funds are in a satisfactory condition, and although the stormy winter, now fairly set in, will prevent me doing much more for some months, I am prepared to start in early spring with renewed vigour to a widely different, and what I now believe to be a better region than the one I have explored. I trust that I shall be able to take the field by the 1st of April with advantage to the Association, and with every prospect of much greater success attending my efforts than during the previous six months. I know of some good species of coniferæ which I could not obtain this year, but I am almost certain of next year—such as *Abies Williamsoni* from "The Three Sisters," in the Cascade Mountains, *Abies bracteata*, *Picea nobilis*, and a new species of *Thuja*.

I have had much to contend against this season, but the difficulties are now, I am glad to say, in a great measure over. I now know the country, and what is just as important, the people, so well, that should there be only half the subscribers of the previous year, I will be able to do tolerably well. Many expenses were incurred last year which will not be required again. These are in addition to the sum paid for travelling to this distant part of the world (Kamtschatka is the furthest sea voyage, and that is only a few days' sail from here). Last night I was awoke from my camp-fire sleep by the "long cry" of the same wolf that "howls from Unalaska shore."

On my return to civilisation, I will send my seeds to England by Wells, Fargo, and Company's expresses, in several air and water-tight boxes, addressed to you, "care of the Hudson Bay Company," prepaying the freight. This will be expensive; but the agent here refuses to receive them on any other terms, a standing order having been issued to that effect. I will at the same time send a fuller account of my procedure, and by later mail a statement of my intromissions will be sent to the treasurer. In order that you may be early advised of the despatch of the seeds, and be ready to make the necessary arrangement with the secretary of the Hudson Bay Company, I will send this and the next letter overland by the pony rider, *via* the Salt Lake City, to St Joseph, in Missouri, and thence *via* New York to Liverpool. It will save nearly three weeks, and I hope will escape the emissaries of President Davis.

I am in good health and strong; rough, ragged, weather-beaten, perhaps a little dirty, and certainly a most unpresentable figure at the meeting where this communication will be read. I am sleeping on a curious Indian blanket, woven from the liber of *Pinus Strobus* (is it the same as the *P. Strobus* of the East?), and to keep the hoar frost off my blanket, there is a mat (clay hulk) of the bark of *Thuja gigantea*. Both will, I hope, at some future day, ornament the museum at the Botanic Garden.

III. *Extracts of Letters received from* Mr WILLIAM MILNE, *Old Calabar.* Communicated by Mr JOHN SADLER.

Creek Town, Old Calabar,
June 29, 1863.

I am fairly settled in the district of Old Calabar, exploring the creeks and corners of this majestic river. Africa is certainly rich in botany and other branches of natural history. Years must roll away before the botany of this vast continent is thoroughly investigated, and that will not be until Christianity is upon a more substantial basis. I will give you one extract from my daily journal to show the superstition which still exists amongst the people in Western Africa. While in the district of Ikorofiong, in passing through a large native town in the Ebebo country, I saw a straggling shrub belonging to Bignoniaceæ. While in the act of pulling down some of the flowers, I was surrounded by some hundreds of men, women, and children, shouting and dancing like so many fiends. At first I was inclined to think they were about to hang me in front of their palaver house or heathen temple. On looking round I could see no way of escape, so I held my ground, determined to have some of the flowers; but they were as determined that I should not get them. At last they put me out of the town. On the following Sunday I accompanied the Rev. Zeruh Baillie to several of the plantation villages, where he preaches once a week. We met the Ebebo chief. I wished to shake hands with him, but he would not come near. He said he was afraid of the strange medicine I was making, and told Mr Baillie that I was not to come to his town again.

About a fortnight ago a man told me that if I went into the bush I would be shot: so you see it is not all plain sailing at Calabar. But I have an extensive field before me, and I am determined to make the best of it, in spite of the natives, as it will not do to let them have it all their own way. I will mention a few of the leading characteristics of the vegetation which have come under my own observation.

There are five species of Melastoma, six species of Dracæna, five species of Amomum, and several others belonging to Zingiberaceæ. There are a number of species belonging to Scrophulariaceæ; and amongst them is a Digitalis, which is scattered over all waste ground. Euphorbiaceæ and Cucurbitaceæ are both extensive orders here. Three species of Amaryllis are abundant—one in the river, and the other two spread all over the plantations. Solanaceous plants are numerous: there are two kinds sold in the market as purgatives and for bathing the sides of their faces when they have a discharge from the ears. Anonaceæ is another extensive order. According to the Rev. Mr Thomson, there are sixteen or eighteen kinds. I have collected a number of Bignoniaceæ and Cinchonaceæ. I have also met with eight or nine species of Convolvulus, but there are more than that. Amongst the Labiatæ is a large species of Salvia, which is used as a medicine. There are three true mints which are used for seasoning; in fact, all this order is made use of as articles of food. A species of Nymphæa is frequent in the inland streams.

I think there are from eighteen to twenty-four distinct Orchids: one fine terrestrial species has a flowering stem 6ft. or 7ft. high. There are two fine species of Strophanthus. One true Verbena and two Clerodendrons are abundant. I have also observed two species of Amaranthus. "Love-lies-bleeding" is one of them, but I am doubtful if this is indigenous, although the natives say so. Both kinds are used as vegetables in Calabar chop. One Pentstemon is found by the margin of a small stream at Ikorofiong, but not plentiful. A Phytolacca and a Polygonum also occur at the same place. I have collected specimens of a Loranthus from trees by the banks of the river. Two species of Lonicera are very common. Leguminous plants are very numerous. Amongst them is a sensitive Mimosa. The poison beans (Physostigma) are often used for deadly purposes. One species is largely cultivated for putting into the streams to poison the fish, and another is sold in the markets for Calabar chop: one kind is very like our scarlet runner. The ripe pods are from 6 to 8 inches long, and the fruit is beautifully spotted. Compositæ are not so numerous as might be expected: however, there is a due proportion. A Tillandsia climbs up the palm trees. There is one fine species of Calophyllum, and a tree belonging to Myrtaceæ; also an Aristolochia, which I think is gigantea. There are five different palms. One large species of Juncus is abundant on the sides of the river at Creek Town; also another smaller species. There are several Cyperaceæ by the river, and amongst the lowland plantations there are a number of Gramineæ sprinkled about. Eighteen varieties of yams and six varieties of Colocasia are cultivated; the flowering stems of the latter, with spathe and spadix, are sold in the markets for putting into Calabar chop. The corms are also boiled, and used by the natives. There are two kinds of Cassava largely cultivated. There is only one true Banana with very small fruit, and eight kinds of plantains, sold in the markets. There is a malvaceous plant cultivated; the fruit is cut up into slices and put into soups. There are also two species of Agaricus sold in the markets. They are said to be very nourishing, and to give a fine flavour to Calabar chop. The larger kind is also put into rum, as it is of an intoxicating nature. Calabar chop is composed of the following ingredients:—Palm oil, yams, mints of all kinds; flowering stems, leaves, and corms of the colocasia; two species of agaricus, the fruit of two leguminous plants, the leaves of two kinds of amaranthus. The rest consists of monkey's flesh, dogs, rats, fish, goats, fowls, parrots, and birds of all kinds; in fact, everything eatable, whether animal or vegetable, is put into this wonderful dish. It is a favourite dish among the natives, and relished by Europeans, only they take nothing but palm oil, fowls, and goats' flesh. I have only seen three species of snakes. There is only one parrot, five species of monkeys, three rats, and three mice, four or five land crabs, and a number of freshwater fish, some sixty or eighty butterflies, and a host of beetles and other insects. Amongst my collection I find twenty-four different ants. Such is a very brief outline of the botany and natural history around Old Calabar.

Creek Town, July 1, 1863.

I am hard at work exploring one of the finest rivers on the west coast of Africa. As you ascend this noble river, the banks and surrounding country are one vast amphitheatre of everlasting green. I am delighted with the appearance of the country and its vegetable products. I have got an extensive field before me, and undoubtedly, if all goes well, I will find many a novelty. This great continent is teeming with animal and vegetable life. Even in my room I have got use for an insect net. It is a great pity that the climate is so unhealthy. Fevers are very prevalent. I have had three attacks of fever since I landed upon the African coast. Fortunately I am blessed with a strong constitution, and I get easy over it; but such is not the case with everybody. The bones of many a blooming youth are bleaching beneath the sun at Calabar.

The natives are not such a murderous class, except amongst themselves. For instance, there is not a day passes but they are killing their twin children instantly after birth, and banishing the unfortunate women to what they call twin villages, where they are left to languish out a life of silent sorrow, and are denied all intercourse with the rest of the world. Mrs Goldie, about seven weeks ago, saved two of those little unfortunates. She remained by the poor woman until she was confined, and then at the dead hour of midnight she entered the mission-house with a little boy and a girl rolled in her lap. The mother followed about six o'clock in the morning. They are all under the protection of the Rev. Mr Goldie. About a month ago the King of Creek Town had a sister whose daughter died of consumption. She sent for a number of her slaves, to give them poison bean. Three women died, three more escaped by vomiting, and one girl took refuge in the mission-house, under Mr Goldie, where she is now attending the school. These poor people were to be servants to her daughter in the future world. As soon as the rainy season is over, I am going to the Qua Mountains, and that will be the time for plants of all kinds. My collections will be upon a grand scale by-and-by. Only the other day I found a splendid climbing lily; it is a true turn-cap lily, and will form a grand show upon the rafters of any stove.

IV. *Extract from Letter from* Dr Meredith, *Georgetown, Demerara, dated 21st November* 1863, *to* Professor Balfour.

Dr Meredith says :—I daresay you know the topography of the large streams in this neighbourhood. The Essequibo is the largest, and lies nearly north and south. The Euyuni, nearly east and west, and the Massaruni, lies between the two, and runs nearly north-west. The two last meet about four miles above the penal settlement, and their joint current unites with the Essequibo about five miles below the penal settlement. The penal settlement is on the west or left bank of the river. The country is nearly a dead level, and at irregular intervals creeks open into the rivers on both sides, and are often navigable for boats. Their mouths are usually entirely hid by foliage. But the Indians know the right gaps as perfectly as did the old smugglers know the caves and holes on the Scotch coasts. I have often been in these creeks for miles, going with two or three hands in a small canoe—dodging under branches, hauling over a prostrate log, and charging through leafy shrubs, lying down nearly flat in the canoe all the time. It was very seldom we came to a place we could see the sky above—such is the dense nature of the bush. In many parts there is never more than dim twilight, and in the rainy season when the sky is dark or cloudy it is a perfect midnight even at mid-day. Never is there a breeze experienced. We have only a rustling sough some 80 or 100 feet over head. I have often been thinking that the usual description given of the luxuriance of the vegetation during the coal epoch might be applied with great effect to the vast forest of Guiana. The most remarkable feature of the bush is perhaps the immense quantity of parasites, but particularly epiphytes, with their long descending cord-like roots of all lengths, up to 90 or 100 feet, and not thicker than an ordinary writing quill. They are usually very tough. In one of our excursions my companion and I set to work to pull at one of these. We could not see from what height it came, nor to what kind of plant it was attached; however, we pulled, and with great success, as it happened, but we brought a tremendous shower of water upon our heads, as well as a host of black ants, which teased us dreadfully for a while by getting under our clothing. We had dislodged one of the air-plants belonging to the pine-apple order, and the ants' nest must have been attached to it. The root was, I think, about thirty feet long. The Indians always use them as *cables* for mooring their canoes. In April last a friend and I started off to see the *Victoria regia* up the Essequibo river. We were away nearly a week. It was a delightful excursion. The scenery all along the river was really magnificent—thoroughly unlike anything I ever saw at home. The clearness of the atmosphere, the rank luxuriancy of the bush all along the river, mirrored in its gently flowing stream, gave to the whole a picture of exquisite beauty. The habitat of the lily in this district is a lake on an island in the river about the 6th parallel of N. lat.; where Schomburgk first saw it was up above Berbice on the Corentyn river. This lake (Essequibo) was surrounded by dense vegetation, which rendered it very difficult to approach. It was covered with the lily in all stages of its existence—the large white flower, the beautiful crimson edges of the young leaf changing into green as it grows. When it has attained its full dimensions the leaf begins to wither and die at its circumference; but as fast as it decays its place is taken up by a new one. Thus the great *struggle* goes on. I managed to get a good view of the place by climbing along the trunk of a tree which partly overhung the lake. A drizzling rain had been falling for an hour or so before we arrived at the lake; but just as I had ascended my perch on the tree it ceased, and a gleam of sunshine fell on the water, with a slight breeze, causing the vast flotilla to roll gracefully to leeward, throwing out various shades of colours. It was certainly the prettiest water scene I ever witnessed. A day or two afterwards I saw the Indians poisoning a small lake with the "Hiari" bush rope. I entered into the sport with right goodwill. The Indians cut the "Hiari" into pieces of about two feet long, then proceed into the water and beat out the juice of it as a blacksmith beats hot iron with a hammer. The fish begin to show signs of uneasiness, very soon get regularly intoxicated, and often jump clean out upon the dry land, unable to bear the poisoned water any longer. When they were in this state of helplessness the Indians, as well as my friend and I, with some half a dozen negroes, speared or caught them with nets by the dozen. They were most extraordinary looking fish. Some of them bite like dogs, and have very sharp teeth. One of the Indians got a severe wound from one of them. The most ferocious fish I have met here is what the Indians call the "piari;" it cuts as clean as a knife. The quantity of fish in these rivers is perfectly surprising, and if I am not mistaken, their natural history has yet to be commenced.

We spent a day at an Indian settlement, where the inhabitants were most scrupulous in painting themselves with red paint all over, but considered themselves amply clothed by wearing a lappet about six inches square. I asked one young man who had painted his body various colours like the belts on a tiger, and had his head dressed out with Macaw feathers, why he did it? His answer was, "because it looked pretty, and the women liked it." I wonder what an Eastern satrap would say if he was transported into the midst of an Indian village, and saw the difference of clothing from that in his own country, where a woman dare hardly show the tips of her fingers. We spent a few nights in the bush, swinging hammocks between trees with a sail stretched over them. We were often disturbed by the noise of the "howling monkeys." They make a most awful and dismal noise. They can be heard several miles away. I see the London Zoological Gardens obtained a pair of them the other day. I have been very much interested in trying to make out the history of the enormous boulders which are to be met with about the penal settlement and above. There is no rock like them within hundreds of miles, and the ground on which they rest is pure alluvial soil. The ground of the penal settlement is covered by a collection of them, forming what, I suppose, would be called a moraine. How came these huge blocks down here—could it be by means of glaciers? The gold discovered up the Euyuni, which is now beginning to excite deserved attention, is found chiefly in boulders of this description. The parent rock is in some of the mountains in the interior—but where, no one knows. There is ample room for explorers in this part of the world. I have not been up to the gold district, although I once tried it. Our boat proved too heavy for hauling over the falls. Descending these rapids is most exciting work. We came down almost at railway-train speed. We have always to get Indians to steer, two in each boat.

V. *Notice of Mosses found near Blair-Athole, Perthshire.* By Miss M'INROY, of Lude. Communicated by Mr JOHN SADLER.

VI. *Principal Plants of the Sutlej Valley with Hill, Botanical, and English Names; together with approximate Elevations, and Remarks.* By Dr HUGH F. C. CLEGHORN.

VII. *Letter from Henry Stephens, Esq., relative to Dry-rot.*

Thursday, 11*th February* 1864.—Professor BALFOUR, President, in the Chair.

The following Communications were read :—

I. *On Diplostemonous Flowers, with some remarks upon the position of the Carpels in the Malvaceæ.*—By Dr ALEXANDER DICKSON. (This paper appears in the present number of the Journal.)

II. *On the Cinchona Plantations in connection with the Botanical Garden at Bath, Jamaica.*—By NATHANIEL WILSON, Curator of the Garden. Communicated by Dr LAUDER LINDSAY.

The most important event in the history of this Botanic Garden for many years past has been the introduction, by seeds, of the quinine-yielding Cinchona in the autumn of 1860. By the month of October 1861, I succeeded in rearing upwards of 400 healthy plants, quite ready for planting out; but unfortunately the selection of a proper site for their final establishment was overlooked, and in consequence one-half of the number perished. Being anxious, however, to test the adaptability of the plants for cultivation in the higher altitudes of this island, I caused the whole of them to be removed in small pots to Cold Spring Coffee Plantation, the elevation of which is about 4000 feet. I soon found the climate and soil of this locality to be all I could desire for the plants; and as it afforded every facility for carrying out so valuable an experiment, I at once availed myself of it, and planted out in the coffee fields on the 16th November several plants of each species, then about two or two and a half inches in height. In twelve months after, a plant of the red bark, *Cinchona succirubra*, had attained to the height of forty-four inches, with leaves measuring thirteen and a half inches in length, by eight and three-quarters in breadth. The same plant, now two years old, measures six feet in height with ten branches, having a circumference of stem at base of four and a half inches. The plants of *Cinchona micrantha*, grey bark, being of more slender habit of growth, have not made such rapid progress; the highest has attained to five feet with three branches. The leaves, however, are larger, and measure fourteen by ten inches. So far the experiment has thus proved eminently successful. It would be difficult to find more healthy trees in the forests of that neighbourhood; and, in about three or four years hence, they may be expected to produce seeds. In the meantime they can be largely increased by cuttings and layers, in the hands of a skilful propaga or. During the months of August and September 1862 the collection was again removed to Bath. The plants were at this time eighteen or twenty months old, a critical period for forest trees in flower pots under artificial treatment, and in a climate too so uncongenial for them as that of Bath, which would have terminated their lives had they not soon afterwards (13th October) been planted out at Mount Essex, near Bath, at an altitude of two thousand feet, or little more. This site, as a temporary one, was had recourse to to save the plants alive until a better could be obtained; and so far it has answered the purpose. and a majority of the plants are healthy, but have not made such rapid progress as could have been desired. The soil is too loamy and by far too stiff to admit of a free and rapid escape of the rains which fall here in torrents during the greater part of the year, and the altitude far too low for the Peruvian barks. The red bark thrives at a much lower altitude, and, being a more hardy tree, the plants are more healthy. A very important fact has now been established—viz., that the climate of our higher, and many of our intermediate, mountains is suited for the growth of the most valuable species of quinine-yielding plants—*Cinchona succirubra*. A knowledge also of the method of increasing the plants, and of the soil best adapted for their full development has been obtained.

III. *Notice of the Occurrence of* Woodsia alpina (hyperborea) *in Gaspé, Canada East.*—By GEORGE LAWSON, LL.D., Professor of Chemistry, Dalhousie College, Halifax, Nova Scotia.

Professor Lawson stated that this rare fern (*Woodsia alpina*) had not been found by any recent collector in Canada, and no one knew where to look for it. He had recently described, however, in connection with *Woodsia glabella*, and under the name of var. *Belli*, a remarkable Woodsia found in Gaspé on the Dartmouth River, 20 miles from its mouth, by his former pupil, Mr John Bell. This plant, on further examination, turned out to be the *Woodsia alpina.* He enclosed a specimen.

IV. *Remarks on* Myrica cerifera, *or Candleberry Myrtle.*—By Professor LAWSON, Dalhousie College, Halifax.

I found *Myrica cerifera* a few days ago in some quantity on a hillside near Halifax, with small birds feeding on the berries. It is variously called Wax Myrtle, Candleberry Tree, and Tallow Shrub, in reference to the wax with which its berries are coated. The wax is yellowish green, emitting a fragrant balsamic odour not so unctuous as bees' wax, nor so brittle as resin, sinking in water, whereas bees' wax swims, sp. gr. 1·015; fusing point, 110°. The wax is obtained by boiling the berries in water. The plant is common all over the hills overlooking the innumerable bays and harbours of the Nova Scotian coast, but I do not know that the product has been made an object of commerce here. Some years ago Professor Simmonds called attention to the desirability of encouraging the production of Myrtle wax, and gave in the " Pharmaceutical Journal, vol. xiii. p. 418," very full details respecting the wax-yielding Myrtles of South Africa, and the mode of manufacturing the product. The species which is cultivated at the Cape as the best South African one is *Myrica cordifolia.* Professor Simmonds speaks of our American Myrtle wax (*M. cerifera*) as identical commercially with that produced at the Cape. A consignment of Cape wax (2561 lbs.) yielded a clear profit of L.54, 4s. 5d., after payment of all expenses, collecting, shipping, &c. Should any manufacturer in Britain feel an interest in the matter, I will be happy to furnish such additional information as may be procurable here. In " Annales des Sciences Naturelles 1855," it is stated that the European *Myrica Gale* yields a little wax. The *Myrica cerifera,* like its African congener, is a coast plant. I never saw it in Upper Canada ; but Professor Gray gives a station on Lake Erie. On the Halifax hills it is a small spreading bush, three or four feet high, forming a close brush. At this season the stems are leafless, but the withered leaves scattered about still retain their fragrance.

V. *Note on the leaves of Ulex (Whin).* By Professor LAWSON, Dalhousie College, Halifax.

The seedling *Ulex* has at first no spines. The young stem is clothed with leaves—from twelve to twenty in number—these are shortly petiolate and trifoliate, consisting each of three small elliptical hairy articulated leaflets. When the stem becomes five or six inches in length (usually) the trifoliate leaves cease to be developed, and spines are then produced. We thus see that in *Ulex* the perfect leaves appear during the early period of the plant's development, while in the Australian *Leguminosæ,* their production is delayed till the maturity of the plant. *Ulex,* however, is truly a plant with compound trifoliate leaves, not simple-leaved, as stated in many works. The above facts were ascertained from observing the development of some seedlings of the whin raised by Mrs Lawson in a flower-pot, from seeds brought from Ireland by Dr Dickson, Dean of the Medical Faculty of Kingston.

VI. *Notice of Mosses found in the neighbourhood of The Burn, near Brechin, Forfarshire.* By Mrs M'INROY. Communicated by Mr JOHN SADLER.

Mrs M'Inroy gave a list of mosses to be met with in the neighbourhood of The Burn, so far as at present known. The greater proportion are to be found within the policies, and principally on the banks of the North Esk, which flows through the grounds. The list contains 144 species, but Mrs M'Inroy states that she believes this by no means exhausts the mosses of the place.

The following note to Professor Balfour from Dr John Kirk was read :—
" I enclose two pods of a tree nearly allied to *Bauhinia* or *Copaifera,* but of a new genus. Perhaps you may succeed in growing it. As yet

the flowers are unknown, for my specimens were lost. This tree has hard black wood, called by the Portuguese *Pao ferro*; by the Bechuanas, *Mopane*. I have got the foliage and fruit, but, being a new genus, the flower is a sad want. It inhabits the driest and most baked clay plains; no heat kills it, the leaves, like those of Australian forests, turning their edge upwards when the sun shines very powerfully. The testa of the seed is full of resin in large cysts, and the cotyledons are convoluted like the surface of the human brain."

A note from Mr W. J. HAIG, Dollarfield, was read, in which he says:—" I enclose a specimen of a plant which I have just received from a cousin who is settled in the Banda Oriental Republic of Uraguay. He writes to me:—' I should be glad if you could get a scientific opinion on the following case:—It had long been a known fact here that sheep removed from the province of Buenos Ayres to the Banda Oriental, or imported from Europe, are liable to die in great numbers during the first fortnight after arriving. I saw an instance myself last week. At an *estancia* about five leagues from here, the owner had just received from Buenos Ayres sixty rams and ewes freshly imported from Europe, which cost L.15 per head: of these, fourteen died in five days. It is believed that the mortality arises from the sheep eating a plant called *min-min*, peculiar to this province. I enclose a twig of it. It grows in patches about the size of one's hand, and the twig enclosed is of average height. The stomach and intestines after death are found much inflamed; and in some cases the under side of the skin is quite red with effused blood. I may add that sheep born here, or acclimatised, are in no way injured by the plant—in fact, they almost entirely avoid it." This seems very analogous to the effect which the twigs of the yew tree have on animals here; and I should think that the only way to avoid the danger would be to put the animals for the first week or two in an enclosure cleared of this plant. By this time desire for green food would be somewhat abated, and they would, like the native born, become more discriminating.

Thursday, 10th March 1864.—Professor BALFOUR, President, in the Chair.

The following Communications were read :—

I. *New Researches on Hybridity in Plants.* By M. CH. NAUDIN. Part II. Translated from the French, and communicated by Mr GEORGE M. LOWE.

(This paper appears in the present number of the Journal.)

II. *On the Chemical and Natural History of Lupuline.* By M. J. PERSONNE. Translated by GEORGE LAWSON, LL.D., Professor of Chemistry and Natural History in Dalhousie College, Nova Scotia.

Drawings of Lupuline were shown, and specimen from the Industrial Museum was exhibited by Professor Archer

(This paper appears in the present number of the Journal.)

III. *Remarks on the Sexuality of the Higher Cryptogams, with a Notice of a Hybrid specimen of the species of* Selaginella. By Mr JOHN SCOTT.

Specimens of the species of Selaginella and the Hybrid in a living state were exhibited.

(This paper appears in the present number of the Journal.)

IV. *Abstract of a paper on the Constitution of Gymnospermous Flowers.*
By A. W. Eichler.　Communicated by Professor Balfour.

V. *List of Fossil Plants found in the Tertiary Strata of the South-
East of France.* By Gaston de Saporta.　Communicated by Professor
Balfour.

(From " Annales des Sciences Naturelles," 4th ser. Bot. tom. xix.)

PUBLICATIONS RECEIVED.

1. Canadian Naturalist and Geologist, October and December 1863.—
From the Editor.

2. Bulletin de l'Académie Royale des Sciences, des Lettres, et des
Beaux Arts de Belgique, Nos. 11 and 12, 1863.—*From the Editor.*

3. Remarks on the Present Condition of Geological Science.　By
Geo. E. Roberts.—*From the Author.*

4. The Story of the Guns.　By Sir James Emerson Tennent.—*From
the Author.*

5. Natural History Review for January 1864.—*From the Publishers.*

6. Functional Diseases of Women　By John Chapman, M.D.—*From
the Publisher.*

7. The Classification of Animals based on the Principle of Cephalization.　By James D. Dana.—*From the Author.*

8. On Fossil Insects from the Carboniferous Formation in Illinois.
By James D. Dana.—*From the Author.*

9. Programme de la Societé Batave de Philosophie Experimentale de
Rotterdam.—*From the Society.*

10. Journal of the Asiatic Society of Bengal, Nos. III. and IV,
1863.—*From the Society.*

11. Handbook of Astronomy.　By George F. Chambers, F.R.G.S.
—*From the Author.*

12. American Journal of Science and Arts, May and November 1863,
and January 1864.—*From the Editors.*

13. Proceedings of the Literary and Philosophical Society of Liverpool, No. XVII.—*From the Society.*

INDEX.

THE END.

PRINTED BY NEILL AND COMPANY, EDINBURGH